Numerical Methods for Scientists and Engineers

Numerical Methods for Scientists and Engineers

Third Edition

K. SANKARA RAO
Formerly, Professor of Mathematics
Anna University, Chennai

PHI Learning Private Limited
Delhi-110092
2014

₹ 275.00

NUMERICAL METHODS FOR SCIENTISTS AND ENGINEERS, 3rd ed.
K. Sankara Rao

ISBN-978-81-203-3217-1

Eleventh Printing (Third Edition) ... ... February, 2014

Published by Asoke K. Ghosh, PHI Learning Private Limited, Rimjhim House, 111, Patparganj Industrial Estate, Delhi-110092 and Printed by Meenakshi Art Printers, Delhi-110006.

To

my wife ***Leela***

Contents

Preface

The objective of this third edition is the same as in previous two editions: to provide a broad coverage of various numerical techniques, that are widely used for solving many important problems in engineering, science and technology with the help of computational tools.

I have updated the previous edition by adding new material as suggested by my old colleagues and students. Two new chapters: Chapter 12, Boundary Value Problems and Chapter 13, Approximation of Functions, have been added.

The text now has thirteen chapters. Chapter 1 is unchanged. Chapter 2 has been updated by adding a new section explaining Bairstow method, to find the complex roots of a polynomial with real coefficients and another section with a method for solving a system of nonlinear equations. Chapter 3 remains unchanged.

In Chapter 4, Gerschgorin's theorem is included to get the bounds for eigenvalues of a square matrix, while Chapter 5 has been repeated as such.

Chapter 6 has been updated by including a new section on Hermite interpolation. In Chapter 7, a new section, where Gaussian quadrature type of formula has been derived for evaluating multiple integrals, has been added. Chapter 8 is unchanged except that the concept of stability is introduced and illustrated. Chapters 9 to 11 remain unchanged.

Chapter 12 is new to this edition. Here, finite difference method, shooting method, and weighted residual methods such as Galerkin and collocation methods are discussed in detail. Finally, Chapter 13 is also new to this edition. Here, approximation of a real continuous function by a polynomial, using least squares, Chebyshev and Pade approximations are discussed extensively. In addition, Fourier series approximation is also given with an introduction to Fast Fourier Transform.

I wish to thank all my old colleagues, friends and students whose feedback has helped me to improve over previous two editions.

I also wish to thank the publisher, Prentice-Hall of India, for their careful processing of the manuscript, both at the editorial and production stages.

K. SANKARA RAO

Preface

Preface to the Second Edition

In the present-day world, computer-based knowledge has become an integral part of the educational system. Any kind of application is amenable to numerical treatment, the only limitation being the reader's imagination. In fact, many practical problems in science and engineering, when formulated mathematically, give rise to several interesting situations, for which numerical answers can be obtained, while analytical methods may fail. Due to the advent of fast and more powerful digital computers, in the last five decades, the developments of numerical techniques to solve scientific problems have been quite phenomenal.

This self-contained text has numerous distinguishing features over the already existing books on the same topic. A comprehensive and balanced treatment of the subject covering many important topics such as computation of eigenvalues and eigenvectors of a matrix by the power method and Jacobi's method (Chapter 4), derivation of the second-order Runge–Kutta method for solving ordinary differential equations (Chapter 8), and numerical solutions of parabolic, elliptic and hyperbolic partial differential equations (Chapters 9–11) are dealt with in great detail. New methodology for finding interpolation in two dimensions and cubic-spline interpolation (Chapter 6) is also discussed in this book. Curve-fitting techniques based on statistical methods, the least square principle and the concepts of stability and truncation error are introduced in the text without loss of rigour, which are not discussed in most of the existing books.

This is a self-contained textbook, primarily designed for senior undergraduate students in engineering and technology. It also meets the needs of postgraduate students of computer science, mathematics and physics. The topics covered in this book are supported by many solved examples to facilitate clear understanding of the numerical methods. Most of the examples presented in this book are taken from B.E., B.Tech. and M.Sc. question papers of various Indian universities. Of course, it is assumed that the reader is familiar with calculus taught at the higher secondary level.

From my own experience at Anna University, Chennai, this entire book can be covered in a single semester with three credits per week. In case of two-credit course, without loss of continuity, Chapters 9–11 may be omitted, and the rest of the book can be covered in one semester.

I would like to express my gratitude to various authors whose books and research papers are referred to and have been listed in the Bibliography. Exercises along with detailed hints are provided in the book to drill the students in problem solving.

I wish to thank many of my colleagues in the Department of Mathematics of Anna University, in particular, Dr. (Mrs.) Prabhamani R. Patil, Dr. N. Venkatesan, and Dr. (late) N. Muthiyalu. I also wish to thank my friends Dr. A. Avudainayagam and Dr. Y.V.S.S.S. Raju of I.I.T. (Madras) for their invaluable suggestions for improving the manuscript. Acknowledgement is also due to Mr. K.V. Ravikumar for efficiently keying the entire manuscript into PC.

I wish to thank my wife Leela Sankar who has been the real source of inspiration during the preparation of the manuscript. I must express my special thanks to my daughter Aruna, granddaughter Sangeetha, and my son-in-law Prof. R. Parthasarathi, for their understanding and constant encouragement while I was busy preparing the manuscript.

Having received a few important suggestions, on the first edition of the book, from the teaching faculty and to take care of the revised syllabi of B.E., B.Tech., M.Sc. (Computer Science) and MCA courses of various universities and I.I.T.'s, I have included sections on 'Gaussion Quadrature Formulae', 'Maxima and Minima of a Tabulated Function' and 'Von-Neumann Stability Concept of Finite Difference Schemes for Solving Partial Differential Equations'. I am also particularly grateful to Dr. Usha (Maths. Dept.) of I.I.T. (Madras), Dr. J. Pandurangan of Anna University and Dr. (Mrs.) Rajashree of Sai–Ram Engineering College (Chennai), for their valuable suggestions.

I hope and trust that both students and teaching fraternity will enjoy reading this new edition of the book.

Finally, I wish to thank the publisher, Prentice-Hall of India, for the careful processing of the manuscript—both at the editorial and production stages.

Suggestions for improving the contents will be warmly appreciated.

K. SANKARA RAO

Chapter 1

Basics in Computing

1.1 INTRODUCTION

We begin this chapter with some of the basic concepts of representation of numbers on computers and errors introduced during computation. Problem-solving using computers and the steps involved are also discussed in brief.

Many of the available digital computing systems fall mainly under four categories: personal computers, workstations, mainframe computers and super computers, based on their speed, cost and facilities. Mainframe and super computers being costly and are used only for research and development purposes. These computers involve large-scale programmes with huge data. In the new millennium, computers with storage capacities of several hundred billion words and capable of making 15 billion calculations a second are made available at select installations over the globe.

With the availability of such powerful digital computers and vastly improved numerical methods, scientists and engineers will be able to develop models that can be used for numerous purposes: weather prediction, effect of solar storms on Earth, the performance of an aircraft as a whole, some aspects of space flight simulations and many more practical problems meant for the welfare of the mankind.

Personal computers with local area networking (LAN) and pentium processors can meet most of the demands of teaching, project evaluations, computer-aided design and almost all business applications. In fact, many computer installations provide the user with the necessary software of routine nature in the name of utility sub-programmes.

1.2 REPRESENTATION OF NUMBERS

It is known that in our daily life, we use numbers based on the decimal system. In this system, we use ten symbols 0, 1, 2, 3, 4, 5, 6, 7, 8, 9 and the number 10 is called the base of the system. Thus, when a base N is given, we need N different symbols 0, 1, 2, . . . , $(N - 1)$ to represent an arbitrary number. The number systems commonly used in computers are shown as in Table 1.1.

Table 1.1 Number Systems

Base, N	Number
2	Binary
8	Octal
10	Decimal
16	Hexadecimal

Thus, if a number system has only two symbols 0 and 1, then, its base is 2 and so on. In general, an arbitrary real number, a can be written as

$$a = a_m N^m + a_{m-1} N^{m-1} + \cdots + a_1 N^1 + a_0 + a_{-1} N^{-1} + \cdots + a_{-m} N^{-m}$$

In binary system, it has the form,

$$a = a_m 2^m + a_{m-1} 2^{m-1} + \cdots + a_1 2^1 + a_0 + a_{-1} 2^{-1} + \cdots + a_{-m} 2^{-m}$$

In hexadecimal system, we write it as

$$a = a_m 16^m + a_{m-1} 16^{m-1} + \cdots + a_1 16^1 + a_0 + a_{-1} 16^{-1} + \cdots + a^{-m} 16^{-m}$$

For example, the value of the decimal number 1729 is represented and calculated as

$$(1729)_{10} = 1 \times 10^3 + 7 \times 10^2 + 2 \times 10^1 + 9 \times 10^0$$

While the decimal equivalent of binary number 1.0011001 is

$$1 \times 2^0 + 0 \times 2^{-1} + 0 \times 2^{-2} + 1 \times 2^{-3} + 1 \times 2^{-4} + 0 \times 2^{-5} + 0 \times 2^{-6} + 1 \times 2^{-7}$$

$$= 1 + \frac{1}{8} + \frac{1}{16} + \frac{1}{128} = (1.1953125)_{10}$$

Electronic computers use binary system whose base is 2. The two symbols used in this system are 0 and 1, which are called *binary digits* or simply *bits*. The internal representation of any data within a computer is in binary form. However, we prefer data input and output of numerical results in decimal system. Within the computer, the arithmetic is carried out in binary form. Infact, there is a built-in circuit design in every computer, which converts decimal input to binary and binary result to decimal output, and carry-out binary addition, subtraction, multiplication and division. For example, the method of converting decimal to binary equivalent can be seen as follows: we divide the given decimal number by 2 and the resulting successive quotients and continue to do so till the quotient becomes zero. Then, the binary equivalent of the decimal number is obtained as a string of remainders and the process can be seen through examples.

Similarly, for converting a decimal fraction into its binary equivalent, we multiply it by 2, the resultant integer part gives the most significant bit of the binary fraction. We keep multiplying by 2 and extract the next significant digit, and the process is continued until the fractional part becomes zero.

Example 1.1 Convert the decimal number 47 into its binary equivalent.

Solution

		Remainder	
2	47	↓	
2	23	1	
2	11	1	
2	5	1	
2	2	1	
2	1	0	
	0	1	⟵ Most significant bit

Thus,

$$(47)_{10} = (101111)_2$$

Example 1.2 Find the binary equivalent of the decimal fraction 0.7625.

Solution

	Product	Integer part	
0.7625 × 2	1.5250	1	← Most significant digit
0.5250 × 2	1.0500	1	
0.05 × 2	0.1	0	
0.1 × 2	0.2	0	
0.2 × 2	0.4	0	
0.4 × 2	0.8	0	
0.8 × 2	1.6	1	
0.6 × 2	1.2	1	
0.2 × 2	0.4	0	Repeated hereafter

Therefore,

$$(0.7625)_{10} = (0.11000011(0011))_2$$

Suppose, we consider 8 bits only, that is, $(0.11000011)_2$; its decimal value is equal to $(0.7617187)_{10}$. While if we take 12 bits, its decimal equivalent is $(0.76245)_{10}$.

Example 1.3 Convert $(59)_{10}$ into binary and then into octal.

Solution

		Remainder	
2	59		
2	29	1	
2	14	1	
2	7	0	
2	3	1	
2	1	1	
	0	1	⟵ Most significant bit

Thus,

$$(59)_{10} = (111011)_2$$

Now, if we group the binary digits such that each group contains three bits each, we can easily go from binary to octal. Thus,

$$(111011)_2 = 111\ 011 = (73)_8$$

1.2.1 Floating-point Representation

In general, two types of arithmetic operations are carried out in computers: integer arithmetic and floating point arithmetic. However, most scientific and engineering calculations are essentially carried out in floating point arithmetic. For example, an n digit floating point number in base b can be represented as

$$a = \pm(.d_1 d_2, \ldots, d_n)_b b^e$$

where, $(.d_1 d_2, \ldots, d_n)_b$ is called *mantissa* and e is the *exponent*. Let us consider a binary number as

$$.10110101 \times 2^{11} = .10110101E01011$$

Here, .10110101 is called the mantissa and 1011 is the exponent. The precision of floating point numbers on any computer is determined by the number of digits used in the mantissa, which in general, varies.

Computers having 48 bits to represent single precision real number, in general, allocate 1 bit for sign, 7 bits for exponent and 40 bits for mantissa. Thus, the range is limited from 2.939E−39 to 1.701E+38. The numerical precision in this case is 11 decimal digits. Similarly, those computers having 32 bits to represent a single precision real number, in general, allocate 1 bit for sign, 7 bits for exponent and 24 bits for mantissa. In this case, the limit ranges from 2.939E−39 to 1.701E+38. The numerical precision in this case is only six decimal digits.

When 64-bit double precision arithmetic is used, the computer output may be accurate even up to 16 decimal digits.

1.3 ERRORS IN COMPUTATIONS

Numerically, computed solutions are subject to certain errors. It may be fruitful to identify the error sources and their growth while classifying the errors in numerical computation. There are essentially three error sources: inherent errors, local round-off errors and local truncation errors.

1.3.1 Inherent Errors

It is that quantity of error which is present in the statement of the problem itself, before finding its solution. It arises due to the simplified assumptions made in the mathematical modelling of a problem. It can also arise when the data is obtained from certain physical measurements of the parameters of the problem.

1.3.2 Local Round-off Errors

Every computer has a finite word length, and therefore, it is possible to store only a fixed number of digits of a given input number. Since computers store information in binary form, storing an exact decimal number in its binary form into the computer memory gives an error. This error is computer-dependent. Also, at the end of computation of a particular problem, the final results in the computer, which is obviously in binary form, should be converted into decimal form — a form understandable to the user — before their print out. Therefore, an additional error is committed at this stage too. This error is called *local round-off error.*

For example, in Section 1.2, we have noted that

$$(0.7625)_{10} = (0.11000011\ (0011))_2$$

If a particular computer system has a word length of 12 bits only, then the decimal number 0.7625 is stored in the computer memory in binary form as 0.110000110011. However, it is equivalent to 0.76245. Thus, in storing the number 0.7625, we have committed an error equal to 0.00005, which is the round-off error; inherent with the computer system considered. Thus, we define the *error* as

$$\text{Error} = \text{True value} - \text{Computed value}$$

Now, in order to determine the accuracy of an approximate solution, errors are measured in different ways. *Absolute error*, denoted by |Error|, while, the *relative error* is defined as

$$\text{Relative error} = \frac{|\text{Error}|}{|\text{True value}|}$$

For example, consider the value of $\sqrt{2} = (1.414213\ .\ .\ .)$ up to four decimal places, then

$$\sqrt{2} = 1.4142 + \text{Error}$$

Hence, we get

$$\text{Absolute error} = |\text{Error}| = 0.00001$$

$$\text{Relative error} = \frac{0.00001}{1.4142}$$

We are aware of the fact that $\sqrt{2}$ is irrational. However, widely used value up to four decimal digits is taken as the true value for the computation of relative error. These error measures are generally used in numerical analysis for measuring the accuracy of the results.

When a number N is written in floating point form with t digits, say, in base 10 as

$$N = (.d_1 d_2\ .\ .\ .\ d_t)10^e$$

we say that the number N has t significant digits. Here, d_1 is called the *most significant digit*. For example, 0.3 agrees with 1/3 to one significant digit, while 0.3333 agrees with 1/3 to four significant digits.

1.3.3 Local Truncation Error

It is generally easier to expand a function into a power series using Taylor series expansion and evaluate it by retaining the first few terms. For example, we may approximate the function $f(x) = \cos x$ by the series

$$\cos x = 1 - \frac{x^2}{2!} + \frac{x^4}{4!} - \cdots + (-1)^n \frac{x^{2n}}{(2n)!} + \cdots \tag{1.1}$$

In fact, it is an infinite series expansion. If we use only the first three terms to compute $\cos x$ for a given x, we get an approximate answer. Here, the error is due to truncating the series. Suppose, we retain the first n terms, the *truncation error* (TE) is given by

$$\text{TE} \leq \frac{x^{2n+2}}{(2n+2)!} \tag{1.2}$$

It may be noted that the TE is independent of the computer used.

If we wish to compute $\cos x$ for $|x| < \pi/2$ accurate with five significant digits, the question is, how many terms in the expansion (1.1) are to be included? In this situation

$$\frac{x^{2n+2}}{(2n+2)!} < .5 \times 10^{-5} = 5 \times 10^{-6}$$

Taking logarithm on both sides, we get

$$(2n + 2) \log x - \log [(2n + 2)!] < \log_{10} 5 - 6 \log_{10} 10 = 0.699 - 6 = -5.3$$

or

$$\log [(2n + 2)!] - (2n + 2) \log x > 5.3$$

We can observe that, for x in the interval $[-\pi/2, \pi/2]$, the above inequality is satisfied for $n = 7$. Hence, seven terms in the expansion (1.1) are required to get the value of $\cos x$, with the prescribed accuracy, in the interval $[-\pi/2, \pi/2]$. In this example, the truncation error is given by

$$\text{TE} \leq \frac{x^{16}}{16!}$$

1.4 PROBLEM-SOLVING USING COMPUTERS

In order to solve a given problem using a computer, the major steps involved are:

(i) Choosing an appropriate numerical method,

(ii) Designing an algorithm,

(iii) Programming and debugging, and

(iv) Computer execution.

These steps are briefly explained as follows.

We define the numerical method as a mathematical formula for finding the solution to a given problem. There may be many methods available to solve the same problem. For example, in Chapter 2, we shall present various computer-based numerical methods, such as, bisection method, regula-falsi method, method of iteration, Newton–Raphson method, Muller's method, Graeffe's root squaring method, etc., for solving an algebraic or transcendental equation. One should choose an appropriate method, which suits best in the given situation, as a first step.

Once we chose a particular method for solving a problem, we should write down the sequence of steps to be followed in order, precisely and unambiguously, to obtain the solution. This is called *designing an algorithm.*

Now, the flow chart for the algorithm is drawn and then translated into a programming language, which we call *computer programme.* This programme should be debugged and free from coding errors. The choice of the languages is for the user to decide. It can be FORTRAN, BASIC, COBOL, Pascal, C, etc.

As a last step, the computer programme is fed to a personal computer or to a mainframe computer, with the necessary data through an input unit. Then, the central processing unit (CPU) of the computing system interprets the programme steps and executes them if the programme is free from coding errors. When it encounters output statement, the numerical answers to the problem are sent to the output unit chosen by the user; it may be a printer. This completes the problem-solving task using a computer.

Chapter 2

Solution of Algebraic and Transcendental Equations

2.1 INTRODUCTION

One of the basic problems in science and engineering is the computation of roots of an equation in the form, $f(x) = 0$. The equation $f(x) = 0$ is called an *algebraic equation*, if it is purely a polynomial in x; it is called a *transcendental equation* if $f(x)$ contains trigonometric, exponential or logarithmic functions. For example,

$$x^3 + 5x^2 - 6x + 3 = 0$$

is an algebraic equation, whereas

$$M = E - e \sin E \qquad \text{and} \qquad ax^2 + \log (x - 3) + e^x \sin x = 0$$

are transcendental equations.

To find the solution of an equation $f(x) = 0$, we find those values of x for which $f(x) = 0$ is satisfied. Such values of x are called the *roots* of $f(x) = 0$. Thus a is a root of an equation $f(x) = 0$, if and only if, $f(a) = 0$.

Before, we develop various numerical methods, we shall list below some of the basic properties of an algebraic equation:

(i) Every algebraic equation of nth degree, where n is a positive integer, has n and only n roots.

(ii) Complex roots occur in pairs. That is, if $(a + ib)$ is a root of $f(x) = 0$, then $(a - ib)$ is also a root of this equation.

(iii) If $x = a$ is a root of $f(x) = 0$, a polynomial of degree n, then $(x - a)$ is a factor of $f(x)$. On dividing $f(x)$ by $(x - a)$ we obtain a polynomial of degree $(n - 1)$.

(iv) Descartes rule of signs: The number of positive roots of an algebraic equation $f(x) = 0$ with real coefficients cannot exceed the number of changes in sign of the coefficients in the polynomial $f(x) = 0$. Similarly, the number of negative roots of $f(x) = 0$ cannot exceed the number of changes in the sign of the coefficients of $f(-x) = 0$. For example, consider an equation

$$x^3 - 3x^2 + 4x - 5 = 0$$

As there are three changes in sign, also, the degree of the equation is three, and hence the given equation will have all the three positive roots.

(v) Intermediate value property: If $f(x)$ is a real valued continuous function in the closed interval $a \le x \le b$. If $f(a)$ and $f(b)$ have opposite signs, then the graph of the function $y = f(x)$ crosses the x-axis at least once; that is $f(x) = 0$ has at least one root ξ such that $a < \xi < b$.

Broadly speaking, all the known numerical methods for solving either a transcendental equation or an algebraic equation can be classified into two groups: *direct methods* and *iterative methods*. Direct methods require no knowledge of the initial approximation of a root of the equation $f(x) = 0$, while iterative methods do require first approximation to initiate iteration. How to get the first approximation? We can find the approximate value of the root of $f(x) = 0$ either by a *graphical method* or by an *analytical method* as explained below:

Graphical method

Often, the equation $f(x) = 0$ can be rewritten as $f_1(x) = f_2(x)$ and the first approximation to a root of $f(x) = 0$ can be taken as the abscissa of the point of intersection of the graphs of $y = f_1(x)$ and $y = f_2(x)$. For example, consider,

$$f(x) = x - \sin x - 1 = 0$$

It can be written as $x - 1 = \sin x$. Now, we shall draw the graphs of

$$y = x - 1 \qquad \text{and} \qquad y = \sin x$$

as shown in Fig. 2.1. The approximate value of the root is found to be 1.9.

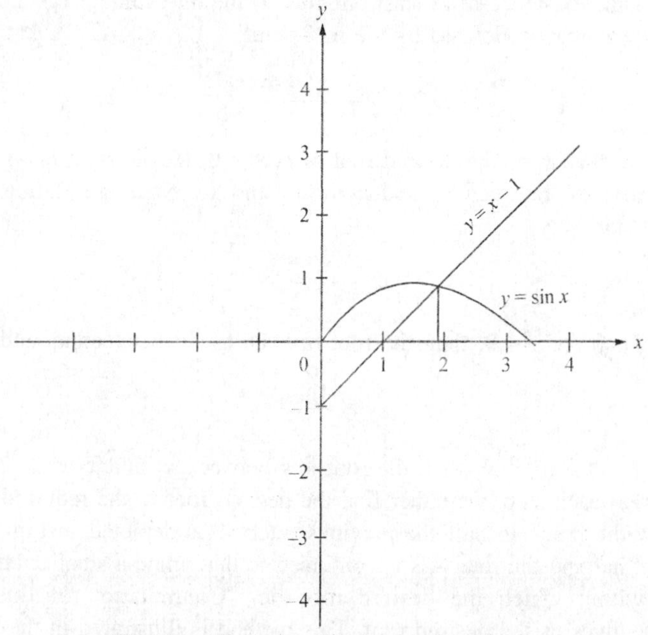

Fig. 2.1 Illustration by graphical method.

Analytical method

This method is based on 'intermediate value property'. We shall illustrate it through an example. Let,

$$f(x) = 3x - \sqrt{1 + \sin x} = 0$$

We can easily verify

$$f(0) = -1$$

$$f(1) = 3 - \sqrt{1 + \sin\left(1 \times \frac{180}{\pi}\right)} = 3 - \sqrt{1 + 0.84147} = 1.64299$$

We observe that $f(0)$ and $f(1)$ are of opposite signs. Therefore, using intermediate value property we infer that there is at least one root between $x = 0$ and $x = 1$. This method is often used to find the first approximation to a root of either transcendental equation or algebraic equation. Hence, in analytical method, we must always start with an initial interval (a, b), so that $f(a)$ and $f(b)$ have opposite signs.

2.2 BISECTION METHOD

This method is due to Bolzano. Suppose, we wish to locate the root cf an equation $f(x) = 0$ in an interval, say (x_0, x_1). Let $f(x_0)$ and $f(x_1)$ are of opposite signs, such that $f(x_0)\, f(x_1) < 0$.

Then the graph of the function crosses the x-axis between x_0 and x_1, which guarantees the existence of at least one root in the interval (x_0, x_1). The desired root is approximately defined by the mid-point

$$x_2 = \frac{x_0 + x_1}{2}$$

If $f(x_2) = 0$, then x_2 is the desired root of $f(x) = 0$. However, if $f(x_2) \neq 0$, then the root may be between x_0 and x_2 or x_2 and x_1. Now, we define the next approximation by

$$x_3 = \frac{x_0 + x_2}{2}$$

provided $f(x_0)\, f(x_2) < 0$, then the root may be found between x_0 and x_2 or by

$$x_3 = \frac{x_1 + x_2}{2}$$

provided $f(x_1)\, f(x_2) < 0$, then the root lies between x_1 and x_2 etc.

Thus, at each step, we either find the desired root to the required accuracy or narrow the range to half the previous interval as depicted in Fig. 2.2. This process of halving the intervals is continued to determine a smaller and smaller interval within which the desired root lies. Continuation of this process eventually gives us the desired root. This method is illustrated in the following example.

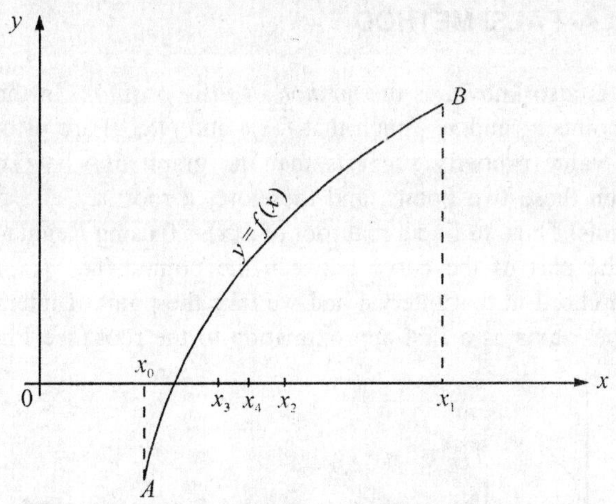

Fig. 2.2 Geometrical illustration of bisection method.

Example 2.1 Solve $x^3 - 9x + 1 = 0$ for the root between $x = 2$ and $x = 4$ by the bisection method.

Solution Given $f(x) = x^3 - 9x + 1$. We can verify $f(2) = -9$, $f(4) = 29$. Therefore, $f(2)\,f(4) < 0$ and hence the root lies between 2 and 4. Let $x_0 = 2$, $x_1 = 4$. Now, we define

$$x_2 = \frac{x_0 + x_1}{2} = \frac{2 + 4}{2} = 3$$

as a first approximation to a root of $f(x) = 0$ and note that $f(3) = 1$, so that $f(2)\,f(3) < 0$. Thus, the root lies between 2 and 3. We further define,

$$x_3 = \frac{x_0 + x_2}{2} = \frac{2 + 3}{2} = 2.5$$

and note that $f(x_3) = f(2.5) < 0$, so that $f(2.5)\,f(3) < 0$. Therefore, we define the mid-point,

$$x_4 = \frac{x_3 + x_2}{2} = \frac{2.5 + 3}{2} = 2.75, \text{ etc.}$$

Similarly, we find that

$$x_5 = 2.875 \qquad \text{and} \qquad x_6 = 2.9375$$

and the process can be continued until the root is obtained to the desired accuracy. These results are presented in the table.

n	x_n	$f(x_n)$
2	3	1.0
3	2.5	–5.875
4	2.75	–2.9531
5	2.875	–1.1113
6	2.9375	–0.0901

2.3 REGULA–FALSI METHOD

This method is also known as the *method of false position.* In this method, we choose two points x_n and x_{n-1} such that $f(x_n)$ and $f(x_{n-1})$ are of opposite signs. Intermediate value property suggests that the graph of $y = f(x)$ crosses the x-axis between these two points, and therefore, a root say $x = \xi$ lies between these two points. Thus, to find a real root of $f(x) = 0$ using Regula-Falsi method, we replace the part of the curve between the points $A[x_n, f(x_n)]$ and $B[x_{n-1}, f(x_{n-1})]$ by a chord in that interval and we take the point of intersection of this chord with the x-axis as a first approximation to the root (see Fig. 2.3).

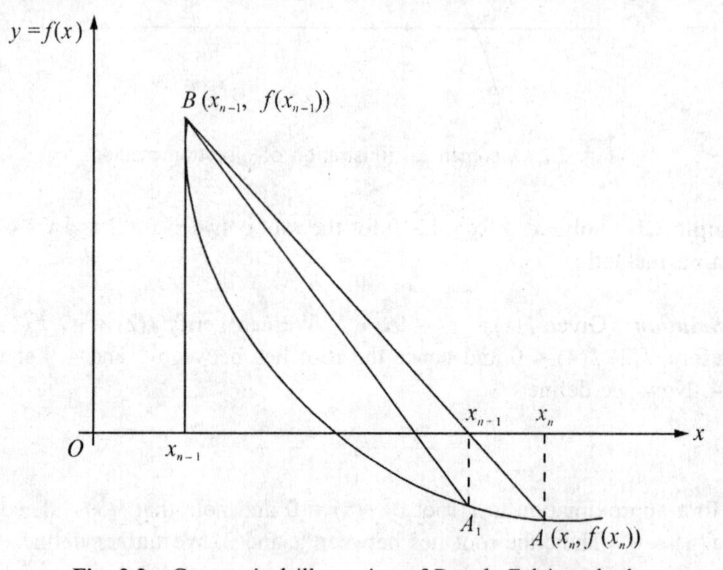

Fig. 2.3 Geometrical illustration of Regula-Falsi method.

Now, the equation of the chord joining the points A and B is

$$\frac{y - f(x_n)}{f(x_{n-1}) - f(x_n)} = \frac{x - x_n}{x_{n-1} - x_n} \tag{2.1}$$

Setting $y = 0$ in Eq. (2.1), we get

$$x = x_n - \frac{x_{n-1} - x_n}{f(x_{n-1}) - f(x_n)} f(x_n)$$

Hence, the first approximation to the root of $f(x) = 0$ is given by

$$x_{n+1} = x_n - \frac{x_n - x_{n-1}}{f(x_n) - f(x_{n-1})} f(x_n) \tag{2.2}$$

From Fig. 2.3, we observe that $f(x_{n-1})$ and $f(x_{n+1})$ are of opposite sign. Thus, it is possible to apply the above procedure, to determine the line through B and A_1 and so on. Hence, the successive approximations to the root of $f(x) = 0$ is given by Eq. (2.2). This method can best be understood through the following examples.

Example 2.2 Use the Regula-Falsi method to compute a real root of the equation $x^3 - 9x + 1 = 0$,

(i) if the root lies between 2 and 4
(ii) if the root lies between 2 and 3.

Comment on the results.

Solution Let $f(x) = x^3 - 9x + 1$.

(i) $f(2) = -9$ and $f(4) = 29$. Since $f(2)$ and $f(4)$ are of opposite signs, the root of $f(x) = 0$ lies between 2 and 4. Taking $x_1 = 2$, $x_2 = 4$ and using Regula-Falsi method, the first approximation is given by

$$x_3 = x_2 - \frac{x_2 - x_1}{f(x_2) - f(x_1)} f(x_2) = 4 - \frac{2 \times 29}{38} = 2.47368$$

and $f(x_3) = -6.12644$. Since $f(x_2)$ and $f(x_3)$ are of opposite signs, the root lies between x_2 and x_3. The second approximation to the root is given as

$$x_4 = x_3 - \frac{x_3 - x_2}{f(x_3) - f(x_2)} f(x_3) = 2.73989$$

and $f(x_4) = -3.090707$. Now, since $f(x_2)$ and $f(x_4)$ are of opposite signs, the third approximation is obtained from

$$x_5 = x_4 - \frac{x_4 - x_2}{f(x_4) - f(x_2)} f(x_4) = 2.86125$$

and $f(x_5) = -1.32686$. This procedure can be continued till we get the desired result. The first three iterations are shown as in the table.

n	x_{n+1}	$f(x_{n+1})$
2	2.47368	–6.12644
3	2.73989	–3.090707
4	2.86125	–1.32686

(ii) $f(2) = -9$ and $f(3) = 1$. Since $f(2)$ and $f(3)$ are of opposite signs, the root of $f(x) = 0$ lies between 2 and 3. Taking $x_1 = 2$, $x_2 = 3$ and using Regula-Falsi method, the first approximation is given by

$$x_3 = x_2 - \frac{x_2 - x_1}{f(x_2) - f(x_1)} f(x_2) = 3 - \frac{1}{10} = 2.9$$

and $f(x_3) = -0.711$. Since $f(x_2)$ and $f(x_3)$ are of opposite signs, the root lies between x_2 and x_3. The second approximation to the root is given as

$$x_4 = x_3 - \frac{x_3 - x_2}{f(x_3) - f(x_2)} f(x_3) = 2.94156$$

and $f(x_4) = -0.0207$. Now, we observe that $f(x_2)$ and $f(x_4)$ are of opposite signs, the third approximation is obtained from

$$x_5 = x_4 - \frac{x_4 - x_2}{f(x_4) - f(x_2)} f(x_4) = 2.94275$$

and $f(x_5) = -0.0011896$. This procedure can be continued till we get the desired result. The first three iterations are shown as in the table.

n	x_{n+1}	$f(x_{n+1})$
2	2.9	−0.711
3	2.94156	−0.0207
4	2.94275	−0.0011896

From the above computations, we observe that the value of the root as a third approximation is evidently different in both the cases, while the value of x_5, when the interval considered is (2, 3), is closer to the root. Hence, an important observation in this method is that the interval (x_1, x_2) chosen initially in which the root of the equation lies must be sufficiently small.

Example 2.3 Use Regula-Falsi method to find a real root of the equation

$$\log x - \cos x = 0$$

accurate to four decimal places after three successive approximations.

Solution Given $f(x) = \log x - \cos x$. We observe that

$$f(1) = 0 - 0.5403 = -0.5403$$

and

$$f(2) = 0.69315 + 0.41615 = 1.1093$$

Since $f(1)$ and $f(2)$ are of opposite signs, the root lies between $x_1 = 1$, $x_2 = 2$. The first approximation is obtained from

$$x_3 = x_2 - \frac{x_2 - x_1}{f(x_2) - f(x_1)} f(x_2) = 2 - \frac{1.1093}{1.6496} = 1.3275$$

and

$$f(x_3) = 0.2833 - 0.2409 = 0.0424$$

Now, since $f(x_1)$ and $f(x_3)$ are of opposite signs, the second approximation is obtained as

$$x_4 = 1.3275 - \frac{(.3275)(.0424)}{0.0424 + 0.5403} = 1.3037$$

and

$$f(x_4) = 1.24816 \times 10^{-3}$$

Similarly, we observe that $f(x_1)$ and $f(x_4)$ are of opposite signs, so, the third approximation is given by

$$x_5 = 1.3037 - \frac{(0.3037)(0.001248)}{0.001248 + 0.5403} = 1.3030$$

and

$$f(x_5) = 0.62045 \times 10^{-4}$$

Hence, the required real root is 1.3030.

Example 2.4 Using Regula-Falsi method, find the real root of the following equation correct to three decimal places:

$$x \log_{10} x = 1.2$$

Solution Let $f(x) = x \log_{10} x - 1.2$. We observe that $f(2) = -0.5979$, $f(3) = 0.2314$. Since $f(2)$ and $f(3)$ are of opposite signs, the real root lies between $x_1 = 2$, $x_2 = 3$. The first approximation is obtained from

$$x_3 = x_2 - \frac{x_2 - x_1}{f(x_2) - f(x_1)} f(x_2) = 3 - \frac{0.2314}{0.8293} = 2.72097$$

and $f(x_3) = -0.01713$. Since $f(x_2)$ and $f(x_3)$ are of opposite signs, the root of $f(x) = 0$ lies between x_2 and x_3. Now, the second approximation is given by

$$x_4 = x_3 - \frac{x_3 - x_2}{f(x_3) - f(x_2)} f(x_3) = 2.7402$$

and $f(x_4) = -3.8905 \times 10^{-4}$. Thus, the root of the given equation correct to three decimal places is 2.740.

2.4 METHOD OF ITERATION

The method of iteration can be applied to find a real root of the equation $f(x) = 0$ by rewriting the same in the form,

$$x = \phi(x) \tag{2.3}$$

For example, $f(x) = \cos x - 2x + 3 = 0$. It can be rewritten as

$$x = \frac{1}{2}(\cos x + 3) = \phi(x)$$

Let $x = \xi$ is the desired root of Eq. (2.3). Suppose x_0 is its initial approximation. The first and successive approximations to the root can be obtained as

$$\left.\begin{aligned} x_1 &= \phi(x_0) \\ x_2 &= \phi(x_1) \\ &\vdots \\ x_{n+1} &= \phi(x_n) \end{aligned}\right\} \tag{2.4}$$

Definition 2.1 Let $\{x_i\}$ be the sequence obtained by a given method and let $x = \xi$ denotes the root of the equation $f(x) = 0$. Then, the method is said to be *convergent*, if and only if

$$\underset{n\to\infty}{\text{Lt}} |x_n - \xi| = 0$$

The convergence of the above sequence to the root is stated as in Theorem 2.1.

Theorem 2.1 Support $x = \xi$ be a root of the equation $f(x) = 0$, which can be rewritten as $x = \phi(x)$, contained in an interval I. Also, let $\phi(x)$ and $\phi'(x)$ be continuous in I. Then, if $|\phi'(x)| < 1$ for all x in I, the iterative process defined by $x_{n+1} = \phi(x_n)$ converges to the root $x = \xi$, if and only if, the initially chosen approximation $x_0 \in I$.

This method is illustrated through the following examples.

Example 2.5 Use the method of iteration to determine the real root of the equation $e^{-x} = 10x$ correct to four decimal places.

Solution Let $f(x) = e^{-x} - 10x = 0$, we observe that $f(0) = 1$ and $f(1) = -9.6321$. Since $f(0) < f(1)$ numerically, the root is near to $x = 0$. Now, we shall rewrite the given equation in the form

$$x = \frac{1}{10} e^{-x} = \phi(x)$$

Therefore,

$$\phi'(x) = -\frac{1}{10} e^{-x}$$

and

$$|\phi'(x)| = \frac{1}{10} e^{-x} = \frac{1}{10\, e^{x}} < 1$$

for all x in (0, 1). Hence, the method of iteration can be applied. Thus, we start with the initial value $x_0 = 0$, then

$$x_1 = \phi(x_0) = \frac{1}{10} = 0.1, \qquad f(x_1) = -0.09516$$

Similarly, the successive approximations are

$$x_2 = \phi(x_1) = \frac{1}{10} e^{-0.1} = \frac{0.904837}{10} = 0.09048, \quad f(x_2) = 0.00869$$

$$x_3 = \phi(x_2) = 0.091349, \qquad f(x_3) = -7.90877 \times 10^{-4}$$

$$x_4 = \phi(x_3) = 0.091274, \qquad f(x_4) = 2.75784 \times 10^{-5}.$$

Hence, the required root is 0.0913.

Example 2.6 Find a real root of the equation

$$f(x) = x^3 + x^2 - 1 = 0$$

by the method of iteration.

Solution We observe that $f(0) = -1$, $f(1) = 1$ which shows that there is a real root between $x = 0$ and $x = 1$. To find the real root, we rewrite the equation in the form

$$x^2 (x + 1) = 1 \qquad \text{or} \qquad x = \frac{1}{\sqrt{x+1}} = \phi(x)$$

Therefore,

$$\phi'(x) = -\frac{1}{2(x+1)^{3/2}}$$

We note that $|\phi'(x)| < 1$, for all x in (0, 1). Hence, the method of iteration is applicable here.

Taking the initial value $x_0 = 1$, we successively obtain the following values:

$$\begin{aligned}
x_1 &= \phi(x_0) = 1/\sqrt{2} = 0.70711, & f(x_1) &= -0.14644\\
x_2 &= \phi(x_1) = 0.76537, & f(x_2) &= 0.03414\\
x_3 &= \phi(x_2) = 0.75263, & f(x_3) &= 7.2213 \times 10^{-3}\\
x_4 &= \phi(x_3) = 0.75536, & f(x_4) &= 1.55658 \times 10^{-3}\\
x_5 &= \phi(x_4) = 0.75477, & f(x_5) &= -3.44323 \times 10^{-4}\\
x_6 &= \phi(x_5) = 0.7549, & f(x_6) &= 7.38295 \times 10^{-5}
\end{aligned}$$

Hence, the required root is 0.7549.

Note: The given equation can be rewritten in many ways. Suppose, we rewrite

$$x^2 = 1 - x^3 \quad \text{or} \quad x = (1 - x^3)^{1/2} = \phi(x)$$

Then

$$|\phi'(x)| = \frac{3x^2}{2(1 - x^3)^{1/2}}$$

if we take $x = 1$, in the interval (0, 1), $|\phi'(x)| = \infty$, then the condition $|\phi'(x)| < 1$ is violated.

2.5 NEWTON–RAPHSON METHOD

This is a very powerful method for finding the real root of an equation in the form, $f(x) = 0$. Suppose, x_0 is an approximate root of $f(x) = 0$. Let $x_1 = x_0 + h$, where h is small, be the exact root of $f(x) = 0$, then $f(x_1) = 0$. Now, expanding $f(x_0 + h)$ by Taylor's theorem, we get

$$f(x_0 + h) = f(x_0) + h\,f'(x_0) + \frac{h^2}{2} f''(x_0) + \cdots = 0 \tag{2.5}$$

Since h is small, we neglect terms containing h^2 and its higher powers, then

$$f(x_0) + h\,f'(x_0) = 0 \quad \text{or} \quad h = \frac{-f(x_0)}{f'(x_0)}$$

Therefore, a better approximation to the root is given by

$$x_1 = x_0 - \frac{f(x_0)}{f'(x_0)}$$

Still better and successive approximations $x_2, x_3, \ldots, x_n$ to the root can obviously be obtained from the iteration formula,

$$x_{n+1} = x_n - \frac{f(x_n)}{f'(x_n)} \tag{2.6}$$

This is known as Newton–Raphson iteration formula, which has the following geometrical interpretation:

Suppose, the graph of the function $y = f(x)$ crosses the x-axis at α (*see* Fig. 2.4), then $x = \alpha$ is the root of the equation $f(x) = 0$.

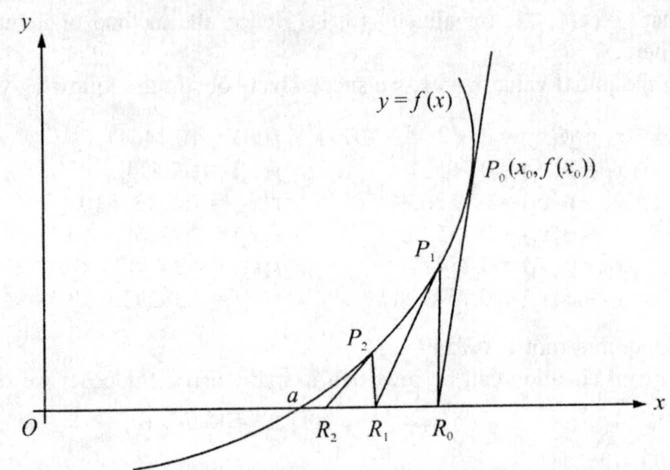

Fig. 2.4 Geometrical interpretation of Newton–Raphson method.

Let x_0 be a point closer to the root α, then the equation of the tangent at $P_0(x_0, f(x_0))$ is

$$y - f(x_0) = f'(x_0)\,(x - x_0) \tag{2.7}$$

This tangent cuts the x-axis at R_0 $(x_1, 0)$. Therefore,

$$x_1 = x_0 - \frac{f(x_0)}{f'(x_0)} \tag{2.8}$$

which is a first approximation to the root α. If P_1 is a point on the curve corresponding to x_1, then the tangent at P_1 cuts the x-axis at $R_1(x_2, 0)$, which is still closer to α, than x_1. Therefore, x_2 is a second approximation to the root. Continuing this process, we arrive at the root α, very rapidly, which is evident from Fig. 2.4. Thus, in this method, we have replaced the part of the curve between the point P_0 and x-axis by a tangent to the curve at P_0 and so on. In order to illustrate this method, we shall consider the following examples.

Example 2.7 Find the real root of the equation $xe^x - 2 = 0$ correct to two decimal places, using Newton–Raphson method.

Solution Given $f(x) = xe^x - 2$, we have

$$f'(x) = xe^x + e^x \text{ and } f''(x) = xe^x + 2e^x$$

clearly, we have

$$f(0) = -2 \quad \text{and} \quad f(1) = e - 2 = 0.71828$$

Hence, the required root lies in the interval (0, 1) and is nearer to 1.

Also, $f'(x)$ and $f''(x)$ do not vanish in (0, 1) and $f(x)$ and $f''(x)$ will have the same sign at $x = 1$. Therefore, we take the first approximation $x_0 = 1$, and using Newton–Raphson method, we get

$$x_1 = x_0 - \frac{f(x_0)}{f'(x_0)} = \frac{e+2}{2e} = 0.867879$$

and

$$f(x_1) = 6.71607 \times 10^{-2}$$

The second approximation is

$$x_2 = x_1 - \frac{f(x_1)}{f'(x_1)} = 0.867879 - \frac{0.06716}{4.44902} = 0.85278$$

and

$$f(x_2) = 7.655 \times 10^{-4}$$

Thus, the required root is 0.853.

Example 2.8 Find a real root of the equation $x^3 - x - 1 = 0$ using Newton–Raphson method, correct to four decimal places.

Solution Let $f(x) = x^3 - x - 1$, then we observe that $f(1) = -1$, $f(2) = 5$. Therefore, the root lies in the interval (1, 2). We also observe

$$f'(x) = 3x^2 - 1, \qquad f''(x) = 6x$$

and

$$f(1) = -1, \qquad f''(1) = 6, \qquad f(2) = 5, \qquad f''(2) = 12$$

Since $f(2)$ and $f''(2)$ are of the same sign, we choose $x_0 = 2$ as the first approximation to the root. The second approximation is computed using Newton–Raphson method as

$$x_1 = x_0 - \frac{f(x_0)}{f'(x_0)} = 2 - \frac{5}{11} = 1.54545 \quad \text{and} \quad f(x_1) = 1.14573$$

The successive approximations are

$$x_2 = 1.54545 - \frac{1.14573}{6.16525} = 1.35961, \qquad f(x_2) = 0.15369$$

$$x_3 = 1.35961 - \frac{0.15369}{4.54562} = 1.32579, \qquad f(x_3) = 4.60959 \times 10^{-3}$$

$$x_4 = 1.32579 - \frac{4.60959 \times 10^{-3}}{4.27316} = 1.32471, \qquad f(x_4) = -3.39345 \times 10^{-5}$$

$$x_5 = 1.32471 + \frac{3.39345 \times 10^{-5}}{4.26457} = 1.324718, \qquad f(x_5) = 1.823 \times 10^{-7}$$

Hence, the required root is 1.3247.

Convergence of Newton–Raphson method

To examine the convergence of Newton–Raphson formula (2.6), that is,

$$x_{n+1} = x_n - \frac{f(x_n)}{f'(x_n)}$$

We compare it with the general iteration formula $x_{n+1} = \phi(x_n)$, and thus obtain

$$\phi(x_n) = x_n - \frac{f(x_n)}{f'(x_n)}$$

In general, we write it as

$$\phi(x) = x - \frac{f(x)}{f'(x)}$$

We have already noted in Theorem 2.1 that the iteration method converges if $|\phi'(x)| < 1$. Therefore, Newton–Raphson formula (2.6) converges, provided

$$|f(x)f''(x)| < |f'(x)|^2 \tag{2.9}$$

in the interval considered. Newton–Raphson formula therefore converges, provided the initial approximation x_0 is chosen sufficiently close to the root and $f(x)$, $f'(x)$ and $f''(x)$ are continuous and bounded in any small interval containing the root.

Definition 2.2 Let

$$x_n = \alpha + \varepsilon_n, \qquad x_{n+1} = \alpha + \varepsilon_{n+1}$$

where α is a root of $f(x) = 0$. If we can prove that $\varepsilon_{n+1} = K\varepsilon_n^p$, where K is a constant and ε_n is the error involved at the n th step, while finding the root by an iterative method, then the rate of convergence of the method is p.

We can now establish that Newton–Raphson method converges quadratically.

Let

$$x_n = \alpha + \varepsilon_n, \qquad x_{n+1} = \alpha + \varepsilon_{n+1}$$

where α is a root of $f(x) = 0$ and ε_n is the error involved at the nth step, while finding the root by Newton–Raphson formula (2.6). Then, Eq. (2.6) gives,

$$\alpha + \varepsilon_{n+1} = \alpha + \varepsilon_n - \frac{f(\alpha + \varepsilon_n)}{f'(\alpha + \varepsilon_n)}$$

i.e.

$$\varepsilon_{n+1} = \varepsilon_n - \frac{f(\alpha + \varepsilon_n)}{f'(\alpha + \varepsilon_n)} = \frac{\varepsilon_n f'(\alpha + \varepsilon_n) - f(\alpha + \varepsilon_n)}{f'(\alpha + \varepsilon_n)}$$

Using Taylor's expansion, we get

$$\varepsilon_{n+1} = \frac{1}{f'(\alpha) + \varepsilon_n f''(\alpha) + \cdots} \left\{ \varepsilon_n \left[f'(\alpha) + \varepsilon_n f''(\alpha) + \cdots \right] - \left[f(\alpha) + \varepsilon_n f'(\alpha) + \frac{\varepsilon_n^2}{2} f''(\alpha) + \cdots \right] \right\}$$

Since α is a root, $f(\alpha) = 0$. Therefore, the above expression simplifies to

$$\varepsilon_{n+1} \simeq \frac{\varepsilon_n^2}{2} f''(\alpha) \frac{1}{f'(\alpha) + \varepsilon_n f''(\alpha)}$$

$$= \frac{\varepsilon_n^2}{2} \frac{f''(\alpha)}{f'(\alpha)} \left[1 + \varepsilon_n \frac{f''(\alpha)}{f'(\alpha)}\right]^{-1}$$

$$\simeq \frac{\varepsilon_n^2}{2} \frac{f''(\alpha)}{f'(\alpha)} \left[1 - \varepsilon_n \frac{f''(\alpha)}{f'(\alpha)}\right]$$

or

$$\varepsilon_{n+1} = \frac{\varepsilon_n^2}{2} \frac{f''(\alpha)}{f'(\alpha)} + O\left(\varepsilon_n^3\right)$$

On neglecting terms of order ε_n^3 and higher powers, we obtain

$$\varepsilon_{n+1} = K\varepsilon_n^2 \tag{2.10}$$

where

$$K = \frac{f''(\alpha)}{2f'(\alpha)} \tag{2.11}$$

It shows that Newton–Raphson method has second order convergence or converges quadratically.

Example 2.9 Set up Newton's scheme of iteration for finding the square root of a positive number N.

Solution The square root of N can be carried out as a root of the equation $x^2 - N = 0$. Let $f(x) = x^2 - N$. By Newton's method, we have

$$x_{n+1} = x_n - \frac{f(x_n)}{f'(x_n)}$$

In this problem, $f(x) = x^2 - N$, $f'(x) = 2x$. Therefore,

$$x_{n+1} = x_n - \frac{x_n^2 - N}{2x_n} = \frac{1}{2}\left(x_n + \frac{N}{x_n}\right) \tag{2.12}$$

Example 2.10 Evaluate $\sqrt{12}$, by Newton's formula.

Solution Since $\sqrt{9} = 3$, $\sqrt{16} = 4$, we take $x_0 = (3 + 4)/2 = 3.5$. Using Eq. (2.12), we have

$$x_1 = \frac{1}{2}\left(x_0 + \frac{N}{x_0}\right) = \frac{1}{2}\left(3.5 + \frac{12}{3.5}\right) = 3.4643$$

$$x_2 = \frac{1}{2}\left(3.4643 + \frac{12}{3.4643}\right) = 3.4641$$

$$x_3 = \frac{1}{2}\left(3.4641 + \frac{12}{3.4641}\right) = 3.4641$$

Hence, $\sqrt{12} = 3.4641$.

Example 2.11 Obtain the Newton–Raphson extended formula

$$x_1 = x_0 - \frac{f(x_0)}{f'(x_0)} - \frac{1}{2}\frac{[f(x_0)]^2}{[f'(x_0)]^3} f''(x_0)$$

for finding the root of the equation $f(x) = 0$.

Solution Expanding $f(x)$ by Taylor's series, in the neighbourhood of x_0, we obtain after retaining the first order term only

$$0 = f(x) = f(x_0) + (x - x_0) f'(x_0) + \cdots$$

Which gives

$$x = x_0 - \frac{f(x_0)}{f'(x_0)}$$

This is the first approximation to the root. Therefore,

$$x_1 = x_0 - \frac{f(x_0)}{f'(x_0)} \tag{2.13}$$

Again, expanding $f(x)$ by Taylor's series and retaining up to second order term, we have

$$0 = f(x) = f(x_0) + (x - x_0) f'(x_0) + \frac{(x - x_0)^2}{2} f''(x_0)$$

Therefore,

$$f(x_1) = f(x_0) + (x_1 - x_0) f'(x_0) + \frac{(x_1 - x_0)^2}{2} f''(x_0) = 0$$

Using Eq. (2.13), the above equation reduces to the form

$$f(x_0) + (x_1 - x_0) f'(x_0) + \frac{1}{2}\frac{[f(x_0)]^2}{[f'(x_0)]^2} f''(x_0) = 0$$

Thus, the Newton–Raphson extended formula is given by

$$x_1 = x_0 - \frac{f(x_0)}{f'(x_0)} - \frac{1}{2}\frac{[f(x_0)]^2}{[f'(x_0)]^3} f''(x_0) \tag{2.14}$$

This is also known as Chebyshev's formula of third order.

2.6 MULLER'S METHOD

In Muller's method, $f(x) = 0$ is approximated by a second degree polynomial; that is by a quadratic equation that fits through three points in the vicinity of a root.

The roots of this quadratic equation are then assumed to be approximated to the roots of the equation $f(x) = 0$. This method is iterative in nature and does not require the evaluation of derivatives as in Newton–Raphson method. This method can also be used to determine both real and complex roots of $f(x) = 0$.

Suppose, x_{i-2}, x_{i-1}, x_i be any three distinct approximations to a root of $f(x) = 0$. Let $f(x_{i-2}) = f_{i-2}$, $f(x_{i-1}) = f_{i-1}$ and $f(x_i) = f_i$. Noting that any three distinct points in the (x, y)-plane uniquely, determine a polynomial of second degree. A general polynomial of second degree is given by

$$f(x) = ax^2 + bx + c \tag{2.15}$$

Suppose, it passes through the points (x_{i-2}, f_{i-2}), (x_{i-1}, f_{i-1}) and (x_i, f_i) as shown in Fig. 2.5, then the following equations will be satisfied:

$$ax_{i-2}^2 + bx_{i-2} + c = f_{i-2} \tag{2.16}$$

$$ax_{i-1}^2 + bx_{i-1} + c = f_{i-1} \tag{2.17}$$

$$ax_i^2 + bx_i + c = f_i \tag{2.18}$$

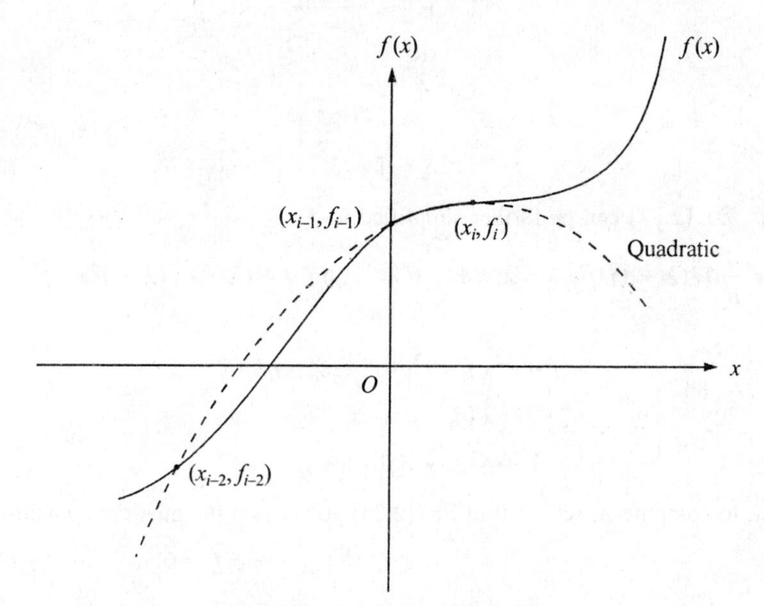

Fig. 2.5 Quadratic polynomial.

Eliminating a, b, c in Eqs. (2.15)–(2.18), we obtain

$$\begin{vmatrix} x^2 & x & 1 & f \\ x_{i-2}^2 & x_{i-2} & 1 & f_{i-2} \\ x_{i-1}^2 & x_{i-1} & 1 & f_{i-1} \\ x_i^2 & x_i & 1 & f_i \end{vmatrix} = 0$$

which can be written conveniently as

$$f = \frac{(x - x_{i-1})(x - x_i)}{(x_{i-2} - x_{i-1})(x_{i-2} - x_i)} f_{i-2} + \frac{(x - x_{i-2})(x - x_i)}{(x_{i-1} - x_{i-2})(x_{i-1} - x_i)} f_{i-1}$$

$$+ \frac{(x - x_{i-2})(x - x_{i-1})}{(x_i - x_{i-2})(x_i - x_{i-1})} f_i \tag{2.19}$$

Equation (2.19), obviously is a second degree polynomial. Now, introducing the notation

$$h = x - x_i, \qquad h_i = x_i - x_{i-1}, \qquad h_{i-1} = x_{i-1} - x_{i-2} \tag{2.20}$$

The above equation can be written as

$$f = \frac{(h + h_i)h}{-h_{i-1}(-h_{i-1} - h_i)} f_{i-2} + \frac{(h + h_i + h_{i-1})h}{(h_{i-1})(-h_i)} f_{i-1}$$

$$+ \frac{(h + h_i + h_{i-1})(h + h_i)}{(h_i + h_{i-1})h_i} f_i \tag{2.21}$$

We further define,

$$\lambda = \frac{h}{h_i} = \frac{x - x_i}{x_i - x_{i-1}} \tag{2.22}$$

$$\lambda_i = \frac{h_i}{h_{i-1}} \tag{2.23}$$

and

$$\delta_i = 1 + \lambda_i \tag{2.24}$$

Thus, Eq. (2.21) can be further simplified to

$$f = \frac{1}{\delta_i}[\lambda(\lambda + 1)\lambda_i^2 f_{i-2} - \lambda(\lambda + 1 + \lambda_i^{-1})\lambda_i\delta_i f_{i-1} + (\lambda + 1)(\lambda + 1 + \lambda_i^{-1})\lambda_i f_i]$$

or

$$f = \lambda^2 (f_{i-2}\lambda_i^2 - f_{i-1}\lambda_i\delta_i + f_i\lambda_i)\delta_i^{-1}$$

$$+ \lambda[f_{i-2}\lambda_i^2 - f_{i-1}\delta_i^2$$

$$+ f_i(\lambda_i + \delta_i)]\delta_i^{-1} + f_i \tag{2.25}$$

Now, to compute λ, set $f = 0$ in Eq. (2.25) and obtain the quadratic equation as

$$\lambda_i(f_{i-2}\lambda_i - f_{i-1}\delta_i + f_i)\lambda^2 + g_i\lambda + \delta_i f_i = 0 \tag{2.26}$$

where

$$g_i = f_{i-2}\lambda_i^2 - f_{i-1}\delta_i^2 + f_i(\lambda_i + \delta_i) \tag{2.27}$$

A direct solution of Eq. (2.26) leads to loss of accuracy, and therefore, to obtain maximum accuracy we rewrite Eq. (2.26) as follows:

$$\frac{f_i\delta_i}{\lambda^2} + \frac{g_i}{\lambda} + \lambda_i(f_{i-2}\lambda_i - f_{i-1}\delta_i + f_i) = 0 \tag{2.28}$$

so that,

$$\frac{1}{\lambda} = \frac{-g_i \pm [g_i^2 - 4f_i\delta_i\lambda_i(f_{i-2}\lambda_i - f_{i-1}\delta_i + f_i)]^{1/2}}{2f_i\delta_i}$$

or

$$\lambda = \frac{-2f_i\delta_i}{g_i \pm [g_i^2 - 4f_i\delta_i\lambda_i(f_{i-2}\lambda_i - f_{i-1}\delta_i + f_i)]^{1/2}} \tag{2.29}$$

Here, the positive sign must be so chosen that the denominator becomes largest in magnitude. Using

$$x_{i+1} = x_i + h_i\lambda \tag{2.30}$$

we can get a better approximation to the root.

Example 2.12 Find the root of the equation $x^3 - x - 1 = 0$ using Muller's method.

Solution Let $f(x) = x^3 - x - 1$, then $f(0) = -1, f(1) = -1, f(2) = 5$. Therefore, a root lies between 1 and 2 and is close to 1. Muller's method can be conveniently started by taking

$$x_{i-2} = 0, \qquad x_{i-1} = 1 \qquad \text{and} \qquad x_i = 2$$

Correspondingly, we get

$$f_{i-2} = -1, \qquad f_{i-1} = -1 \qquad \text{and} \qquad f_i = 5$$

We define

$$h = x - x_i = x - 2, \qquad h_i = x_i - x_{i-1} = 2 - 1 = 1$$

$$h_{i-1} = x_{i-1} - x_{i-2} = 1 - 0 = 1, \qquad \lambda = \frac{h}{h_i} = x - 2$$

$$\lambda_i = \frac{h_i}{h_{i-1}} = 1, \qquad \delta_i = 1 + \lambda_i = 2$$

From Eq. (2.27), we note that

$$g_i = f_{i-2}\ \lambda_i^2 - f_{i-1}\ \delta_i^2 + f_i\ (\lambda_i + \delta_i)$$

which gives,

$$g_i = (-1)\ (1) - (-1)\ (2)^2 + (5)\ (3) = 18$$

Now, the quadratic equation for $1/\lambda$ is given by

$$\frac{f_i\delta_i}{\lambda^2} + \frac{g_i}{\lambda} + \lambda_i\ (f_{i-2}\lambda_i - f_{i-1}\ \delta_i + f_i) = 0$$

That is,

$$\frac{5}{\lambda^2} + \frac{9}{\lambda} + 3 = 0$$

which gives

$$\frac{1}{\lambda} = \frac{-9 \pm \sqrt{21}}{10}$$

Taking negative sign, so that the numerator is the largest value in magnitude, we get $\lambda = -0.736236$. Therefore, the first approximation to the required root is given by

$$x_{i+1} = x_i + \lambda h_i = 2 - 0.736236 = 1.26376$$

2.7 GRAEFFE'S ROOT SQUARING METHOD

This method is particularly attractive for finding all the roots of a polynomial equation. We shall illustrate this method by considering a polynomial of third degree. However, it may be noted that this method is applicable for higher degree polynomials too.

Consider a polynomial of third degree

$$f(x) = a_0 + a_1 x + a_2 x^2 + a_3 x^3 \tag{2.31}$$

On putting $x = -x$, we get

$$f(-x) = a_0 - a_1 x + a_2 x^2 - a_3 x^3 \tag{2.32}$$

On multiplying Eq. (2.31) with Eq. (2.32), we get

$$f(x)\, f(-x) = a_3^2 t^3 - (a_2^2 - 2a_1 a_3)\, t^2 + (a_1^2 - 2a_0 a_2)\, t - a_0^2 \tag{2.33}$$

Here, we have replaced x^2 by t. It may be observed that, the roots of Eq. (2.33) are indeed the squares or 2^i $(i = 1)$ powers of the original roots. Here $i = 1$ indicates that squaring is done once.

Now, Eq. (2.33) can again be squared and this squaring process is repeated as many times as required. After each squaring, the coefficients become large and overflow is possible as i increases.

Suppose, we have squared the given polynomial i times, then we can estimate the value of the roots by evaluating 2^i root of

$$\left| \frac{a_i}{a_{i-1}} \right|, \qquad i = 1, 2, \ldots, n$$

where n is the degree of the given polynomial. The proper sign of each root can be determined by recalling the original equation. This method of course fails, when the roots of the given polynomial are repeated. This technique could be understood better from the following example.

Example 2.13 Using Graeffe root squaring method, find all the roots of the equation

$$x^3 - 6x^2 + 11x - 6 = 0$$

Solution Using Graeffe root squaring method, the first three squared polynomials are as under:

For $i = 1$, the polynomial is

$$x^3 - (36 - 22)\, x^2 + (121 - 72)\, x - 36 = x^3 - 14x^2 + 49x - 36 \tag{1}$$

For $i = 2$, the polynomial is

$$x^3 - (196 - 98)\, x^2 + (2401 - 1008)\, x - 1296 = x^3 - 98x^2 + 1393x - 1296 \tag{2}$$

For $i = 3$, the polynomial is

$$x^3 - (9604 - 2786)x^2 + (1940449 - 254016)x - 1679616$$

$$= x^3 - 6818x^2 + 1686433x - 1679616 \quad (3)$$

The roots of polynomial (1) are

$$\sqrt{\frac{36}{49}} = 0.85714, \quad \sqrt{\frac{49}{14}} = 1.8708, \quad \sqrt{\frac{14}{1}} = 3.7417$$

Similarly, the roots of polynomial (2) are

$$\sqrt[4]{\frac{1296}{1393}} = 0.9821, \quad \sqrt[4]{\frac{1393}{98}} = 1.9417, \quad \sqrt[4]{\frac{98}{1}} = 3.1464$$

Still better estimates of the roots obtained from polynomial (3) are

$$\sqrt[8]{\frac{1679616}{1686433}} = 0.99949, \quad \sqrt[8]{\frac{1686433}{6818}} = 1.99143, \quad \sqrt[8]{\frac{6818}{1}} = 3.0144$$

It may be observed that the exact values of the roots of the given polynomial are 1, 2 and 3.

2.8 BAIRSTOW METHOD

To find the complex roots of a polynomial

$$f(x) = a_0x^n + a_1x^{n-1} + a_2x^{n-2} + \cdots + a_{n-1}x + a_n = 0 \quad (2.34)$$

where the coefficients a_i are real, we present below a more convenient iterative method which is due to Bairstow. Since complex roots always occur in pairs as $\alpha \pm i\beta$, the corresponding real quadratic polynomial is

$$[x - (\alpha + i\beta)][x - (\alpha - i\beta)] = x^2 - 2\alpha x + (\alpha^2 + \beta^2)$$

which is of the form $x^2 - px - l$. Thus, if we can find quadratic factors, we can avoid complex arithmetic in view of the fact that quadratic factors will always have real coefficients. Also, it may be noted that in any hand computation, performing arithmetic operations of complex numbers is a tedious job. Now, the division of $f(x)$ by the above quadratic factor can be expressed in the form

$$f(x) = (x^2 - px - l)\, Q_{n-2}(x) + r(x) \quad (2.35)$$

where $Q_{n-2}(x)$ is a polynomial of degree $(n - 2)$ and $r(x)$ is the remainder. Thus, we may write

$$f(x) = (x^2 - px - l)(b_0x^{n-2} + b_1x^{n-3} + \cdots + b_{n-3}x + b_{n-2}) + b_{n-1}(x - p) + b_n \quad (2.36)$$

If $(x^2 - px - l)$ were an exact factor of $f(x)$, the remainder $b_{n-1}(x - p) + b_n = 0$. This particular form is chosen only for convenience of later simplifications. This form imply that $b_{n-1} = 0$ and $b_n = 0$, which means that both b_{n-1} and b_n depend on p and l. In other words, the problem is to determine p and l such that

$$b_{n-1}(p, l) = 0, \qquad b_n(p, l) = 0 \tag{2.37}$$

Using Taylor series expansion in terms of $(p^* - p)$ and $(l^* - l)$ and retaining only the first order terms, we get from Eq. (2.37), the following pair of equations:

$$b_{n-1}(p^*, l^*) = b_{n-1}(p, l) + \frac{\partial b_{n-1}}{\partial p}(p^* - p) + \frac{\partial b_{n-1}}{\partial l}(l^* - l)$$

and

$$b_n(p^*, l^*) = b_n(p, l) + \frac{\partial b_n}{\partial p}(p^* - p) + \frac{\partial b_n}{\partial l}(l^* - l) \tag{2.38}$$

Here, we consider (p^*, l^*) as a point at which the remainder is zero and take

$$p^* - p = \Delta p, \quad l^* - l = \Delta l$$

Then, the above pair becomes

$$b_{n-1}(p^*, l^*) = 0 = b_{n-1} + \frac{\partial b_{n-1}}{\partial p}\Delta p + \frac{\partial b_{n-1}}{\partial l}\Delta l$$

$$b_n(p^*, l^*) = 0 = b_n + \frac{\partial b_n}{\partial p}\Delta p + \frac{\partial b_n}{\partial l}\Delta l \tag{2.39}$$

where, all the terms on the right hand side are to be evaluated at (p, l). The solution of these equations gives us the corrections Δp and Δl, which of course require the evaluation of partial derivatives. This procedure is repeated with the updated values of p and l.

In order to compute the coefficients b_i, p and l we equate the coefficients of like powers of x in Eq. (2.36) and get

$$\left.\begin{aligned}
&a_0 = b_0, && \text{that is } b_0 = a_0\\
&a_1 = b_1 - pb_0, && \text{that is } b_1 = a_1 + pb_0\\
&a_2 = b_2 - pb_1 - lb_0, && \text{that is } b_2 = a_2 + pb_1 + lb_0\\
&\vdots\\
&a_i = b_i - pb_{i-1} - lb_{i-2}, && \text{that is } b_i = a_i + pb_{i-1} + lb_{i-2}\\
&\vdots\\
&\text{Coefficient of } x \text{ gives } b_{n-1} - pb_{n-2} - lb_{n-3} = a_{n-1}, \text{ that is}\\
&\qquad\qquad b_{n-1} = a_{n-1} + pb_{n-2} + lb_{n-3}\\
&\text{Coefficient of constant gives } b_n - pb_{n-1} - lb_{n-2} = a_n, \text{ that is}\\
&\qquad\qquad b_n = a_n + pb_{n-1} + lb_{n-2}
\end{aligned}\right\} \tag{2.40}$$

This set of equations is equivalent to the recurrence relation

$$b_i = a_i + pb_{i-1} + lb_{i-2}, \quad i = 2, 3, \ldots, n \tag{2.41}$$

If we set $b_{-1} = 0$, $b_{-2} = 0$, the above recurrence relation holds for $i = 0, 1, 2, \ldots n$. Of course, the coefficient b_i depend on the numbers p and l.

Bairstow showed that the required partial derivatives in Eq. (2.38) can be obtained from the b's in just the same way that the b's are obtained from a's as in Eq. (2.40). Taking the partial derivatives of Eqs. (2.40) with respect to p and l and letting

$$\frac{\partial b_{i+1}}{\partial p} = C_i \tag{2.42}$$

we arrive at the following relations

$$\frac{\partial b_0}{\partial p} = \frac{\partial a_0}{\partial p} = 0 \qquad \frac{\partial b_0}{\partial l} = \frac{\partial a_0}{\partial l} = 0$$

$$\frac{\partial b_1}{\partial p} = p\frac{\partial b_0}{\partial p} + b_0 = b_0 = C_0 \qquad \frac{\partial b_1}{\partial l} = p\frac{\partial b_0}{\partial l} = 0$$

$$\frac{\partial b_2}{\partial p} = p\frac{\partial b_1}{\partial p} + b_1 + l\frac{\partial b_0}{\partial p} = C_1 \qquad \frac{\partial b_2}{\partial l} = p\frac{\partial b_1}{\partial l} + l\frac{\partial b_0}{\partial l} + b_0$$

Therefore, $C_1 = b_1 + pc_0$. $\qquad = b_0 = C_0$

$$\frac{\partial b_3}{\partial p} = p\frac{\partial b_2}{\partial p} + b_2 + l\frac{\partial b_1}{\partial p} = C_2 \qquad \frac{\partial b_3}{\partial l} = p\frac{\partial b_2}{\partial l} + l\frac{\partial b_1}{\partial l} + b_1$$

Therefore, $C_2 = b_2 + pC_1 + lC_0$ $\qquad = b_1 + pC_0 = C_1$

$$\frac{\partial b_{n-1}}{\partial p} = p\frac{\partial_{n-2}}{\partial p} + b_{n-2} + l\frac{\partial b_{n-3}}{\partial p} \qquad \frac{\partial b_{n-1}}{\partial l} = p\frac{\partial b_{n-2}}{\partial l} + l\frac{\partial b_{n-3}}{\partial l} + b_{n-3}$$

$$= C_{n-2} \qquad = b_{n-3} + pC_{n-4} + lC_{n-5}$$

$$= C_{n-3}$$

Therefore,

$$C_{n-2} = b_{n-2} + pC_{n-3} + lC_{n-4} \tag{2.43}$$

Utilizing these results, Eqs. (2.39) can now be written as

$$\left.\begin{aligned} -b_{n-1} &= C_{n-2}\,\Delta p + C_{n-3}\,\Delta l \\ -b_n &= C_{n-1}\,\Delta p + C_{n-2}\,\Delta l \end{aligned}\right\} \tag{2.44}$$

These are the central equations for Newton's iteration. Solving these equations for Δp and Δl, we find

$$\left.\begin{aligned} \Delta p &= \frac{b_n C_{n-3} - b_{n-1} C_{n-2}}{C_{n-2}^2 - C_{n-1}\,C_{n-3}}, \\ \Delta l &= \frac{b_{n-1} C_{n-1} - b_n C_{n-2}}{C_{n-2}^2 - C_{n-1}\,C_{n-3}}. \end{aligned}\right\} \tag{2.45}$$

For illustration of the method, we consider the following example

Example 2.14 Find the quadratic factors of

$$x^4 - 1.1x^3 + 2.3x^2 + 0.5x + 3.3 = 0$$

Using $(x^2 + x + 1)$ as a starting factor (Gerald et. al)

Solution: Comparing the given starting factor with the standard factor $(x^2 - px - l)$, we note that $p = -1$, $l = -1$. We are also given the data as $a_0 = 1$, $a_1 = -1.1$, $a_2 = 2.3$, $a_3 = 0.5$, $a_4 = 3.3$. Now, using Eq. (2.40). we compute b's as

$$b_0 = a_0 = 1,\ b_1 = a_1 + pb_0 = -1.1 + (-1)(1) = -2.1 (= b_{n-3})$$

$$b_2 = a_2 + pb_1 + lb_0 = 2.3 + (-1)(-2.1) + (-1)(1) = 3.4\ (= b_{n-2})$$

$$b_3 = a_3 + pb_2 + lb_1 = 0.5 + (-1)(3.4) + (-1)(-2.1) = -0.8\ (= b_{n-1})$$

$$b_4 = a_4 + pb_3 + lb_2 = 3.3 + (-1)(-0.8) + (-1)(3.4) = 0.7\ (= b_n)$$

Finally, using Eq. (2.43), we compute C's as

$$C_0 = b_0 = 1$$

$$C_1 = b_1 + pC_0 = (-2.1) + (-1)(1) = -3.1 \qquad (= C_{n-3})$$

$$C_2 = b_2 + pC_1 + lC_0 = 3.4 + (-1)(-3.1) + (-1)(1) = 5.5 \qquad (= C_{n-2})$$

$$C_3 = b_3 + pC_2 + lC_1 = -0.8 + (-1)(5.5) + (-1)(-3.1) = -3.2 \qquad (= C_{n-1})$$

Substituting these values of b's and C's in Eq. (2.45), we get after first iteration that

$$\Delta p = \frac{(0.7)(-3.1) - (-0.8)(5.5)}{(5.5)^2 - (-3.2)(-3.1)} = \frac{2.23}{20.33} = 0.11$$

$$\Delta l = \frac{(-0.8)(-3.2) - (0.7)(5.5)}{(5.5)^2 - (-3.2)(-3.1)} = \frac{-1.29}{20.33} = -0.06$$

Thus, as a first approximation, we have

$$p^* = (-1) + \Delta p = (-1) + (0.11) = -0.89$$

$$l^* = (-1) + \Delta l = (-1) + (-0.06) = -1.06$$

Now, starting with $p = -0.89$, $l = -1.06$ and repeating the above steps, we get

$$\Delta p = -0.01, \qquad \Delta l = -0.04 \quad \text{and}$$

$$p^* = -0.89 - 0.01 = -0.9, \quad l^* = -1.06 - 0.04 = -1.1$$

Hence, the factor after the second iteration is $(x^2 + 0.9x + 1.1)$. Finally, the quadratic factors of the given equation is found to be

$$(x^2 + 0.9x + 1.1)(x^2 - 2x + 3)$$

2.9 SYSTEM OF NON-LINEAR EQUATIONS

The general problem is to solve a system of n non-linear equations in n unknowns. That is, to solve

$$\left.\begin{aligned} f_1(x_1, x_2, \ldots, x_n) &= 0 \\ f_2(x_1, x_2, \ldots, x_n) &= 0 \\ &\vdots \\ f_n(x_1, x_2, \ldots, x_n) &= 0 \end{aligned}\right\} \tag{2.46}$$

Most of the known methods are of iterative type, where we start with initial guess and improve the solution iteratively, until a required tolerance is achieved. By taking the gradients of all the variables, we get a function matrix called the Jacobian defined as

$$J = \begin{bmatrix} \dfrac{\partial f_1}{\partial x_1} & \dfrac{\partial f_1}{\partial x_2} & \cdots & \dfrac{\partial f_1}{\partial x_n} \\ \vdots & \vdots & & \vdots \\ \dfrac{\partial f_n}{\partial x_1} & \dfrac{\partial f_n}{\partial x_2} & \cdots & \dfrac{\partial f_n}{\partial x_n} \end{bmatrix}$$

It is essential that this Jacobian be non-singular for the existence of a solution. It may also be noted that the value of the Jacobian changes from iteration to iteration. Suppose, we denote the vector of components $(x_1, x_2, \ldots, x_n)$ by X and the vector $(f_1, f_2, \ldots, f_n)$ by F, then the system (2.46) can be written as

$$F(X) = 0 \tag{2.47}$$

The task is to solve the system (2.47). We present below a simple Newton's method by considering a system in two variables such as

$$\begin{aligned} f(x, y) &= 0 \\ g(x, y) &= 0 \end{aligned} \tag{2.48}$$

Suppose, we choose (x_0, y_0) as the initial (guess) approximation and let h, k are the quantities to be determined such that

$$\begin{aligned} f(x_0 + h,\ y_0 + k) &= 0 \\ g(x_0 + h,\ y_0 + k) &= 0 \end{aligned} \tag{2.49}$$

Using Taylor series expansion, we get

$$f(x_0 + h, y_0 + k) = f(x_0, y_0) + \left(h\frac{\partial f}{\partial x} + k\frac{\partial f}{\partial y} \right)\Bigg|_{(x_o, y_0)} + \text{H.O.T} = 0$$

$$g(x_0 + h, y_0 + k) = g(x_0, y_0) + \left(h\frac{\partial g}{\partial x} + k\frac{\partial g}{\partial y} \right)\Bigg|_{(x_o, y_0)} + \text{H.O.T} = 0$$

Neglecting the higher order terms (H.O.T) and solving the above system, we get the approximate values of h and k as

$$h = \left. \frac{-f\dfrac{\partial g}{\partial y} + g\dfrac{\partial f}{\partial y}}{J} \right|_{(x_0, y_0)} \tag{2.50}$$

$$k = \left. \frac{-g\dfrac{\partial f}{\partial x} + f\dfrac{\partial g}{\partial x}}{J} \right|_{(x_0, y_0)}$$

where J is the Jacobian defined by

$$J = \begin{vmatrix} f_x & f_y \\ g_x & g_y \end{vmatrix} \tag{2.51}$$

Once h and k are computed, we get the first approximate solution in the form

$$\left. \begin{aligned} x_1 &= x_0 + h \\ y_1 &= y_0 + k \end{aligned} \right\} \tag{2.52}$$

which suggests an iteration

$$\left. \begin{aligned} x_{n+1} &= x_n + h \\ y_{n+1} &= y_n + k \end{aligned} \right\} \tag{2.53}$$

Observe that we need to guess a good initial approximation (x_0, y_0) for convergence of iteration, which we may get graphically. The generalisation to a system of equations in n variables is of course straight forward. Here, follows an example for illustration.

Example 2.15 Solve the following system of equations

$$f(x, y) = x^3 - 3xy^2 - 2x + 2 = 0$$

$$g(x, y) = 3x^2y - y^3 - 2y = 0$$

taking $x_0 = y_0 = 1$, as the initial approximation.

Solution In the present example, we have

$$\begin{aligned} f(x, y) &= x^3 - 3xy^2 - 2x + 2 \\ g(x, y) &= 3x^2y - y^3 - 2y \end{aligned} \tag{1}$$

Therefore,

$$\begin{aligned} f_x &= 3x^2 - 3y^2 - 2, \qquad f_y = -6xy \\ g_x &= 6xy, \qquad g_y = 3x^2 - 3y^2 - 2 \end{aligned} \tag{2}$$

Assume the initial approximation $(x_0, y_0) = (1, 1)$ so that

$$f(x_0, y_0) = -2, \qquad g(x_0, y_0) = 0$$

$$f_x\big|_{(x_0, y_0)} = -2, \quad f_y\big|_{(x_0, y_0)} = -6$$

$$g_x\big|_{(x_0, y_0)} = 6, \quad g_y\big|_{(x_0, y_0)} = -2$$

then the Jacobian

$$J\big|_{(x_0, y_0)} = \begin{vmatrix} -2 & -6 \\ 6 & -2 \end{vmatrix} = 40 \neq 0$$

Hence, the solution exists. Now, from Eq. (2.50) we compute

$$h = \frac{[(2)(-2) + (0)(-6)]}{40} = -\frac{1}{10} = -0.1$$

$$k = \frac{[(0)(-2) + (-2)(6)]}{40} = -\frac{3}{10} = -0.3$$

Thus, the first approximation is

$$x_1 = x_0 + h = 1 - 0.1 = 0.9, \; y_1 = y_0 + k = 1 - 0.3 = 0.7 \qquad (3)$$

Similarly, we find

$$f(x_1, y_1) = 0.729 - 1.323 - 1.8 + 2 = -0.394$$

$$g(x_1, y_1) = 1.701 - 0.343 - 1.4 = -0.042$$

$$f_x = 2.43 - 1.47 - 2 = -1.04, \quad f_y = -3.78$$

$$g_x = 3.78, \; g_y = 2.43 - 1.47 - 2 = -1.04$$

Therefore,

$$J = \begin{vmatrix} -1.04 & -3.78 \\ 3.78 & -1.04 \end{vmatrix} = 15.37 \neq 0$$

$$h = \frac{(0.394)(-1.04) + (-0.042)(-3.78)}{15.37} = -0.0163$$

$$k = \frac{(0.042)(-1.04) + (-0.394)(3.78)}{15.37} = -0.0997$$

Hence, the second approximation is

$$x_2 = x_1 + h = 0.9 - 0.0163 = 0.8837$$

$$y_2 = y_1 + k = 0.7 - 0.0997 = 0.6003$$

This is the solution of the given system after two iterations. At this point, we may note that

$$f(x_2, y_2) = -0.03265$$

and

$$g(x_2, y_2) = -0.01055.$$

Further continuation gives

$$f_x\Big|_{(x_2,y_2)} = 1.2617, \quad f_y\Big|_{(x_2,y_2)} = -3.1829$$

$$g_x\Big|_{(x_2,y_2)} = 3.1829, \quad g_y\Big|_{(x_2,y_2)} = 1.2617$$

and

$$J = \begin{vmatrix} f_x & f_y \\ g_x & g_y \end{vmatrix} = 11.7227 \neq 0$$

which yield

$$h = 0.006379, \qquad k = -0.0773$$

Thus, the third approximation is

$$x_3 = x_2 + h = 0.8901, \qquad y_3 = y_2 + k = 0.5926$$

and

$$f(x_3, y_3) = -0.0127, \quad g(x_3, y_3) = 0.0152$$

The three iterations are tabulated as

i	x_i	y_i	$f(x_i, y_i)$	$g(x_i, y_i)$
1	0.9	0.7	–0.394	–0.042
2	0.8837	0.6003	–0.0327	–0.0106
3	0.8901	0.5926	–0.0127	0.0152

EXERCISES

2.1 Find the real root of the equation, $x^3 - 3x - 5 = 0$ by the bisection method.

2.2 Find the real root of the equation, $x^3 + x - 3 = 0$, using Regula-Falsi method, correct to four places of decimal.

2.3 Find the real root of the equation $x^6 - x^4 - x^3 - 1 = 0$, which lies between 1.4 and 1.5, correct to four places of decimal, by the method of false position, obtained after three successive approximations.

2.4 Use Regula-Falsi method to find the real roots of the equation $x^3 - \sin x + 1 = 0$ correct to four decimal places after three successive approximations between (–2, –1).

2.5 Explain the method of false position for finding a real root of the equation $f(x) = 0$, and hence derive the general formula.

2.6 Use Regula-Falsi method to compute the root of the equation $\cos x - xe^x = 0$.

2.7 Find the root of the equation $2x = \cos x + 3$, correct to three decimal places using iteration method.

2.8 Find a root of the equation $x \log_{10} x = 4.77$ by Newton–Raphson method, correct to two decimal places.

2.9 Explain the Newton–Raphson method to find a root of the equation $f(x) = 0$, and hence derive its iteration formula.

2.10 Geometrically explain Newton–Raphson method to find a root of the equation $f(x) = 0$ and hence derive the general formula.

2.11 Obtain the real root of the equation $x^3 - 3x - 5 = 0$ using Newton–Raphson method, after third iteration.

2.12 Find a real root of the equation, $x^4 - x - 10 = 0$ using Newton–Raphson method correct to four decimal places.

2.13 Apply Newton–Raphson method to determine a root of the equation $\cos x = x\,e^x$ correct to three decimal places, using the initial approximation, $x_0 = 1$.

2.14 Set up the Newton's scheme of iteration for finding the p-th root of a positive number N.

2.15 Obtain the cube root of 12 using Newton–Raphson iteration.

2.16 Find the first approximation of the root of the equation $x^3 - 3x - 5 = 0$ using Muller's method, which lies between 2 and 3.

2.17 Find the first approximation to the root of the equation

$$f(x) = \sin x - \frac{x}{2} = 0$$

near $x = 2.0$, using Muller's method.

2.18 Using Graeffe's root squaring method, find the roots of the equation $x^3 - 4x^2 + 3x + 1 = 0$ with the help of a calculator.

2.19 Using the method of false position, find the root of $x \sin x - 1 = 0$ which lies in the interval (0, 2).

2.20 Find the quadratic factors of

$$x^4 - 5.7x^3 + 26.7x^2 - 42.21x + 69$$

Using Bairstows method with $(x^2 - 1.5x + 4.3)$ as a starting factor.

2.21 Using Bairstows method, find the quadratic factors of the polynomial

$$2x^4 + 7x^3 - 4x^2 + 29x + 14$$

with $(x^2 + 5x + 2)$ as a starting factor.

2.22 Find the solution of

$$f(x, y) = x^3 - 3xy^2 + 1 = 0$$

$$g(x, y) = 3x^2y - y^3 = 0$$

taking (1, 1) as the initial approximation using Newtons method.

2.23 Using Newtons method, find the solution of

$$f(x, y) = 4x^2 + y^2 + 2xy - y - 2 = 0$$

$$g(x, y) = 2x^2 + 3xy + y^2 - 3 = 0$$

taking (0.4, 0.9) as the initial approximation.

Chapter 3

Solution of Linear System of Equations and Matrix Inversion

3.1 INTRODUCTION

Many real-life problems in engineering give rise to a system of linear equations. For example, such systems occur in certain applications of statistical analysis and in finding the numerical solution of partial differential equations and so on. It is therefore, natural to seek efficient methods for solving these equations numerically.

The general form of a system of m linear equations in n unknowns x_1, x_2, x_3, ..., x_n can be represented in matrix form as under:

$$\begin{bmatrix} a_{11} & a_{12} & a_{13} & \cdots & a_{1n} \\ a_{21} & a_{22} & a_{23} & \cdots & a_{2n} \\ \vdots & \vdots & \vdots & & \vdots \\ a_{m1} & a_{m2} & a_{m3} & \cdots & a_{mn} \end{bmatrix} \begin{pmatrix} x_1 \\ x_2 \\ \vdots \\ x_n \end{pmatrix} = \begin{pmatrix} b_1 \\ b_2 \\ \vdots \\ b_m \end{pmatrix} \tag{3.1}$$

Using matrix notation, the above system can be written in compact form as

$$[A]\,(X) = (B) \tag{3.2}$$

The solution of the system of equations (3.2) gives n unknown values x_1, x_2, ..., x_n, which satisfy the system simultaneously. If $m > n$, we may not be able to find a solution, in principle, which satisfy all the equations. If $m < n$, the system usually will have an infinite number of solutions. However, in this chapter, we shall restrict to the case $m = n$. In this case, if $|A| \neq 0$, then the system will have a unique solution, while, if $|A| = 0$, then there exists no solution.

Various numerical methods are available for finding the solution of the system of equations (3.2), and they are classified as *direct* and *iterative methods*. In direct methods, we get the solution of the system after performing all the steps involved in the procedure. The direct methods consist of *elimination methods* and *decomposition methods*. In this chapter, under elimination methods, we consider, *Gaussian elimination* and *Gauss–Jordan elimination methods*. *Crout's reduction* also known as *Cholesky's reduction* is considered under decomposition methods. Under iterative methods, the initial approximate solution is assumed to be known and is improved towards the exact solution in an iterative way. We consider *Jacobi, Gauss–Seidel* and *relaxation methods* under iterative methods. All these methods are easily adoptable to computers and can be used to solve even hundred or more simultaneous linear equations.

3.2 GAUSSIAN ELIMINATION METHOD

In the Gaussian elimination method, the solution to the system of Eqs. (3.2) is obtained in two stages. In the first stage, the given system of equations is reduced to an equivalent upper triangular form using elementary transformations. In the second stage, the upper triangular system is solved using back substitution procedure by which we obtain the solution in the order x_n, x_{n-1}, x_{n-2}, ..., x_2, x_1.

This method is explained by considering a system of n equations in n unknowns in the form as follows

$$\left.\begin{aligned} a_{11}x_1 + a_{12}x_2 + \cdots + a_{1n}x_n &= b_1 \\ a_{21}x_1 + a_{22}x_2 + \cdots + a_{2n}x_n &= b_2 \\ \vdots \qquad \vdots \qquad\qquad \vdots \quad &\quad \vdots \\ a_{n1}x_1 + a_{n2}x_2 + \cdots + a_{nn}x_n &= b_n \end{aligned}\right\} \tag{3.3}$$

Stage I: We divide the first equation by a_{11} and then subtract this equation multiplied by a_{21}, a_{31}, ..., a_{n1} from the 2nd, 3rd, ..., n th equation. Then the system (3.3) reduces to the following form:

$$\left.\begin{aligned} x_1 + a'_{12}x_2 + \cdots + a'_{1n}x_n &= b'_1 \\ a'_{22}x_2 + \cdots + a'_{2n}x_n &= b'_2 \\ \vdots \qquad\qquad \vdots \quad &\quad \vdots \\ a'_{n2}x_2 + \cdots + a'_{nn}x_n &= b'_n \end{aligned}\right\} \tag{3.4}$$

Here, we can observe that the last $(n - 1)$ equations are independent of x_1, that is, x_1 is eliminated from the last $(n - 1)$ equations.

This procedure is repeated with the second equation of (3.4), that is, we divide the second equation by a'_{22} and then x_2 is eliminated from 3rd, 4th, ..., nth equations of (3.4). The same procedure is repeated again and again till the given system assumes the following upper triangular form:

$$\left.\begin{aligned} c_{11}x_1 + c_{12}x_2 + \cdots + c_{1n}x_n &= d_1 \\ c_{22}x_2 + \cdots + c_{2n}x_n &= d_2 \\ \vdots \quad &\quad \vdots \\ c_{nn}x_n &= d_n \end{aligned}\right\} \tag{3.5}$$

Stage II: Now, the values of the unknowns are determined by back substitution procedure, in which we obtain x_n from the last equation of (3.5) and then substituting this value of x_n in the preceding equation, we get the value of x_{n-1}. Continuing this way, we can find the values of all other unknowns in the order x_n, x_{n-1}, ..., x_2, x_1.

In this method, we observe that the determinant of the coefficient matrix is obtained as a by-product, that is,

$$|A| = c_{11}c_{22} \ldots c_{nn} \tag{3.6}$$

To familiarize with the method, we consider the following example:

Example 3.1 Solve the following system of equations using Gaussian elimination method

$$2x + 3y - z = 5$$

$$4x + 4y - 3z = 3$$

$$-2x + 3y - z = 1$$

Solution The given system of equations is solved in two stages.

Stage I (Reduction to upper-triangular form): We divide the first equation by 2 and then subtract the resulting equation (multiplied by 4 and –2) from the second and third equations respectively. Thus, we eliminate x from the 2nd and 3rd equations. The resulting new system is given by

$$\left.\begin{aligned} x + \frac{3}{2}y - \frac{z}{2} &= \frac{5}{2} \\ -2y - z &= -7 \\ 6y - 2z &= 6 \end{aligned}\right\} \tag{1}$$

Now, we divide the second equation of (1) by –2 and eliminate y from the last equation and the modified system is given by

$$\left.\begin{aligned} x + \frac{3}{2}y - \frac{z}{2} &= \frac{5}{2} \\ y + \frac{z}{2} &= \frac{7}{2} \\ -5z &= -15 \end{aligned}\right\} \tag{2}$$

Stage II (Back substitution): From the last equation of (2), we immediately get

$$z = 3 \tag{3}$$

using this value of z, the second equation of (2) gives

$$y = \frac{7}{2} - \frac{3}{2} = 2 \tag{4}$$

Using these values of y and z in the first equation of (2), we get

$$x = \frac{5}{2} + \frac{3}{2} - 3 = 1 \tag{5}$$

Thus, the solution of the given system is given by Eqs. (3) – (5).

Partial and full pivoting

The Gaussian elimination method fails if any one of the pivot elements becomes zero. In such a situation, we rewrite the equations in a different order to avoid zero pivots. Changing the order of equations is called *pivoting*.

We now introduce the concept of partial pivoting. In this technique, if the pivot a_{ii} happens to be zero, then the ith column elements are searched for the numerically largest element. Let the jth row ($j > i$) contains this element, then we interchange the ith equation with the jth equation and proceed for elimination. This process is continued whenever pivots become zero during elimination. For example, let us examine the solution of the following simple system

$$10^{-5}x_1 + x_2 = 1$$
$$x_1 + x_2 = 2$$

Using Gaussian elimination method with and without partial pivoting, assuming that we require the solution accurate to only four decimal places. The solution by Gaussian elimination gives $x_1 = 0$, $x_2 = 1$. If we use partial pivoting, the system takes the form

$$x_1 + x_2 = 2$$
$$10^{-5}x_1 + x_2 = 1$$

Using Gaussian elimination method, the solution is found to be $x_1 = 1$, $x_2 = 1$, which is a meaningful and perfect result.

In full pivoting which is also known as *complete pivoting*, we interchange rows as well as columns, such that the largest element in the matrix of the system becomes the pivot element. In this process, the position of the unknown variables also get changed. Full pivoting, in fact, is more complicated than the partial pivoting. Partial pivoting is preferred for hand computation.

Example 3.2 Solve the system of equations

$$x + y + z = 7$$
$$3x + 3y + 4z = 24$$
$$2x + y + 3z = 16$$

by Gaussian elimination method with partial pivoting.

Solution In matrix notation, the given system can be written as

$$\begin{bmatrix} 1 & 1 & 1 \\ 3 & 3 & 4 \\ 2 & 1 & 3 \end{bmatrix} \begin{pmatrix} x \\ y \\ z \end{pmatrix} = \begin{pmatrix} 7 \\ 24 \\ 16 \end{pmatrix} \tag{1}$$

To start with, we observe that the pivot element $a_{11} = 1$ ($\neq 0$). However, a glance at the first column reveals that the numerically largest element is 3 which is in the second row. Hence, we interchange the first row with the second row and then proceed for elimination. Thus, Eq. (1) takes the form

$$\begin{bmatrix} 3 & 3 & 4 \\ 1 & 1 & 1 \\ 2 & 1 & 3 \end{bmatrix} \begin{pmatrix} x \\ y \\ z \end{pmatrix} = \begin{pmatrix} 24 \\ 7 \\ 16 \end{pmatrix} \tag{2}$$

after partial pivoting.

Stage I (Reduction to upper triangular form): By dividing the first row of the system (2) by 3 and then subtracting the resulting row, multiplied by 1 and 2 from the second and third rows of the system (2), we get

$$\begin{bmatrix} 1 & 1 & \frac{4}{3} \\ & & -\frac{1}{3} \\ & -1 & \frac{1}{3} \end{bmatrix} \begin{pmatrix} x \\ y \\ z \end{pmatrix} = \begin{pmatrix} 8 \\ -1 \\ 0 \end{pmatrix} \tag{3}$$

The second row in Eq. (3) cannot be used as the pivot row, as $a_{22} = 0$. Interchanging the second and third rows, we obtain

$$\begin{bmatrix} 1 & 1 & \frac{4}{3} \\ & -1 & \frac{1}{3} \\ & & -\frac{1}{3} \end{bmatrix} \begin{pmatrix} x \\ y \\ z \end{pmatrix} = \begin{pmatrix} 8 \\ 0 \\ -1 \end{pmatrix} \tag{4}$$

which is in the upper triangular form.

Stage II (Back substitution): From the last row of Eq. (4), we at once get

$$z = 3 \tag{5}$$

The second row of Eq. (4) with this value of z gives

$$-y + 1 = 0 \quad \text{or} \quad y = 1 \tag{6}$$

Using these values of y and z, the first row of Eq. (4) gives

$$x + 1 + 4 = 8 \quad \text{or} \quad x = 3 \tag{7}$$

Thus, Eqs. (5)–(7) constitute the solution to the given system of equations.

Example 3.3 Solve by Gaussian elimination method with partial pivoting, the following system of equations:

$$0x_1 + 4x_2 + 2x_3 + 8x_4 = 24$$
$$4x_1 + 10x_2 + 5x_3 + 4x_4 = 32$$
$$4x_1 + 5x_2 + 6.5x_3 + 2x_4 = 26$$
$$9x_1 + 4x_2 + 4x_3 + 0x_4 = 21$$

Solution In matrix notation, the given system can be written as

$$\begin{bmatrix} 0 & 4 & 2 & 8 \\ 4 & 10 & 5 & 4 \\ 4 & 5 & 6.5 & 2 \\ 9 & 4 & 4 & 0 \end{bmatrix} \begin{pmatrix} x_1 \\ x_2 \\ x_3 \\ x_4 \end{pmatrix} = \begin{pmatrix} 24 \\ 32 \\ 26 \\ 21 \end{pmatrix} \tag{1}$$

To start with, we observe that the pivot row, that is, the first row has a zero pivot element ($a_{11} = 0$). This row should be interchanged with any row following it, which on becoming a pivot row should not have a zero pivot element. While interchanging rows it is better to interchange with a row having largest pivotal element. Thus, we interchange the first and fourth rows, which is called partial pivoting and get,

$$\begin{bmatrix} 9 & 4 & 4 & 0 \\ 4 & 10 & 5 & 4 \\ 4 & 5 & 6.5 & 2 \\ 0 & 4 & 2 & 8 \end{bmatrix} \begin{pmatrix} x_1 \\ x_2 \\ x_3 \\ x_4 \end{pmatrix} = \begin{pmatrix} 21 \\ 32 \\ 26 \\ 24 \end{pmatrix} \tag{2}$$

We observe that, in partial pivoting, the unknown vector remains unaltered, while the right-hand side vector gets changed.

Now, we shall carry out Gaussian elimination process in two stages.

Stage I (Reduction to upper-triangular form): In this stage, by dividing the first row of the system (2) by 9 and then subtracting this resulting row, multiplied by 4 and 4 from the second and third rows of Eq. (2), we get

$$\begin{bmatrix} 1 & \frac{4}{9} & \frac{4}{9} & 0 \\ 0 & 8.2222 & 3.2222 & 4 \\ 0 & 3.2222 & 4.7222 & 2 \\ 0 & 4 & 2 & 8 \end{bmatrix} \begin{pmatrix} x_1 \\ x_2 \\ x_3 \\ x_4 \end{pmatrix} = \begin{pmatrix} 2.3333 \\ 22.6666 \\ 16.6666 \\ 24 \end{pmatrix} \tag{3}$$

Now, we divide the second pivot row by 8.2222 and subtract the resultant row multiplied by 3.2222 and 4 from the third and fourth rows of Eq. (3) to get

$$\begin{bmatrix} 1 & \frac{4}{9} & \frac{4}{9} & 0 \\ 0 & 1 & 0.3919 & 0.4865 \\ 0 & 0 & 3.4594 & 0.4324 \\ 0 & 0 & 0.4324 & 6.0540 \end{bmatrix} \begin{pmatrix} x_1 \\ x_2 \\ x_3 \\ x_4 \end{pmatrix} = \begin{pmatrix} 2.3333 \\ 2.7568 \\ 7.7836 \\ 12.9728 \end{pmatrix} \tag{4}$$

Finally, we divide the third pivot row by 3.4594 and subtract the resultant row multiplied by 0.4324 from fourth row of Eq. (4), thus getting the upper triangular form

$$\begin{bmatrix} 1 & \frac{4}{9} & \frac{4}{9} & 0 \\ 0 & 1 & 0.3919 & 0.4865 \\ 0 & 0 & 1 & 0.1250 \\ 0 & 0 & 0 & 5.9999 \end{bmatrix} \begin{pmatrix} x_1 \\ x_2 \\ x_3 \\ x_4 \end{pmatrix} = \begin{pmatrix} 2.3333 \\ 2.7568 \\ 2.2500 \\ 11.9999 \end{pmatrix} \tag{5}$$

Stage II (Back substitution): From the last row of Eq. (5), we immediately get $x_4 = 2.0000$. Using this value of x_4 into the third row of Eq. (5), we obtain

$$x_3 + 0.25 = 2.25 \quad \text{or} \quad x_3 = 2.0000 \tag{6}$$

Similarly, we get

$$x_2 = 1.0000, \quad x_1 = 1.0000$$

Thus, the solution of the given system is given by

$$x_1 = 1.0, \quad x_2 = 1.0, \quad x_3 = 2.0, \quad x_4 = 2.0$$

3.3 GAUSS–JORDAN ELIMINATION METHOD

This method is a variation of Gaussian elimination method. In this method, the elements above and below the diagonal are simultaneously made zero and thereby the given system is reduced to an equivalent diagonal form using elementary transformations. Then the solution of the resulting diagonal system can be readily obtained.

Sometimes, we normalize the pivot row with respect to the pivot element, before elimination. Partial pivoting is also used whenever the pivot element becomes zero.

This method is illustrated through the following examples.

Example 3.4 Solve the system of equations

$$\left.\begin{aligned} x + 2y + z &= 8 \\ 2x + 3y + 4z &= 20 \\ 4x + 3y + 2z &= 16 \end{aligned}\right\} \tag{1}$$

using Gauss–Jordan elimination method.

Solution In matrix notation, the given system (1) can be written as

$$\begin{bmatrix} 1 & 2 & 1 \\ 2 & 3 & 4 \\ 4 & 3 & 2 \end{bmatrix} \begin{pmatrix} x \\ y \\ z \end{pmatrix} = \begin{pmatrix} 8 \\ 20 \\ 16 \end{pmatrix} \tag{2}$$

We subtract the first row multiplied by 2 and 4 from the second and third rows respectively of Eq. (2), and eliminate x

$$\begin{bmatrix} 1 & 2 & 1 \\ 0 & -1 & 2 \\ 0 & -5 & -2 \end{bmatrix} \begin{pmatrix} x \\ y \\ z \end{pmatrix} = \begin{pmatrix} 8 \\ 4 \\ -16 \end{pmatrix} \tag{3}$$

Now, we eliminate y from the first and third rows using the second row. Thus, we get

$$\begin{bmatrix} 1 & 0 & 5 \\ 0 & -1 & 2 \\ 0 & 0 & -12 \end{bmatrix} \begin{pmatrix} x \\ y \\ z \end{pmatrix} = \begin{pmatrix} 16 \\ 4 \\ -36 \end{pmatrix} \tag{4}$$

Before, eliminating z from the first and second row, normalizing the third row with respect to the pivot element, we get

$$\begin{bmatrix} 1 & 0 & 5 \\ 0 & -1 & 2 \\ 0 & 0 & 1 \end{bmatrix} \begin{pmatrix} x \\ y \\ z \end{pmatrix} = \begin{pmatrix} 16 \\ 4 \\ 3 \end{pmatrix} \tag{5}$$

Using the third row of Eq. (5), eliminating z from the first and second rows of Eq. (5), we obtain

$$\begin{bmatrix} 1 & 0 & 0 \\ 0 & -1 & 0 \\ 0 & 0 & 1 \end{bmatrix} \begin{pmatrix} x \\ y \\ z \end{pmatrix} = \begin{pmatrix} 1 \\ -2 \\ 3 \end{pmatrix} \tag{6}$$

From Eq. (16), we get the solution directly as $x = 1$, $y = 2$, $z = 3$.

3.4 CROUT'S REDUCTION METHOD

This method is based on the fact that the coefficient matrix $[A]$ of the system of equations (3.3) can be decomposed into the product of two matrices $[L]$ and $[U]$, where $[L]$ is a lower-triangular matrix and $[U]$ is an upper-triangular matrix with 1's on its main diagonal. The rules for getting $[L]$ and $[U]$ can be obtained from the fact

$$[L]\,[U] = [A] \tag{3.7}$$

For the purpose of illustration, let us consider a (3×3) general matrix in the form

$$\begin{bmatrix} l_{11} & 0 & 0 \\ l_{21} & l_{22} & 0 \\ l_{31} & l_{32} & l_{33} \end{bmatrix} \begin{bmatrix} 1 & u_{12} & u_{13} \\ 0 & 1 & u_{23} \\ 0 & 0 & 1 \end{bmatrix} = \begin{bmatrix} a_{11} & a_{12} & a_{13} \\ a_{21} & a_{22} & a_{23} \\ a_{31} & a_{32} & a_{33} \end{bmatrix} \tag{3.8}$$

The sequence of steps for getting $[L]$ and $[U]$ are given below:

Step I: Multiplying all the rows of $[L]$ by the first column of $[U]$, we get

$$l_{11} = a_{11}, \qquad l_{21} = a_{21}, \qquad l_{31} = a_{31} \tag{3.9}$$

Thus, we observe that the first column of $[L]$ is same as the first column of $[A]$.

Step II: Now, multiplying the first row of $[L]$ by the second and third columns of $[U]$, we obtain

$$l_{11}u_{12} = a_{12}, \qquad l_{11}u_{13} = a_{13}$$

or

$$u_{12} = \frac{a_{12}}{l_{11}}, \qquad u_{13} = \frac{a_{13}}{l_{11}} \tag{3.10}$$

Thus, the first row of $[U]$ is obtained. Now, we continue this process, thus getting alternately the column of $[L]$ and a row of $[U]$.

Step III: Multiply the second and third rows of $[L]$ by the second column of $[U]$ to get

$$l_{21}u_{12} + l_{22} = a_{22}, \qquad l_{31}u_{12} + l_{32} = a_{32}$$

which gives

$$l_{22} = a_{22} - l_{21}u_{12}, \qquad l_{32} = a_{32} - l_{31}u_{12} \tag{3.11}$$

Thus, the second column of $[L]$ is obtained.

Step IV: Now, multiply the second row of $[L]$ by the third column of $[U]$ which yields

$$l_{21}u_{13} + l_{22}u_{23} = a_{23} \qquad \text{or} \qquad u_{23} = \frac{a_{23} - l_{21}u_{13}}{l_{22}} \tag{3.12}$$

Step V: Lastly, we multiply the third row of $[L]$ by the third column of $[U]$ and get

$$l_{31}u_{13} + l_{32}u_{23} + l_{33} = a_{33}$$

which gives

$$l_{33} = a_{33} - l_{31}u_{13} - l_{32}u_{23} \tag{3.13}$$

Thus, the above five steps determine $[L]$ and $[U]$. This algorithm can be generalized to any linear system of order n.

Now, to obtain the solution of the linear system

$$\left.\begin{aligned} a_{11}x_1 + a_{12}x_2 + a_{13}x_3 &= b_1 \\ a_{21}x_1 + a_{22}x_2 + a_{23}x_3 &= b_2 \\ a_{31}x_1 + a_{32}x_2 + a_{33}x_3 &= b_3 \end{aligned}\right\} \tag{3.14}$$

in matrix notation as $[A]\,(X) = (B)$. Let $[A] = [L]\,[U]$, then we get,

$$[L]\,[U]\,(X) = (B) \tag{3.15}$$

Substituting $[U]\,(X) = (Z)$ in Eq. (3.15), we obtain

$$[L]\,(Z) = (B) \tag{3.16}$$

Now, Eq. (3.16) is equivalent to

$$\left.\begin{aligned} l_{11}z_1 &= b_1 \\ l_{21}z_1 + l_{22}z_2 &= b_2 \\ l_{31}z_1 + l_{32}z_2 + l_{33}z_3 &= b_3 \end{aligned}\right\} \tag{3.17}$$

The first of these equations gives z_1. Knowing z_1, the second equation of (3.17) gives z_2; then the third equation of (3.17) can be solved and z_3 is obtained. Having computed z_1, z_2 and z_3, we can compute x_1, x_2 and x_3 from equation $[U]\,(X) = (Z)$ or from

$$\begin{bmatrix} 1 & u_{12} & u_{13} \\ 0 & 1 & u_{23} \\ 0 & 0 & 1 \end{bmatrix} \begin{pmatrix} x_1 \\ x_2 \\ x_3 \end{pmatrix} = \begin{pmatrix} z_1 \\ z_2 \\ z_3 \end{pmatrix}$$

This method is also known as *Cholesky reduction method*. This technique is widely used in the numerical solutions of partial differential equations.

This method is very popular from computer programming point of view, since the storage space reserved for matrix $[A]$ can be used to store the elements of $[L]$ and $[U]$ at the end of computation.

It may be noted that this method fails if any $a_{ii} = 0$. In that case, the system is singular. In order to familiarize with the Crout's reduction method, we consider the following examples.

Example 3.5 Solve the following system of equations

$$5x_1 - 2x_2 + x_3 = 4$$

$$7x_1 + x_2 - 5x_3 = 8$$

$$3x_1 + 7x_2 + 4x_3 = 10$$

by Crout's reduction method using hand computation.

Solution Let the coefficient matrix $[A]$ be written as $[L]\,[U]$. Thus,

$$\begin{bmatrix} l_{11} & 0 & 0 \\ l_{21} & l_{22} & 0 \\ l_{31} & l_{32} & l_{33} \end{bmatrix} \begin{bmatrix} 1 & u_{12} & u_{13} \\ 0 & 1 & u_{23} \\ 0 & 0 & 1 \end{bmatrix} = \begin{bmatrix} 5 & -2 & 1 \\ 7 & 1 & -5 \\ 3 & 7 & 4 \end{bmatrix} \tag{1}$$

Step I: Multiply all the rows of $[L]$ by the first column of $[U]$, we get

$$l_{11} = 5, \qquad l_{21} = 7, \qquad l_{31} = 3 \tag{2}$$

Step II: Multiply the first row of $[L]$ by the second and third columns of $[U]$, we have

$$l_{11}u_{12} = -2, \qquad l_{11}u_{13} = 1$$

Using Eq. (2), we get

$$u_{12} = -\frac{2}{5}, \qquad u_{13} = \frac{1}{5} \tag{3}$$

Step III: Multiply the second and third rows of [L] by the second column of [U]. Using Eqs. (2) and (3), we get

$$\left.\begin{aligned} l_{21}u_{12} + l_{22} = 1 \quad &\text{or} \quad l_{22} = 1 + \frac{14}{5} = \frac{19}{5} \\ l_{31}u_{12} + l_{32} = 7 \quad &\text{or} \quad l_{32} = 7 + \frac{6}{5} = \frac{41}{5} \end{aligned}\right\} \tag{4}$$

Step IV: Multiply the second row of [L] by the third column of [U] which yields $l_{21}u_{13} + l_{22}u_{23} = -5$. Using Eqs. (2)–(4), we get

$$\frac{19}{5}u_{23} = -5 - \frac{7}{5}$$

Therefore,

$$u_{23} = -\frac{32}{19} \tag{5}$$

Step V: Finally, multiply the third row of [L] with the third column of [U], we obtain

$$l_{31}u_{13} + l_{32}u_{23} + l_{33} = 4$$

Using Eqs. (2)–(5), we get

$$l_{33} = \frac{327}{19} \tag{6}$$

Thus, the given system of equations takes the form [L] [U] [X] = (B). That is,

$$\begin{bmatrix} 5 & 0 & 0 \\ 7 & \frac{19}{5} & 0 \\ 3 & \frac{41}{5} & \frac{327}{19} \end{bmatrix} \begin{bmatrix} 1 & -\frac{2}{5} & \frac{1}{5} \\ 0 & 1 & -\frac{32}{19} \\ 0 & 0 & 1 \end{bmatrix} \begin{pmatrix} x_1 \\ x_2 \\ x_3 \end{pmatrix} = \begin{pmatrix} 4 \\ 8 \\ 10 \end{pmatrix} \tag{7}$$

Let [U](X) = (Z), then

$$[L](Z) = (4\ 8\ 10)^{\mathrm{T}}$$

or

$$\begin{bmatrix} 5 & 0 & 0 \\ 7 & \frac{19}{5} & 0 \\ 3 & \frac{41}{5} & \frac{327}{19} \end{bmatrix} \begin{pmatrix} z_1 \\ z_2 \\ z_3 \end{pmatrix} = \begin{pmatrix} 4 \\ 8 \\ 10 \end{pmatrix} \tag{8}$$

which gives

$$z_1 = \frac{4}{5}, \quad z_2 = \frac{12}{19}, \quad z_3 = \frac{46}{327} \tag{9}$$

utilizing these values of z, Eq. (7) becomes

$$\begin{bmatrix} 1 & -\frac{2}{5} & \frac{1}{5} \\ 0 & 1 & -\frac{32}{19} \\ 0 & 0 & 1 \end{bmatrix} \begin{pmatrix} x_1 \\ x_2 \\ x_3 \end{pmatrix} = \begin{pmatrix} \frac{4}{5} \\ \frac{12}{19} \\ \frac{46}{327} \end{pmatrix} \tag{10}$$

By back substitution method, we obtain

$$x_3 = \frac{46}{327}, \quad x_2 = \frac{284}{327}, \quad x_1 = \frac{366}{327}$$

This is the required solution.

3.5 JACOBI'S METHOD

Jacobi's method is an iterative method, where initial approximate solution to a given system of equations is assumed and is improved towards the exact solution in an iterative way. In general, when the coefficient matrix of the system of equations is a sparse matrix (many elements are zero), iterative methods have definite advantage over direct methods in respect of economy in computer memory. Such spare matrices arise in computing the numerical solution of partial differential equations.

To illustrate Jacobi's method, let us consider a linear system given by

$$\left.\begin{aligned} a_{11}x_1 + a_{12}x_2 + \cdots + a_{1n}x_n &= b_1 \\ a_{21}x_1 + a_{22}x_2 + \cdots + a_{2n}x_n &= b_2 \\ \vdots \quad\quad \vdots \quad\quad\quad\quad \vdots \quad &\quad \vdots \\ a_{n1}x_1 + a_{n2}x_2 + \cdots + a_{nn}x_n &= b_n \end{aligned}\right\} \tag{3.18}$$

In this method, we assume that the coefficient matrix $[A]$ is strictly diagonally dominant, that is, in each row of $[A]$ the modulus of the diagonal element exceeds the sum of the off-diagonal elements. We also assume that the diagonal element a_{ii} do not vanish. If any diagonal element vanishes, the equations can always be rearranged to satisfy this condition. Now the system (3.18) can be written as

$$\left.\begin{aligned} x_1 &= \frac{b_1}{a_{11}} - \frac{a_{12}}{a_{11}}x_2 - \cdots - \frac{a_{1n}}{a_{11}}x_n \\ x_2 &= \frac{b_2}{a_{22}} - \frac{a_{21}}{a_{22}}x_1 - \cdots - \frac{a_{2n}}{a_{22}}x_n \\ \vdots \quad & \quad \vdots \quad\quad \vdots \quad\quad\quad\quad \vdots \\ x_n &= \frac{b_n}{a_{nn}} - \frac{a_{n1}}{a_{nn}}x_1 - \cdots - \frac{a_{n(n-1)}}{a_{nn}}x_{n-1} \end{aligned}\right\} \tag{3.19}$$

We shall take this solution vector $(x_1, x_2, \ldots, x_n)^T$ as a first approximation to the exact solution of system (3.18). For convenience, let us denote the first approximation vector by $(x_1^{(1)}, x_2^{(1)}, \ldots, x_n^{(1)})$ got after taking (0, 0, ..., 0) as an initial starting vector. Substituting this first approximation in the right-hand size of system (3.19), we obtain the second approximation to the given system in the form

$$\left.\begin{aligned} x_1^{(2)} &= \frac{b_1}{a_{11}} - \frac{a_{12}}{a_{11}}x_2^{(1)} - \cdots - \frac{a_{1n}}{a_{11}}x_n^{(1)} \\ x_2^{(2)} &= \frac{b_2}{a_{22}} - \frac{a_{21}}{a_{22}}x_1^{(1)} - \cdots - \frac{a_{2n}}{a_{22}}x_n^{(1)} \\ &\vdots \\ x_n^{(2)} &= \frac{b_n}{a_{nn}} - \frac{a_{n1}}{a_{nn}}x_1^{(1)} - \cdots - \frac{a_{n(n-1)}}{a_{nn}}x_{n-1}^{(1)} \end{aligned}\right\} \quad (3.20)$$

This second approximation is substituted into the right-hand side of Eqs. (3.20) and obtain the third approximation and so on. This process is repeated and $(r + 1)$th approximation is calculated from

$$\left.\begin{aligned} x_1^{(r+1)} &= \frac{b_1}{a_{11}} - \frac{a_{12}}{a_{11}}x_2^{(r)} - \cdots - \frac{a_{1n}}{a_{11}}x_n^{(r)} \\ x_2^{(r+1)} &= \frac{b_2}{a_{22}} - \frac{a_{21}}{a_{22}}x_1^{(r)} - \cdots - \frac{a_{2n}}{a_{22}}x_n^{(r)} \\ &\vdots \\ x_n^{(r+1)} &= \frac{b_n}{a_{nn}} - \frac{a_{n1}}{a_{nn}}x_1^{(r)} - \cdots - \frac{a_{n(n-1)}}{a_{nn}}x_{n-1}^{(r)} \end{aligned}\right\} \quad (3.21)$$

Briefly, we can rewrite Eqs. (3.21) as

$$x_i^{(r+1)} = \frac{b_i}{a_{ii}} - \sum_{\substack{j=1 \\ j\neq i}}^{n} \frac{a_{ij}}{a_{ii}} x_j^{(r)}, \qquad r = 1, 2, \cdots, \quad i = 1, 2, \cdots, n \quad (3.22)$$

This method is due to Jacobi and is called *Jacobi's iterative method.* It is also known as method of *simultaneous displacements*, since no element of $x_i^{(r+1)}$ is used in this iteration until every element is computed.

A sufficient condition for convergence of the iterative solution to the exact solution is

$$|a_{ii}| > \sum_{\substack{j=1 \\ j\neq i}}^{n} |a_{ij}|, \qquad i = 1, 2, \cdots, n \quad (3.23)$$

When this condition (diagonal dominance) is true, Jacobi's method converges.

Example 3.6 Find the solution to the following system of equations

$$\begin{aligned} 83x + 11y - 4z &= 95 \\ 7x + 52y + 13z &= 104 \\ 3x + 8y + 29z &= 71 \end{aligned}$$

using Jacobi's iterative method for the first five iterations.

Solution At first, we rewrite the given system in the form

$$\left.\begin{aligned} x &= \frac{95}{83} - \frac{11}{83}y + \frac{4}{83}z \\ y &= \frac{104}{52} - \frac{7}{52}x - \frac{13}{52}z \\ z &= \frac{71}{29} - \frac{3}{29}x - \frac{8}{29}y \end{aligned}\right\} \tag{1}$$

Taking the initial starting of solution vector as $(0, 0, 0)^{\mathrm{T}}$, from Eq. (1), we have the first approximation as

$$\begin{pmatrix} x^{(1)} \\ y^{(1)} \\ z^{(1)} \end{pmatrix} = \begin{pmatrix} 1.1446 \\ 2.0000 \\ 2.4483 \end{pmatrix} \tag{2}$$

Now, using Eq. (1), the second approximation is computed from the equations

$$\left.\begin{aligned} x^{(2)} &= 1.1446 - 0.1325y^{(1)} + 0.0482z^{(1)} \\ y^{(2)} &= 2.0 - 0.1346x^{(1)} - 0.25z^{(1)} \\ z^{(2)} &= 2.4483 - 0.1035x^{(1)} - 0.2759y^{(1)} \end{aligned}\right\} \tag{3}$$

Substituting Eq. (2) into Eq. (3), we get the second approximation as

$$\begin{pmatrix} x^{(2)} \\ y^{(2)} \\ z^{(2)} \end{pmatrix} = \begin{pmatrix} 0.9976 \\ 1.2339 \\ 1.7424 \end{pmatrix} \tag{4}$$

Similar procedure yields the third, fourth and fifth approximations to the required solution and they are tabulated as below:

Iteration	Variables		
number, r	x	y	z
1	1.1446	2.0000	2.4483
2	0.9976	1.2339	1.7424
3	1.0651	1.4301	2.0046
4	1.0517	1.3555	1.9435
5	1.0587	1.3726	1.9655

3.6 GAUSS–SEIDEL ITERATION METHOD

It is another well-known iterative method for solving a system of linear equations of the form of system (3.18). In Jacobi method, the $(r + 1)$th approximation to the system (3.18) is given by Eqs. (3.21), from which we can observe that no element of $x_i^{(r+1)}$ replaces $x_i^{(r)}$ entirely for the next cycle of computation.

However, in Gauss–Seidel method, the corresponding elements of $x_i^{(r+1)}$ replaces those of $x_i^{(r)}$ as soon as they become available. Hence, it is called the method of *successive displacements*. For illustration, consider the system (3.18). In Gauss–Seidel iteration, the $(r + 1)$th approximation or iteration is computed from

$$\left.\begin{aligned}
x_1^{(r+1)} &= \frac{b_1}{a_{11}} - \frac{a_{12}}{a_{11}}x_2^{(r)} - \cdots - \frac{a_{1n}}{a_{11}}x_n^{(r)} \\
x_2^{(r+1)} &= \frac{b_2}{a_{22}} - \frac{a_{21}}{a_{22}}x_1^{(r+1)} - \cdots - \frac{a_{2n}}{a_{22}}x_n^{(r)} \\
&\vdots \\
x_n^{(r+1)} &= \frac{b_n}{a_{nn}} - \frac{a_{n1}}{a_{nn}}x_1^{(r+1)} - \cdots - \frac{a_{n,(n-1)}}{a_{nn}}x_{n-1}^{(r+1)}
\end{aligned}\right\} \tag{3.24}$$

Thus, the general procedure can be written in the following compact form

$$x_i^{(r+1)} = \frac{b_i}{a_{ii}} - \sum_{j=1}^{i-1} \frac{a_{ij}}{a_{ii}} x_j^{(r+1)} - \sum_{j=i+1}^{n} \frac{a_{ij}}{a_{ii}} x_j^{(r)} \tag{3.25}$$

for all $i = 1, 2, \ldots, n$ and $r = 1, 2, \ldots$

To describe system (3.24); in the first equation, we substitute the rth approximation into the right-hand side and denote the result by $x_1^{(r+1)}$. In the second equation, we substitute $(x_1^{(r+1)}, x_3^{(r)}, \ldots, x_n^{(r)})$ and denote the result by $x_2^{(r+1)}$. In the third equation, we substitute $(x_1^{(r+1)}, x_2^{(r+1)}, x_4^{(r)}, \ldots, x_n^{(r)})$ and denote the result by $x_3^{(r+1)}$, and so on. This process is continued till we arrive at the desired result. For illustration, we consider the following example.

Example 3.7 Find the solution of the following system of equations

$$x_1 - \frac{1}{4}x_2 - \frac{1}{4}x_3 = \frac{1}{2}$$

$$-\frac{1}{4}x_1 + x_2 - \frac{1}{4}x_4 = \frac{1}{2}$$

$$-\frac{1}{4}x_1 + x_3 - \frac{1}{4}x_4 = \frac{1}{4}$$

$$-\frac{1}{4}x_2 - \frac{1}{4}x_3 + x_4 = \frac{1}{4}$$

using Gauss–Seidel method and perform the first-five iterations.

Solution The given system of equations can be rewritten as

$$\left.\begin{aligned} x_1 &= 0.5 + 0.25x_2 + 0.25x_3 \\ x_2 &= 0.5 + 0.25x_1 + 0.25x_4 \\ x_3 &= 0.25 + 0.25x_1 + 0.25x_4 \\ x_4 &= 0.25 + 0.25x_2 + 0.25x_3 \end{aligned}\right\} \qquad (1)$$

Taking $x_2 = x_3 = x_4 = 0$ on the right-hand side of the first equation of system (1), we get $x_1^{(1)} = 0.5$. Taking $x_3 = x_4 = 0$ and the current value of x_1, we get

$$x_2^{(1)} = 0.5 + (0.25)\,(0.5) + 0 = 0.625$$

from the second equation of system (1). Further, we take $x_4 = 0$ and the current value of x_1, we obtain

$$x_3^{(1)} = 0.25 + (0.25)\,(0.5) + 0 = 0.375$$

from the third equation of system (1). Now, using the current values of x_2 and x_3, the fourth equation of system (1) gives

$$x_4^{(1)} = 0.25 + (0.25)\,(0.625) + (0.25)\,(0.375) = 0.5$$

The Gauss–Seidel iterations for the given set of equations can be written as

$$\begin{aligned} x_1^{(r+1)} &= 0.5 + 0.25x_2^{(r)} + 0.25x_3^{(r)} \\ x_2^{(r+1)} &= 0.5 + 0.25x_1^{(r+1)} + 0.25x_4^{(r)} \\ x_3^{(r+1)} &= 0.25 + 0.25x_1^{(r+1)} + 0.25x_4^{(r)} \\ x_4^{(r+1)} &= 0.25 + 0.25x_2^{(r+1)} + 0.25x_3^{(r+1)} \end{aligned}$$

Now, by Gauss–Seidel procedure, the second and subsequent approximations can be obtained and the sequence of the first-five approximations are tabulated as below:

Iteration	Variables			
number r	x_1	x_2	x_3	x_4
1	0.5	0.625	0.375	0.5
2	0.75	0.8125	0.5625	0.59375
3	0.84375	0.85938	0.60938	0.61719
4	0.86719	0.87110	0.62110	0.62305
5	0.87305	0.87402	0.62402	0.62451

3.7 THE RELAXATION METHOD

This method is an iterative method and is due to Southwell. To explain the details, consider again the system of equations (3.18). Let

$$X^{(p)} = (x_1^{(p)}, x_2^{(p)}, \ldots, x_n^{(p)})^{\mathrm{T}}$$

be the solution vector obtained iteratively after pth iteration. If $R_i^{(p)}$ denotes the residual of the ith equation of system (3.18), that is of

$$a_{i1}x_1 + a_{i2}x_2 + \cdots + a_{in}x_n = b_i$$

defined by

$$R_i^{(p)} = b_i - a_{i1}x_1^{(p)} - a_{i2}x_2^{(p)} - \cdots - a_{in}x_n^{(p)}$$

then, we can improve the solution vector successively by reducing the largest residual to zero at that iteration. This is the basic idea of *relaxation method.*

To achieve the fast convergence of the procedure, we take all terms to one side and then reorder the equations so that the largest negative coefficients in the equations appear on the diagonal. Now, if at any iteration, R_i is the largest residual in magnitude, then we give an increment

$$dx_i = -\frac{R_i}{a_{ii}} \tag{3.26}$$

to x_i; a_{ii} being the coefficient of x_i. In other words, we change x_i to $(x_i + dx_i)$ to relax R_i; that is to reduce R_i to zero. We shall illustrate this procedure through an example.

Example 3.8 Solve the system of equations

$$6x_1 - 3x_2 + x_3 = 11$$
$$2x_1 + x_2 - 8x_3 = -15$$
$$x_1 - 7x_2 + x_3 = 10$$

by the relaxation method, starting with the vector (0, 0, 0).

Solution At first, we transfer all the terms to the right-hand side and reorder the equations, so that the largest coefficients in the equations appear on the diagonal. Thus, we get

$$\left.\begin{aligned} 0 &= 11 - 6x_1 + 3x_2 - x_3 \\ 0 &= 10 - x_1 + 7x_2 - x_3 \\ 0 &= -15 - 2x_1 - x_2 + 8x_3 \end{aligned}\right\} \tag{1}$$

after interchanging the second and third rows. Starting with the initial solution vector (0, 0, 0), that is taking $x_1 = x_2 = x_3 = 0$, we find the residuals $R_1 = 11$, $R_2 = 10$, $R_3 = -15$ of which the largest residual in magnitude is R_3. It indicates that the third equation has more error and needs immediate attention for improvement. Thus, we introduce a change, dx_3, in x_3 which is obtained from the formula

$$dx_3 = -\frac{R_3}{a_{33}} = \frac{15}{8} = 1.875$$

similarly, we find the new residuals of large magnitude and relax it to zero, and so on. We shall continue this process, until all the residuals are zero or very small.

The detailed calculations are shown as in the table.

Iteration number	Residuals R_1	R_2	R_3	Maximum R_i	Difference dx_i	Variables x_1	x_2	x_3
0	11	10	−15	−15	15/8 = 1.875	0	0	0
1	9.125	8.125	0	9.125	−9.125/(−6) = 1.5288	0	0	1.875
2	0.0478	6.5962	−3.0576	6.5962	−6.5962/7 = −0.9423	1.5288	0	1.875
3	−2.8747	0.0001	−2.1153	−2.8747	2.8747/(−6) = −0.4791	1.5288	−0.9423	1.875
4	−0.0031	0.4792	−1.1571	−1.1571	1.1571/8 = 0.1446	1.0497	−0.9423	1.875
5	−0.1447	0.3346	0.0003	.3346	−.3346/7 = −0.0478	1.0497	−0.9423	2.0196
6	0.2881	0.0000	0.0475	.2881	−.2881/(−6) = 0.0480	1.0497	−0.9901	2.0196
7	−0.0001	0.048	0.1435	0.1435	−0.1435/8 = −0.0179	1.0017	−0.9901	2.0196
8	0.0178	0.0659	0.0003	–	–	1.0017	−0.9901	2.0017

At this stage, we observe that all the residuals R_1, R_2 and R_3 are small enough, and therefore, we may take the corresponding values of x_i at this iteration as the solution. Hence, the numerical solution to the given system is given by

$$x_1 = 1.0017, \quad x_2 = -0.9901, \quad x_3 = 2.0017,$$

The exact solution is found to be

$$x_1 = 1.0, \quad x_2 = -1.0, \quad x_3 = 2.0$$

3.8 MATRIX INVERSION

Consider a system of equations in the form

$$[A]\,(X) = (B) \tag{3.27}$$

One way of writing its solution is in the form

$$(X) = [A]^{-1}\,(B) \tag{3.28}$$

Thus, the solution to the system (3.27) can also be obtained if the inverse of the coefficient matrix $[A]$ is known. Alternatively, if the product of two square matrices is an identity matrix, that is, if

$$[A]\,[B] = [I] \tag{3.29}$$

then,

$$[B] = [A]^{-1} \quad \text{and} \quad [A] = [B]^{-1}$$

Every square non-singular matrix will have an inverse. Gauss elimination and Gauss–Jordan methods are popular among many methods available for finding the inverse of a matrix.

3.8.1 Gaussian Elimination Method

In this method, if A is a given matrix, for which we have to find the inverse; at first, we place an identity matrix, whose order is same as that of A, adjacent to A which we call an *augmented matrix*. Then the inverse of A is computed in two stages. In the first stage, A is converted into an upper triangular form, using Gaussian elimination method as discussed in Section 3.2. In the second stage, the above upper triangular matrix is reduced to an identity matrix by row transformations. All these operations are also performed on the adjacently placed identity matrix. Finally, when A is transformed into an identity matrix, the adjacent matrix gives the inverse of A. In order to increase the accuracy of the result, it is essential to employ partial pivoting. To understand the sequence of the steps involved, we consider an example.

Example 3.9 Use the Gaussian elimination method to find the inverse of the matrix

$$A = \begin{bmatrix} 1 & 1 & 1 \\ 4 & 3 & -1 \\ 3 & 5 & 3 \end{bmatrix}$$

Solution At first, we place an identity matrix of the same order adjacent to the given matrix. Thus, the augmented matrix can be written as

$$\left[\begin{array}{ccc|ccc} 1 & 1 & 1 & 1 & 0 & 0 \\ 4 & 3 & -1 & 0 & 1 & 0 \\ 3 & 5 & 3 & 0 & 0 & 1 \end{array}\right] \tag{1}$$

Stage I (Reduction to upper triangular form): Let R_1, R_2 and R_3 denote the first, second and third rows of a matrix. In the first column of Eq. (1), 4 is the largest element, thus interchanging R_1 and R_2 to bring the pivot element 4 to the place of a_{11}, we have the augmented matrix in the form

$$\left[\begin{array}{ccc|ccc} 4 & 3 & -1 & 0 & 1 & 0 \\ 1 & 1 & 1 & 1 & 0 & 0 \\ 3 & 5 & 3 & 0 & 0 & 1 \end{array}\right] \tag{2}$$

Divide R_1 by 4 to get

$$\left[\begin{array}{ccc|ccc} 1 & \frac{3}{4} & -\frac{1}{4} & 0 & \frac{1}{4} & 0 \\ 1 & 1 & 1 & 1 & 0 & 0 \\ 3 & 5 & 3 & 0 & 0 & 1 \end{array}\right] \tag{3}$$

Perform $R_2 - R_1 \longrightarrow R_2$, which gives

$$\left[\begin{array}{ccc|ccc} 1 & \frac{3}{4} & -\frac{1}{4} & 0 & \frac{1}{4} & 0 \\ 0 & \frac{1}{4} & \frac{5}{4} & 1 & -\frac{1}{4} & 0 \\ 3 & 5 & 3 & 0 & 0 & 1 \end{array}\right] \tag{4}$$

Perform $R_3 - 3R_1 \longrightarrow R_3$ in Eq. (4), which yields

$$\left[\begin{array}{ccc|ccc} 1 & \frac{3}{4} & -\frac{1}{4} & 0 & \frac{1}{4} & 0 \\ 0 & \frac{1}{4} & \frac{5}{4} & 1 & -\frac{1}{4} & 0 \\ 0 & \frac{11}{4} & \frac{15}{4} & 0 & -\frac{3}{4} & 1 \end{array}\right] \tag{5}$$

Now, looking at the second column for the pivot, the max (1/4, 11/4) is 11/4. Therefore, we interchange R_2 and R_3 in Eq. (5) and get

$$\left[\begin{array}{ccc|ccc} 1 & \frac{3}{4} & -\frac{1}{4} & 0 & \frac{1}{4} & 0 \\ 0 & \frac{11}{4} & \frac{15}{4} & 0 & -\frac{3}{4} & 1 \\ 0 & \frac{1}{4} & \frac{5}{4} & 1 & -\frac{1}{4} & 0 \end{array}\right] \tag{6}$$

Now, divide R_2 by the pivot a_{22} = 11/4, and obtain

$$\left[\begin{array}{ccc|ccc} 1 & \frac{3}{4} & -\frac{1}{4} & 0 & \frac{1}{4} & 0 \\ 0 & 1 & \frac{15}{11} & 0 & -\frac{3}{11} & \frac{4}{11} \\ 0 & \frac{1}{4} & \frac{5}{4} & 1 & -\frac{1}{4} & 0 \end{array}\right] \tag{7}$$

Performing $R_3 - (1/4)R_2 \longrightarrow R_3$ in (7) yields

$$\left[\begin{array}{ccc|ccc} 1 & \frac{3}{4} & -\frac{1}{4} & 0 & \frac{1}{4} & 0 \\ 0 & 1 & \frac{15}{11} & 0 & -\frac{3}{11} & \frac{4}{11} \\ 0 & 0 & \frac{10}{11} & 1 & -\frac{2}{11} & -\frac{1}{11} \end{array}\right] \tag{8}$$

Finally, we divide R_3 by (10/11), thus getting an upper triangular form

$$\left[\begin{array}{ccc|ccc} 1 & \frac{3}{4} & -\frac{1}{4} & 0 & \frac{1}{4} & 0 \\ 0 & 1 & \frac{15}{11} & 0 & -\frac{3}{11} & \frac{4}{11} \\ 0 & 0 & 1 & \frac{11}{10} & -\frac{1}{5} & -\frac{1}{10} \end{array}\right] \tag{9}$$

Stage II (Reduction to an identity matrix): Multiply R_3 by $-1/4$ and $15/11$ respectively and subtract it from R_1 and R_2 of Eq. (9), we get

$$\left[\begin{array}{ccc|ccc} 1 & \frac{3}{4} & 0 & \frac{11}{40} & \frac{1}{5} & -\frac{1}{40} \\ 0 & 1 & 0 & -\frac{3}{2} & 0 & \frac{1}{2} \\ 0 & 0 & 1 & \frac{11}{10} & -\frac{1}{5} & -\frac{1}{10} \end{array}\right] \tag{10}$$

Finally, performing $R_1 - (3/4)\, R_2 \longrightarrow R_1$ in Eq. (10), we obtain

$$\left[\begin{array}{ccc|ccc} 1 & 0 & 0 & \frac{7}{5} & \frac{1}{5} & -\frac{2}{5} \\ 0 & 1 & 0 & -\frac{3}{2} & 0 & \frac{1}{2} \\ 0 & 0 & 1 & \frac{11}{10} & -\frac{1}{5} & -\frac{1}{10} \end{array}\right]$$

Thus, we have

$$A^{-1} = \begin{bmatrix} \frac{7}{5} & \frac{1}{5} & -\frac{2}{5} \\ -\frac{3}{2} & 0 & \frac{1}{2} \\ \frac{11}{10} & -\frac{1}{5} & -\frac{1}{10} \end{bmatrix} \tag{11}$$

We can easily cheque $[A]\,[A^{-1}] = [I]$.

3.8.2 Gauss–Jordan Method

This method is similar to Gaussian elimination method, with the essential difference that the stage *I* of reducing the given matrix to an upper triangular form is not needed. However, the given matrix can be directly reduced to an identity matrix using elementary row transformations. This technique is illustrated in the following example.

Example 3.10 Find the inverse of

$$A = \begin{bmatrix} 1 & 1 & 1 \\ 4 & 3 & -1 \\ 3 & 5 & 3 \end{bmatrix}$$

by Gauss–Jordan method.

Solution Let R_1, R_2 and R_3 denote the first, second and third rows of a matrix. We place an identity matrix adjacent to the given matrix as a first step and the resulting augmented matrix is given by

$$\left[\begin{array}{ccc|ccc} 1 & 1 & 1 & 1 & 0 & 0 \\ 4 & 3 & -1 & 0 & 1 & 0 \\ 3 & 5 & 3 & 0 & 0 & 1 \end{array}\right] \tag{1}$$

Performing $R_2 - 4R_1 \longrightarrow R_2$, we get

$$\left[\begin{array}{ccc|ccc} 1 & 1 & 1 & 1 & 0 & 0 \\ 0 & -1 & -5 & -4 & 1 & 0 \\ 3 & 5 & 3 & 0 & 0 & 1 \end{array}\right] \tag{2}$$

Now, performing $R_3 - 3R_1 \longrightarrow R_3$, we obtain

$$\left[\begin{array}{ccc|ccc} 1 & 1 & 1 & 1 & 0 & 0 \\ 0 & -1 & -5 & -4 & 1 & 0 \\ 0 & 2 & 0 & -3 & 0 & 1 \end{array}\right] \tag{3}$$

Carrying out further operations $R_2 + R_1 \longrightarrow R_1$ and $R_3 + 2R_2 \longrightarrow R_3$, we arrive at

$$\left[\begin{array}{ccc|ccc} 1 & 0 & -4 & -3 & 1 & 0 \\ 0 & -1 & -5 & -4 & 1 & 0 \\ 0 & 0 & -10 & -11 & 2 & 1 \end{array}\right] \tag{4}$$

Now, dividing the third row by -10, we get

$$\left[\begin{array}{ccc|ccc} 1 & 0 & -4 & -3 & 1 & 0 \\ 0 & -1 & -5 & -4 & 1 & 0 \\ 0 & 0 & 1 & \dfrac{11}{10} & -\dfrac{1}{5} & -\dfrac{1}{10} \end{array}\right] \tag{5}$$

Further, we perform $R_1 + 4R_3 \longrightarrow R_1$, and $R_2 + 5R_3 \longrightarrow R_2$ to get

$$\left[\begin{array}{ccc|ccc} 1 & 0 & 0 & \frac{7}{5} & \frac{1}{5} & -\frac{2}{5} \\ 0 & -1 & 0 & \frac{3}{2} & 0 & -\frac{1}{2} \\ 0 & 0 & 1 & \frac{11}{10} & -\frac{1}{5} & -\frac{1}{10} \end{array}\right] \quad (6)$$

Finally, multiplying R_2 by -1, we obtain

$$\left[\begin{array}{ccc|ccc} 1 & 0 & 0 & \frac{7}{5} & \frac{1}{5} & -\frac{2}{5} \\ 0 & 1 & 0 & -\frac{3}{2} & 0 & \frac{1}{2} \\ 0 & 0 & 1 & \frac{11}{10} & -\frac{1}{5} & -\frac{1}{10} \end{array}\right] \quad (7)$$

Hence, we have

$$A^{-1} = \begin{bmatrix} \frac{7}{5} & \frac{1}{5} & -\frac{2}{5} \\ -\frac{3}{2} & 0 & \frac{1}{2} \\ \frac{11}{10} & -\frac{1}{5} & -\frac{1}{10} \end{bmatrix} \quad (8)$$

It can be easily verified that $[A]\ [A^{-1}] = [I]$.

EXERCISES

3.1 Solve the following systems of equations

(i) $x_1 + \frac{1}{2}x_2 + \frac{1}{3}x_3 = 1$

$\frac{1}{2}x_1 + \frac{1}{3}x_2 + \frac{1}{4}x_3 = 0$

$\frac{1}{3}x_1 + \frac{1}{4}x_2 + \frac{1}{5}x_3 = 0$

(ii) $4x_1 + x_2 + x_3 = 4$

$x_1 + 4x_2 - 2x_3 = 4$

$3x_1 + 2x_2 - 4x_3 = 6$

(iii) $10x - 7y + 3z + 5w = 6$

$-6x + 8y - z - 4w = 5$

$3x + y + 4z + 11w = 2$

$5x - 9y - 2z + 4w = 7$

by Gaussian elimination method.

3.2 Using Gaussian elimination method with partial pivoting, solve the following systems of equations

(i)
$$\begin{aligned} x_1 + x_2 - 2x_3 &= 3 \\ 4x_1 - 2x_2 + x_3 &= 5 \\ 3x_1 - x_2 + 3x_3 &= 8 \end{aligned}$$

(ii)
$$\begin{aligned} 2.5x - 3y + 4.6z &= -1.05 \\ -3.5x + 2.6y + 1.5z &= -14.46 \\ -6.5x - 3.5y + 7.3z &= -17.735 \end{aligned}$$

3.3 Using Gauss–Jordan elimination method, solve the systems of equations

(i)
$$\begin{aligned} 10x_1 + x_2 + x_3 &= 12 \\ x_1 + 10x_2 + x_3 &= 12 \\ x_1 + x_2 + 10x_3 &= 12 \end{aligned}$$

(ii)
$$\begin{aligned} x + y + z &= 7 \\ 3x + 3y + 4z &= 24 \\ 2x + y + 3z &= 16 \end{aligned}$$

3.4 Using Crout's reduction method, solve the following systems of equations

(i)
$$\begin{aligned} x + y + z &= 3 \\ 2x - y + 3z &= 16 \\ 3x + y - z &= -3 \end{aligned}$$

(ii)
$$\begin{aligned} 6x_1 - x_2 &= 3 \\ -x_1 + 6x_2 - x_3 &= 4 \\ -x_2 + 6x_3 &= 3 \end{aligned}$$

3.5 Solve the following system of equations

$$\begin{aligned} 4x_1 - 3x_2 + 2x_3 &= 11 \\ 2x_1 + x_2 + 7x_3 &= 2 \\ 3x_1 - x_2 + 5x_3 &= 8 \end{aligned}$$

using Cholesky's reduction method.

3.6 Using Crout's reduction, decompose the matrix

$$[A] = \begin{bmatrix} 5 & -2 & 1 \\ 7 & 1 & -5 \\ 3 & 7 & 4 \end{bmatrix}$$

into $[L]$ $[U]$ form and hence solve the system of equations

$$\begin{aligned} 5x - 2y + z &= 4 \\ 7x + y - 5z &= 8 \\ 3x + 7y + 4z &= 10 \end{aligned}$$

3.7 Explain how Jacobi's method is used to obtain numerical solution of a system of linear equations. What is the condition of convergence for any choice for the first approximation.

3.8 Find the solution of the following system of equations

$$\begin{aligned} x_1 - \frac{1}{4}x_2 - \frac{1}{4}x_3 &= \frac{1}{2} \\ -\frac{1}{4}x_1 + x_2 - \frac{1}{4}x_4 &= \frac{1}{2} \\ -\frac{1}{4}x_1 + x_3 - \frac{1}{4}x_4 &= \frac{1}{4} \\ -\frac{1}{4}x_2 - \frac{1}{4}x_3 + x_4 &= \frac{1}{4} \end{aligned}$$

using Jacobi method of iteration. Carry computation up to the seventh iteration.

3.9 Solve the following system of equations

$$\begin{aligned} 2x_1 - x_2 &= 7 \\ -x_1 + 2x_2 - x_3 &= 1 \\ -x_2 + 2x_3 &= 1 \end{aligned}$$

using Gauss–Seidel method of iteration and perform the first-five iterations. The exact solution is $(6\ 5\ 3)^T$.

3.10 Solve the system of equations

$$\begin{aligned} 20x + y - 2z &= 17 \\ 3x + 20y - z &= -18 \\ 2x - 3y + 20z &= 25 \end{aligned}$$

by Gauss–Seidel iterative method and perform the first-three iterations.

3.11 Solve the following system of equations

$$\begin{aligned} 5x - 2y + z &= 13 \\ 3x + 7y - 11z &= 2 \\ x + 20y - 2z &= 8 \end{aligned}$$

by relaxation method.

3.12 Using relaxation method, find the solution to the system of equations

$$\begin{aligned} 8x_1 + x_2 - x_3 &= 8 \\ 2x_1 + x_2 + 9x_3 &= 12 \\ x_1 - 7x_2 + 2x_3 &= -4 \end{aligned}$$

taking the initial solution vector (0, 0, 0).

3.13 Using Gaussian elimination method, find the inverse of the matrix

$$A = \begin{bmatrix} 0 & 1 & 2 \\ 1 & 2 & 3 \\ 3 & 1 & 1 \end{bmatrix}$$

3.14 Find the inverse of the matrix

$$A = \begin{bmatrix} 2 & 0 & 1 \\ 3 & 2 & 5 \\ 1 & -1 & 0 \end{bmatrix}$$

3.15 Find the inverse of the following matrices

$$\text{(i) } A = \begin{bmatrix} 1 & 1 & 3 \\ 1 & 3 & -3 \\ -2 & -4 & -4 \end{bmatrix} \qquad \text{(ii) } A = \begin{bmatrix} 1 & 1 & 2 \\ 1 & 2 & 4 \\ 2 & 4 & 7 \end{bmatrix}$$

by Gauss–Jordan method.

Chapter 4

Eigenvalue Problems

4.1 INTRODUCTION

Computation of eigenvalues and the corresponding eigenvectors of a matrix is of practical importance. For example, in solid mechanics, where we consider an element in a continuum, subjected to normal and shear stresses, usually one will be interested in finding the principal stresses, which are the maximum and minimum stresses in an element.

Consider the wedge of unit thickness, subjected to normal and shear stresses. If it has to be in equilibrium, a system of equations written in matrix notation as

$$\begin{bmatrix} \sigma_x - \sigma_\theta & \sigma_{xy} \\ \sigma_{xy} & \sigma_y - \sigma_\theta \end{bmatrix} \begin{pmatrix} \cos\theta \\ \sin\theta \end{pmatrix} = \begin{pmatrix} 0 \\ 0 \end{pmatrix} \tag{4.1}$$

has to be satisfied. For non-trivial solution to exist, the determinant of the matrix must be zero. Its characteristic equation is

$$\sigma_\theta^2 - \sigma_\theta(\sigma_x + \sigma_y) + (\sigma_x\sigma_y - \sigma_{xy}^2) = 0 \tag{4.2}$$

This is a quadratic equation whose roots give two eigenvalues corresponding to principal stresses.

In general, let $[A]$ be an $n \times n$ square matrix. Suppose, there exists a scalar λ and a vector $X = (x_1\ x_2\ \dots\ x_n)^T$ such that

$$[A]\,(X) = \lambda\,(X) \tag{4.3}$$

then λ is the eigenvalue and X is the corresponding eigenvector of the matrix $[A]$. Equation (4.3) can also be written as

$$[A - \lambda I]\,(X) = (O) \tag{4.4}$$

This represents a set of n homogeneous equations possessing non-trivial solution, provided

$$|A - \lambda I| = 0 \tag{4.5}$$

This determinant, on expansion, gives an nth degree polynomial in λ, which is called *characteristic* polynomial of $[A]$, which has n roots. Corresponding to each root, we can solve Eq. (4.4) in principle, and determine a vector called *eigenvector*. However, finding the roots of the characteristic equation is laborious. Hence, we look for better methods suitable from the point of view of computation. Depending upon the type of matrix [A] and on what one is looking for, various numerical methods are available. Power method, Jacobi's method, Given's method, Householder, Lanczos method, Ruthishauser and Francis method are well known in the literature.

In this chapter, we shall consider only real and real-symmetric matrices and discuss power method and Jacobi's method in detail. For further study, one can consult Wilkinson (1965).

4.2 POWER METHOD

To compute the largest eigenvalue and the corresponding eigenvector of the system

$$[A](X) = \lambda (X)$$

where [A] is a real, symmetric or unsymmetric matrix, the power method is widely used in practice. It is an iterative technique. Let $\lambda_1, \lambda_2, \ldots, \lambda_n$ be the distinct eigenvalues of an $(n \times n)$ matrix [A], such that

$$|\lambda_1| > |\lambda_2| > \cdots > |\lambda_n| \tag{4.6}$$

and suppose $v_1, v_2, \ldots, v_n$ are the corresponding eigenvectors. Power method is applicable if the above eigenvalues are real and distinct, and hence, the corresponding eigenvectors are linearly independent. Then, any eigenvector v in the space spanned by the eigenvectors $v_1, v_2, \ldots, v_n$ can be written as their linear combination. Therefore,

$$v = c_1 v_1 + c_2 v_2 + \cdots + c_n v_n \tag{4.7}$$

Pre-multiplying Eq. (4.7) by A and substituting

$$Av_1 = \lambda_1 v_1, \qquad Av_2 = \lambda_2 v_2, \ldots, \qquad Av_n = \lambda_n v_n$$

We get

$$Av = \lambda_1 \left(c_1 v_1 + c_2 \frac{\lambda_2}{\lambda_1} v_2 + \cdots + c_n \frac{\lambda_n}{\lambda_1} v_n \right) \tag{4.8}$$

Again, pre-multiplying by A and simplifying, we obtain

$$A^2 v = \lambda_1^2 \left[c_1 v_1 + c_2 \left(\frac{\lambda_2}{\lambda_1} \right)^2 v_2 + \cdots + c_n \left(\frac{\lambda_n}{\lambda_1} \right)^2 v_n \right]$$

Similarly, we have

$$A^r v = \lambda_1^r \left[c_1 v_1 + c_2 \left(\frac{\lambda_2}{\lambda_1} \right)^r v_2 + \cdots + c_n \left(\frac{\lambda_n}{\lambda_1} \right)^r v_n \right] \tag{4.9}$$

and

$$A^{r+1} v = (\lambda_1)^{r+1} \left[c_1 v_1 + c_2 \left(\frac{\lambda_2}{\lambda_1} \right)^{r+1} v_2 + \cdots + c_n \left(\frac{\lambda_n}{\lambda_1} \right)^{r+1} v_n \right] \tag{4.10}$$

Since $\lambda_i/\lambda_1 < 1$ for $i = 2, 3, \ldots, n$ and as $r \to \infty$, the right-hand sides of Eqs. (4.9) and (4.10) tend to $\lambda_1^r c_1 v_1$ and $\lambda_1^{r+1} c_1 v_1$ respectively. Now, the eigenvalue λ_1 can be computed as the limit of the ratio of the corresponding components of $A^r v$ and $A^{r+1} v$. That is,

$$\lambda_1 = \frac{\lambda_1^{r+1}}{\lambda_1^r} = \underset{r\to\infty}{\text{Lt}} \frac{(A^{r+1}v)_p}{(A^r v)_p}, \qquad p = 1, 2, \ldots, n \tag{4.11}$$

Here, the index p stands for the pth component in the corresponding vector.

Sometimes, we may be interested in finding the least eigenvalue and the corresponding eigenvector. In that case, we proceed as follows. We note that $[A]\,(X) = \lambda\,(X)$. Pre-multiplying by $[A^{-1}]$, we get

$$[A^{-1}]\,[A]\,(X) = [A^{-1}]\,\lambda\,(X) = \lambda\,[A^{-1}]\,(X)$$

or

$$(X) = \lambda\,[A^{-1}]\,(X)$$

which can be rewritten as

$$[A^{-1}]\,(X) = \frac{1}{\lambda}(X) \tag{4.12}$$

which shows that the inverse matrix has a set of eigenvalues which are the reciprocals of the eigenvalues of $[A]$. Thus, for finding the eigenvalue of the least magnitude of the matrix $[A]$, we have to apply power method to the inverse of $[A]$. In order to see that the power method converges fast, the following numerical algorithm is adopted, particularly when working with numerical examples.

Step 1: Choose the initial vector such that the largest element is unity.

Step 2: This normalized vector $v^{(0)}$ is pre-multiplied by the given matrix $[A]$.

Step 3: The resultant vector is again normalized.

Step 4: This process of iteration is continued and the new normalized vector is repeatedly pre-multiplied by the matrix $[A]$ until the required accuracy is obtained. At this point, the result looks like

$$u^{(k)} = [A]v^{(k-1)} = q_k v^{(k)}$$

Here, q_k is the desired largest eigenvalue and $v^{(k)}$ is the corresponding eigenvector.

Example 4.1 Find the eigenvalue of largest modulus, and the associated eigenvector of the matrix

$$[A] = \begin{bmatrix} 2 & 3 & 2 \\ 4 & 3 & 5 \\ 3 & 2 & 9 \end{bmatrix}$$

by power method.

Solution We choose an initial vector $v^{(0)}$ as $(1, 1, 1)^T$. Then, compute the first iteration

$$u^{(1)} = [A]v^{(0)} = \begin{bmatrix} 2 & 3 & 2 \\ 4 & 3 & 5 \\ 3 & 2 & 9 \end{bmatrix} \begin{pmatrix} 1 \\ 1 \\ 1 \end{pmatrix} = \begin{pmatrix} 7 \\ 12 \\ 14 \end{pmatrix}$$

and normalize the resultant vector to get

$$u^{(1)} = 14 \begin{pmatrix} \frac{1}{2} \\ \frac{6}{7} \\ 1 \end{pmatrix} = q_1 v^{(1)}$$

The second iteration gives,

$$u^{(2)} = [A]v^{(1)} = \begin{bmatrix} 2 & 3 & 2 \\ 4 & 3 & 5 \\ 3 & 2 & 9 \end{bmatrix} \begin{pmatrix} \frac{1}{2} \\ \frac{6}{7} \\ 1 \end{pmatrix} = \begin{pmatrix} \frac{39}{7} \\ \frac{67}{7} \\ \frac{171}{14} \end{pmatrix} = 12.2143 \begin{pmatrix} 0.456140 \\ 0.783626 \\ 1.0 \end{pmatrix} = q_2 v^{(2)}$$

Similarly, continuing this procedure, the third and subsequent iterations are given as

$$u^{(3)} = [A]v^{(2)} = \begin{bmatrix} 2 & 3 & 2 \\ 4 & 3 & 5 \\ 3 & 2 & 9 \end{bmatrix} \begin{pmatrix} 0.456140 \\ 0.783626 \\ 1.0 \end{pmatrix} = \begin{pmatrix} 5.263158 \\ 9.175438 \\ 11.935672 \end{pmatrix}$$

$$= 11.935672 \begin{pmatrix} 0.44096 \\ 0.76874 \\ 1.0 \end{pmatrix} = q_3 v^{(3)}$$

$$u^{(4)} = [A]v^{(3)} = \begin{pmatrix} 5.18814 \\ 9.07006 \\ 11.86036 \end{pmatrix} = 11.86036 \begin{pmatrix} 0.437435 \\ 0.764737 \\ 1.0 \end{pmatrix} = q_4 v^{(4)}$$

$$u^{(5)} = [A]v^{(4)} = \begin{pmatrix} 5.16908 \\ 9.04395 \\ 11.84178 \end{pmatrix} = 11.84178 \begin{pmatrix} 0.436512 \\ 0.763732 \\ 1.0 \end{pmatrix} = q_5 v^{(5)}$$

After rounding-off, we take the largest eigenvalue as $\lambda = 11.84$ and the corresponding eigenvector as

$$(X) = \begin{pmatrix} 0.44 \\ 0.76 \\ 1.00 \end{pmatrix}$$

accurate to two decimals.

4.3 JACOBI'S METHOD

Definition 4.1 An $(n \times n)$ matrix $[A]$ is said to be *orthogonal* if

$$[A]^T [A] = [I], \qquad \text{i.e. } [A]^T = [A]^{-1}$$

In order to compute all the eigenvalues and the corresponding eigenvectors of a real symmetric matrix, Jacobi's method is highly recommended. It is based on an important property from matrix theory, which states that, if $[A]$ is an $(n \times n)$ real symmetric matrix, its eigenvalues are real, and there exists an orthogonal matrix $[S]$ such that $[S^{-1}]$ $[A]$ $[S]$ is a diagonal matrix $[D]$. This diagonalization can be carried out by applying a series of orthogonal transformations $S_1, S_2, \ldots, S_n$, as explained below.

Let A be an $(n \times n)$ real symmetric matrix. Suppose $|a_{ij}|$ be numerically the largest element amongst the off-diagonal elements of A. We construct an orthogonal matrix S_1 defined as

$$s_{ij} = -\sin\theta, \quad s_{ji} = \sin\theta, \quad s_{ii} = \cos\theta, \quad s_{jj} = \cos\theta \tag{4.13}$$

while each of the remaining off-diagonal elements are zero, the remaining diagonal elements are assumed to be unity. Thus, we construct S_1 as under

$$S_1 = \begin{bmatrix} 1 & 0 & \cdots & 0 & \cdots & 0 & \cdots & 0 \\ 0 & 1 & \cdots & 0 & \cdots & 0 & \cdots & 0 \\ \vdots & \vdots & & \vdots & & \vdots & & \vdots \\ 0 & 0 & \cdots & \cos\theta & \cdots & -\sin\theta & \cdots & 0 \\ \vdots & \vdots & & \vdots & & \vdots & & \vdots \\ 0 & 0 & \cdots & \sin\theta & \cdots & \cos\theta & \cdots & 0 \\ \vdots & \vdots & & \vdots & & \vdots & & \vdots \\ 0 & 0 & \cdots & 0 & \cdots & 0 & \cdots & 1 \end{bmatrix} \begin{matrix} \\ \\ \\ \leftarrow i\text{th row} \\ \\ \leftarrow j\text{th row} \\ \\ \\ \end{matrix} \tag{4.14}$$

(with $\cos\theta$ and $\sin\theta$ in the ith column ↓ and $-\sin\theta$ and $\cos\theta$ in the jth column ↓)

where $\cos\theta$, $-\sin\theta$, $\sin\theta$ and $\cos\theta$ are inserted in (i, i), (i, j), (j, i), (j, j)th positions respectively, and elsewhere it is identical with a unit matrix. Now, we compute

$$D_1 = S_1^{-1} A S_1 = S_1^T A S_1$$

since S_1 is an orthogonal matrix, such that $S_1^{-1} = S_1^T$. After the transformation, the elements at the positions (i, j), (j, i) get annihilated, that is, d_{ij} and d_{ji} reduce to zero, which is seen as follows:

$$\begin{bmatrix} d_{ii} & d_{ij} \\ d_{ji} & d_{jj} \end{bmatrix} = \begin{bmatrix} \cos\theta & \sin\theta \\ -\sin\theta & \cos\theta \end{bmatrix} \begin{bmatrix} a_{ii} & a_{ij} \\ a_{ij} & a_{jj} \end{bmatrix} \begin{bmatrix} \cos\theta & -\sin\theta \\ \sin\theta & \cos\theta \end{bmatrix}$$

$$\begin{bmatrix} a_{ii}\cos^2\theta + 2a_{ij}\sin\theta\cos\theta + a_{jj}\sin^2\theta & (a_{jj} - a_{ii})\sin\theta\cos\theta + a_{ij}\cos 2\theta \\ (a_{jj} - a_{ii})\sin\theta\cos\theta + a_{ij}\cos 2\theta & a_{ii}\sin^2\theta + a_{jj}\cos^2\theta - 2a_{ij}\sin\theta\cos\theta \end{bmatrix}$$

Therefore, $d_{ij} = 0$, only if,

$$a_{ij}\cos 2\theta + \frac{a_{jj} - a_{ii}}{2}\sin 2\theta = 0$$

That is, if

$$\tan 2\theta = \frac{2a_{ij}}{a_{ii} - a_{jj}} \tag{4.15}$$

Thus, we choose θ such that, Eq. (4.15) is satisfied, thereby, the pair of off-diagonal elements d_{ij} and d_{ji} reduces to zero.

However, though it creates a new pair of zeros, it also introduces non-zero contributions at formerly zero positions. Also, Eq. (4.15) gives four values of θ, but to get the least possible rotation, we choose $-\pi/4 \le \theta \le \pi/4$.

As a next step, the numerically largest off-diagonal element in the newly obtained rotated matrix D_1 is identified and the above procedure is repeated using another orthogonal matrix S_2 to get D_2. That is, we obtain

$$D_2 = S_2^{-1} D_1 S_2 = S_2^{\mathrm{T}}(S_1^{\mathrm{T}} A S_1) S_2$$

Similarly, we perform a series of such two-dimensional rotations or orthogonal transformations. After making r transformations, we obtain

$$\begin{aligned} D_r &= S_r^{-1} S_{r-1}^{-1} \ldots S_2^{-1} S_1^{-1} A S_1 S_2 \ldots S_{r-1} S_r \\ &= (S_1 S_2 \ldots S_{r-1} S_r)^{-1} A (S_1 S_2 \ldots S_{r-1} S_r) \\ &= S^{-1} A S \end{aligned} \tag{4.16}$$

where $S = S_1 S_2 \ldots S_{r-1} S_r$. Now, as $r \to \infty$, D_r approaches to a diagonal matrix, with the eigenvalues on the main diagonal. The corresponding eigenvectors are the columns of S.

It is estimated that the minimum number of rotations required to transform the given $(n \times n)$ real symmetric matrix $[A]$ into a diagonal form is $n(n-1)/2$.

Example 4.2 Find all the eigenvalues and the corresponding eigenvectors of the matrix

$$A = \begin{bmatrix} 1 & \sqrt{2} & 2 \\ \sqrt{2} & 3 & \sqrt{2} \\ 2 & \sqrt{2} & 1 \end{bmatrix}$$

by Jacobi's method.

Solution The given matrix is real and symmetric. The largest off-diagonal element is found to be $a_{13} = a_{31} = 2$. Now, we compute

$$\tan 2\theta = \frac{2a_{ij}}{a_{ii} - a_{jj}} = \frac{2a_{13}}{a_{11} - a_{33}} = \frac{4}{0} = \infty$$

which gives, $\theta = \pi/4$. Thus, we construct an orthogonal matrix S_1 as

$$S_1 = \begin{bmatrix} \cos\frac{\pi}{4} & 0 & -\sin\frac{\pi}{4} \\ 0 & 1 & 0 \\ \sin\frac{\pi}{4} & 0 & \cos\frac{\pi}{4} \end{bmatrix} = \begin{bmatrix} \frac{1}{\sqrt{2}} & 0 & -\frac{1}{\sqrt{2}} \\ 0 & 1 & 0 \\ \frac{1}{\sqrt{2}} & 0 & \frac{1}{\sqrt{2}} \end{bmatrix}$$

The first rotation gives,

$$D_1 = S_1^{-1} A S_1 = \begin{bmatrix} \frac{1}{\sqrt{2}} & 0 & \frac{1}{\sqrt{2}} \\ 0 & 1 & 0 \\ -\frac{1}{\sqrt{2}} & 0 & \frac{1}{\sqrt{2}} \end{bmatrix} \begin{bmatrix} 1 & \sqrt{2} & 2 \\ \sqrt{2} & 3 & \sqrt{2} \\ 2 & \sqrt{2} & 1 \end{bmatrix} \begin{bmatrix} \frac{1}{\sqrt{2}} & 0 & -\frac{1}{\sqrt{2}} \\ 0 & 1 & 0 \\ \frac{1}{\sqrt{2}} & 0 & \frac{1}{\sqrt{2}} \end{bmatrix}$$

$$= \begin{bmatrix} 3 & 2 & 0 \\ 2 & 3 & 0 \\ 0 & 0 & -1 \end{bmatrix}$$

we may observe that the elements d_{13} and d_{31} got annihilated. To make sure that our calculations are correct up to this step, we may also observe that the sum of the diagonal elements of D_1 is same as the sum of the diagonal elements of the original matrix A.

As a second step, we choose the largest off-diagonal element of D_1 and is found to be $d_{12} = d_{21} = 2$, and compute

$$\tan 2\theta = \frac{2d_{12}}{d_{11} - d_{22}} = \frac{4}{0} = \infty$$

which again gives $\theta = \pi/4$. Thus, we construct the second rotation matrix as

$$S_2 = \begin{bmatrix} \frac{1}{\sqrt{2}} & -\frac{1}{\sqrt{2}} & 0 \\ \frac{1}{\sqrt{2}} & \frac{1}{\sqrt{2}} & 0 \\ 0 & 0 & 1 \end{bmatrix}$$

At the end of second rotation, we get

$$D_2 = S_2^{-1} D_1 S_2 = \begin{bmatrix} \frac{1}{\sqrt{2}} & \frac{1}{\sqrt{2}} & 0 \\ -\frac{1}{\sqrt{2}} & \frac{1}{\sqrt{2}} & 0 \\ 0 & 0 & 1 \end{bmatrix} \begin{bmatrix} 3 & 2 & 0 \\ 2 & 3 & 0 \\ 0 & 0 & -1 \end{bmatrix} \begin{bmatrix} \frac{1}{\sqrt{2}} & \frac{1}{\sqrt{2}} & 0 \\ -\frac{1}{\sqrt{2}} & \frac{1}{\sqrt{2}} & 0 \\ 0 & 0 & 1 \end{bmatrix}$$

$$= \begin{bmatrix} 5 & 0 & 0 \\ 0 & 1 & 0 \\ 0 & 0 & -1 \end{bmatrix} \qquad (1)$$

which turned out to be a diagonal matrix, and therefore, we stop the computation. From (1) we notice that the eigenvalues of the given matrix are 5, 1 and –1. The eigenvectors are the column vectors of $S = S_1 S_2$. Therefore,

$$S = \begin{bmatrix} \frac{1}{\sqrt{2}} & 0 & -\frac{1}{\sqrt{2}} \\ 0 & 1 & 0 \\ \frac{1}{\sqrt{2}} & 0 & \frac{1}{\sqrt{2}} \end{bmatrix} \begin{bmatrix} \frac{1}{\sqrt{2}} & -\frac{1}{\sqrt{2}} & 0 \\ \frac{1}{\sqrt{2}} & \frac{1}{\sqrt{2}} & 0 \\ 0 & 0 & 1 \end{bmatrix} = \begin{bmatrix} \frac{1}{2} & -\frac{1}{2} & -\frac{1}{\sqrt{2}} \\ \frac{1}{\sqrt{2}} & \frac{1}{\sqrt{2}} & 0 \\ \frac{1}{2} & -\frac{1}{2} & \frac{1}{\sqrt{2}} \end{bmatrix}$$

Example 4.3 Find all the eigenvalues of the matrix

$$A = \begin{bmatrix} 2 & -1 & 0 \\ -1 & 2 & -1 \\ 0 & -1 & 2 \end{bmatrix}$$

by Jacobi's method.

Solution In this example, we find that all the off-diagonal elements are of the same order of magnitude. Therefore, we can choose any one of them. Suppose, we choose a_{12} as the largest element and compute

$$\tan 2\theta = \frac{-1}{0} = \infty$$

which gives, $\theta = \pi/4$. Then $\cos\theta = \sin\theta = 1/\sqrt{2}$ and we construct an orthogonal matrix S_1 such that

$$S_1 = \begin{bmatrix} \frac{1}{\sqrt{2}} & -\frac{1}{\sqrt{2}} & 0 \\ \frac{1}{\sqrt{2}} & \frac{1}{\sqrt{2}} & 0 \\ 0 & 0 & 1 \end{bmatrix}$$

The first rotation gives

$$D_1 = S_1^{-1}AS_1 = \begin{bmatrix} \frac{1}{\sqrt{2}} & \frac{1}{\sqrt{2}} & 0 \\ -\frac{1}{\sqrt{2}} & \frac{1}{\sqrt{2}} & 0 \\ 0 & 0 & 1 \end{bmatrix} \begin{bmatrix} 2 & -1 & 0 \\ -1 & 2 & -1 \\ 0 & -1 & 2 \end{bmatrix} \begin{bmatrix} \frac{1}{\sqrt{2}} & -\frac{1}{\sqrt{2}} & 0 \\ \frac{1}{\sqrt{2}} & \frac{1}{\sqrt{2}} & 0 \\ 0 & 0 & 1 \end{bmatrix}$$

$$= \begin{bmatrix} 1 & 0 & -\frac{1}{\sqrt{2}} \\ 0 & 3 & -\frac{1}{\sqrt{2}} \\ -\frac{1}{\sqrt{2}} & -\frac{1}{\sqrt{2}} & 2 \end{bmatrix}$$

Now, we choose $d_{13} = -1/\sqrt{2}$ as the largest element of D_1 and compute

$$\tan 2\theta = \frac{2d_{13}}{d_{11} - d_{33}} = \frac{-\sqrt{2}}{1 - 2}$$

which gives, $\theta = 27° \, 22' \, 41''$.

Now we construct another orthogonal matrix S_2, such that

$$S_2 = \begin{bmatrix} 0.888 & 0 & -0.459 \\ 0 & 1 & 0 \\ 0.459 & 0 & 0.888 \end{bmatrix}$$

At the end of second rotation, we obtain

$$D_2 = S_2^{-1}D_1S_2 = \begin{bmatrix} 0.634 & -0.325 & 0 \\ 0.325 & 3 & -0.628 \\ 0 & -0.628 & 2.365 \end{bmatrix}$$

Now, the numerically largest off-diagonal element of D_2 is found to be $d_{23} = -0.628$ and compute

$$\tan 2\theta = \frac{-2 \times 0.628}{3 - 2.365}$$

we get, $\theta = -31° \, 35' \, 24''$. Thus, the orthogonal matrix S_3 is seen to be

$$S_3 = \begin{bmatrix} 1 & 0 & 0 \\ 0 & 0.852 & 0.524 \\ 0 & -0.524 & 0.852 \end{bmatrix}$$

At the end of third rotation, we get

$$D_3 = S_3^{-1} D_2 S_3 = \begin{bmatrix} 0.634 & -0.277 & 0 \\ 0.277 & 3.386 & 0 \\ 0 & 0 & 1.979 \end{bmatrix}$$

To reduce D_3 to a diagonal form, some more rotations are required. However, we may take 0.634, 3.386 and 1.979 as eigenvalues of the given matrix.

Example 4.4 Find all the eigenvalues and eigenvectors of the matrix:

$$A = \begin{bmatrix} 5 & 0 & 1 \\ 0 & -2 & 0 \\ 1 & 0 & 5 \end{bmatrix}$$

by Jacobi's method.

Solution The given matrix is

$$A = \begin{bmatrix} 5 & 0 & 1 \\ 0 & -2 & 0 \\ 1 & 0 & 5 \end{bmatrix}$$

In this example, the largest off-diagonal element is found to be $a_{13} = a_{31} = 1$. Now, we compute

$$\tan 2\theta = \frac{2a_{13}}{a_{11} - a_{33}} = \frac{2}{5-5} = \frac{2}{0} = \infty$$

which gives $\theta = \pi/4$. Following Jacobi's method, we construct an orthogonal matrix S_1 as

$$S_1 = \begin{bmatrix} \cos(\pi/4) & 0 & -\sin(\pi/4) \\ 0 & 1 & 0 \\ \sin(\pi/4) & 0 & \cos(\pi/4) \end{bmatrix} = \begin{bmatrix} 1/\sqrt{2} & 0 & -1/\sqrt{2} \\ 0 & 1 & 0 \\ 1/\sqrt{2} & 0 & 1/\sqrt{2} \end{bmatrix}$$

The first rotation gives

$$D_1 = S_1^{-1} A S_1 = \begin{bmatrix} 1/\sqrt{2} & 0 & 1/\sqrt{2} \\ 0 & 1 & 0 \\ -1/\sqrt{2} & 0 & 1/\sqrt{2} \end{bmatrix} \begin{bmatrix} 5 & 0 & 1 \\ 0 & -2 & 0 \\ 1 & 0 & 5 \end{bmatrix} \begin{bmatrix} 1/\sqrt{2} & 0 & -1/\sqrt{2} \\ 0 & 1 & 0 \\ 1/\sqrt{2} & 0 & 1/\sqrt{2} \end{bmatrix}$$

$$= \begin{bmatrix} 6 & 0 & 0 \\ 0 & -2 & 0 \\ 0 & 0 & 4 \end{bmatrix} \quad (1)$$

Which is a diagonal matrix and hence we stop further computation. From (1), we observe that 6, –2 and 4 are the eigenvalues of the given matrix and the corresponding eigenvectors are respectively the column vectors of

$$S_1 = \begin{bmatrix} 1/\sqrt{2} & 0 & -1/\sqrt{2} \\ 0 & 1 & 0 \\ 1/\sqrt{2} & 0 & 1/\sqrt{2} \end{bmatrix}$$

4.4 GERSCHGORIN'S THEOREM

This is one of the useful theorems on the bounds for eigenvalues of a square matrix.

Let λ_i be an eigenvalue of the $n \times n$ matrix [A] and let x_i be the corresponding eigenvector. Suppose R_s be the sum of the moduli of the terms along sth row, excluding the diagonal element a_{ss}. Then, every eigenvalue of $[A]$ lies inside or on the boundary of atleast one of the circles $|\lambda - a_{ss}| = R_s$

Proof: Given that

$$[A]x_i = \lambda_i x_i \tag{4.17}$$

Let $v_1, v_2, \ldots, v_n$ are the components of x_i, then the above equation can be expanded as

$$\left.\begin{aligned} a_{11}v_1 + a_{12}v_2 + \cdots + a_{1n}v_n &= \lambda_i v_1 \\ a_{21}v_1 + a_{22}v_2 + \cdots + a_{2n}v_n &= \lambda_i v_2 \\ \vdots \qquad\qquad\qquad & \qquad \vdots \\ a_{s1}v_1 + a_{s2}v_2 + \cdots + a_{sn}v_n &= \lambda_i v_s \\ \vdots \qquad\qquad\qquad & \qquad \vdots \\ a_{n1}v_1 + a_{n2}v_2 + \cdots + a_{nn}v_n &= \lambda_i v_n \end{aligned}\right\} \tag{4.18}$$

Suppose v_s be the largest in modulus of $v_1, v_2, \ldots, v_n$. Now, let us divide the sth equation by v_s and get

$$\lambda_i = a_{s1}\left(\frac{v_1}{v_s}\right) + a_{s2}\left(\frac{v_2}{v_s}\right) + \cdots + a_{ss} + \cdots + a_{sn}\left(\frac{v_n}{v_s}\right) \tag{4.19}$$

Since $\left|\frac{v_i}{v_s}\right| \le 1,\ i = 1, 2, \ldots, n$, it follows that

$$|\lambda_i| \le |a_{s1}| + |a_{s2}| + \cdots + |a_{ss}| + \cdots + |a_{sn}| \tag{4.20}$$

Eq. (4.19) can also be written as

$$|\lambda_i - a_{ss}| \le |a_{s1}| + |a_{s2}| + \cdots + |a_{sn}|$$

or

$$|\lambda_i - a_{ss}| \le \sum_{\substack{j=1 \\ j \ne s}}^{n} |a_{sj}| = R_s \tag{4.21}$$

Hence the proof. This theorem also holds for any column.

An immediate consequence of Gerschgorin's theorem, when applied to identity matrix or permutation matrix is that, its eigenvalues lie within a circle having center at 1 and radius 0. Here follows an example.

Example 4.4 Apply Gerschgorin's theorem to the matrix

$$[A] = \begin{bmatrix} 4 & -1 & -1 & 0 \\ -1 & 4 & -1 & -1 \\ -1 & -1 & 4 & -1 \\ 0 & -1 & -1 & 4 \end{bmatrix}$$

Solution This matrix has a dominant diagonal.

In this example $a_{ss} = 4$, Max. $R_s = 3$. Thus, Gerschgorin's theorem states that all the eigenvalues of the given matrix lie inside the circle with center at 4 and radius 3. In view of its symmetry, the eigenvalues are also real.

Definition 4.2 (Spectral Norm). Let λ_i be the largest eigenvalue of AA^* or A^*A, where A^* is the conjugate transpose of A, then the spectral norm of the matrix A, denoted by $\sigma(A)$ is defined as

$$\sigma(A) = (\lambda_i)^{1/2} \tag{4.22}$$

Definition 4.3 (Determinant). The determinant of an $n \times n$ matrix A is the product of its eigenvalues.

Definition 4.4 (Trace of a Matrix). The sum of the diagonal elements of an $n \times n$ matrix A is called the trace of the matrix A. Trace of the matrix A is also defined as the sum of its eignvalues.

EXERCISES

4.1 Find the largest eigenvalue of the matrix

$$\begin{bmatrix} 1 & -3 & 2 \\ 4 & 4 & -1 \\ 6 & 3 & 5 \end{bmatrix}$$

and the corresponding eigenvector, by power method after sixth iteration.

4.2 Find the largest eigenvalue of the matrix

$$\begin{bmatrix} 4 & 1 & 0 \\ 1 & 20 & 1 \\ 0 & 0 & 4 \end{bmatrix}$$

and the corresponding eigenvector, by power method after fourth iteration starting with the initial vector $v^{(0)} = (0, 0, 1)^T$.

4.3 Find the dominant eigenvalue and the corresponding eigevector of the matrix

$$\begin{bmatrix} 8 & 1 & 2 \\ 0 & 10 & -1 \\ 6 & 2 & 15 \end{bmatrix}$$

by power method with unit vector as the initial vector.

4.4 Find the largest eigenvalue and the corresponding eigenvector of the matrix

$$\begin{bmatrix} 3 & 1 & 4 \\ 0 & 2 & 6 \\ 0 & 0 & 5 \end{bmatrix}$$

by power method at the end of sixth iteration, taking unit vector as the initial vector.

4.5 Using Jacobi's method, find all the eigenvalues and eigenvectors of the Hilbert matrix

$$A = \begin{bmatrix} 1 & \frac{1}{2} & \frac{1}{3} \\ \frac{1}{2} & \frac{1}{3} & \frac{1}{4} \\ \frac{1}{3} & \frac{1}{4} & \frac{1}{5} \end{bmatrix}$$

Give result after two rotations.

4.6 Use Jacobi's method to find all the eigenvalues and the corresponding eigenvectors of the matrix

$$A = \begin{bmatrix} 4 & 3 & 2 & 1 \\ 3 & 4 & 3 & 2 \\ 2 & 3 & 4 & 3 \\ 1 & 2 & 3 & 4 \end{bmatrix}$$

4.7 Find all the eigenvalues and the corresponding eigenvectors of the matrix

(i) $A = \begin{bmatrix} 1 & 2 & 2 \\ 2 & 1 & 2 \\ 2 & 2 & 1 \end{bmatrix}$ (ii) $A = \begin{bmatrix} 3 & 2 & 1 \\ 2 & 3 & 2 \\ 1 & 2 & 3 \end{bmatrix}$

by Jacobi's method. Give results at the end of third rotation.

4.8 Find the dominant eigenvalue of

(i) $A = \begin{bmatrix} 1 & 2 \\ 3 & 2 \end{bmatrix}$ (ii) $A = \begin{bmatrix} 1 & 2 \\ 3 & 4 \end{bmatrix}$

and the corresponding eigenvector.

Chapter 5

Curve Fitting

5.1 INTRODUCTION

In science and engineering, we often come across experimental observations involving two variables such as, concentration and time, pressure and temperature, load and deflection, etc. (in general x and y). The important question is: can we find a law or a relation connecting them. Suppose, we wish to find a relation connecting load P kg and the deflection of a beam y cm; given the maximum deflection y for various loads P as shown in Fig. 5.1. Typically, the Figure indicates, a linear relation of the form $y = mx + c$. In another

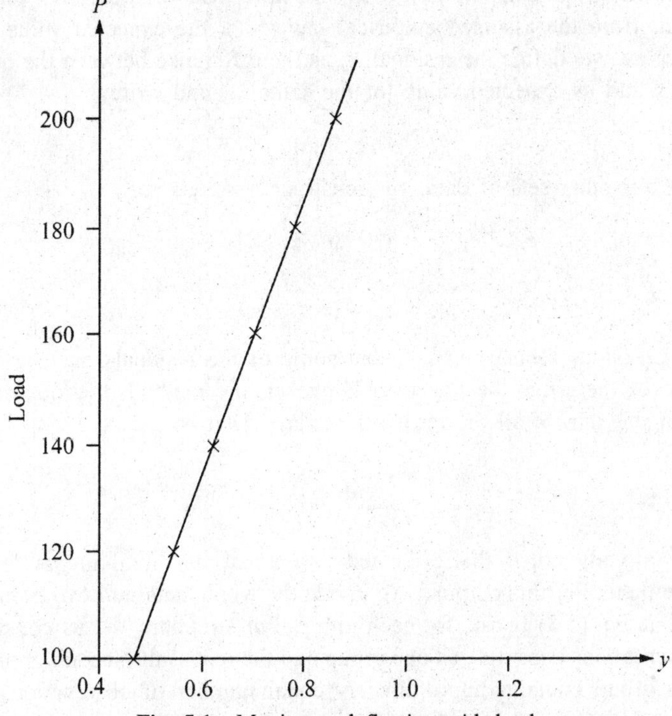

Fig. 5.1 Maximum deflection with load.

situation, the graph may indicate a parabolic or an exponential curve. In fact, Kepler formulated the third law of planetary motion, $T = cx^{3/2}$, from observed data pairs (x, T), of various planets, where T is the period, and x is the mean distance of the planet from the sun.

In general, the problem of finding an equation of an approximating curve, which passes through as many points as possible, is called *curve fitting*. Hence, the basic problem is to find an equation of the curve of the best fit.

The *method* of *group averages*, the *least square method* and the *method of moments* are some of the well-known methods based on the principles of elementary statistics. Amongst them, the least square method gives a unique best fit and is highly recommended.

5.2 METHOD OF GROUP AVERAGES

Suppose we have n sets of observations in an experiment, such as (x_1, y_1), (x_2, y_2), ..., (x_n, y_n). Assume that a straight line

$$y = mx + c \tag{5.1}$$

fits this data very closely. Since Eq. (5.1) contains two constants m and c, we require two equations in m and c to determine them. We shall explain below, how to determine these constants.

From the given data, when $x = x_1$, we note that, the observed value of y is y_1. But, from the assumed empirical law (5.1), the expected value of y is $mx_1 + c$. Now, we define the residual, e, as the difference between the observed value of y and its expected value for the same x_1, and write

$$e_1 = y_1 - (mx_1 + c)$$

Similarly, for other sets of data, we construct residuals as

$$\begin{aligned} e_2 &= y_2 - (mx_2 + c) \\ &\vdots \\ e_n &= y_n - (mx_n + c) \end{aligned} \tag{5.2}$$

These expressions clearly indicate that some of the residuals may be positive and some of them may be negative. However, the method of group averages states that the sum of all the residuals is zero. That is,

$$\sum_{i=1}^{n} e_i = 0 \tag{5.3}$$

We have already noted that we need two equations involving m and c to determine them. For this purpose, we divide the whole data into two groups, and assume that Eq. (5.3) is true for each group. For simplicity, let us consider the first j observations constitute as one group and the rest as the second group, such that each group contains approximately equal number of observations. Thus, using Eq. (5.3) for each group, we have

$$\begin{aligned} \sum_{i=1}^{j} e_i = [y_1 - (mx_1 + c)] + [y_2 - (mx_2 + c)] + \cdots \\ + [y_j - (mx_j + c)] = 0 \end{aligned} \tag{5.4}$$

and

$$\sum_{i=j+1}^{n} e_i = [y_{j+1} - (mx_{j+1} + c)] + [y_{j+2} - (mx_{j+2} + c)] + \cdots$$

$$+ [y_n - (mx_n + c)] = 0 \tag{5.5}$$

On simplification, Eqs. (5.4) and (5.5) become

$$(y_1 + y_2 + \cdots + y_j) - jc - m(x_1 + x_2 + \cdots + x_j) = 0$$

$$(y_{j+1} + y_{j+2} + \cdots + y_n) - (n - j)c - m\,(x_{j+1} + x_{j+2} + \cdots + x_n) = 0$$

which can be rewritten as

$$\frac{y_1 + y_2 + \cdots + y_j}{j} = \frac{m(x_1 + x_2 + \cdots + x_j)}{j} + c \tag{5.6}$$

and

$$\frac{y_{j+1} + y_{j+2} + \cdots + y_n}{n-j} = \frac{m(x_{j+1} + x_{j+2} + \cdots + x_n)}{n-j} + c \tag{5.7}$$

These equations are sufficient to determine the unknown constants m and c.

Obviously, there is no unique way of dividing the observations into two groups and hence different choices will give different values for m and c. This, of course, is the major drawback of the method of group averages. We shall illustrate this method through the following examples.

Example 5.1 Use the method of group averages and find a curve of the form $y = mx^n$, that fits the following data:

x	10	20	30	40	50	60	70	80
y	1.06	1.33	1.52	1.68	1.81	1.91	2.01	2.11

Solution The required curve is of the form

$$y = mx^n \tag{1}$$

Taking logarithms on both sides, we get

$$\log_{10}y = \log_{10}m + n\,\log_{10}x \tag{2}$$

Let $\log_{10}y = Y$, $\log_{10}m = c$ and $\log_{10}x = X$, then Eq. (2) becomes

$$Y = nX + c \tag{3}$$

Now, we take logarithms of the given pairs of data and divide them into two groups, such that the first group contains the first four values and the remaining constitutes the second group, as follows:

Group I

x	y	$Y = \log_{10} y$	$X = \log_{10} x$
10	1.06	0.0253	1.0000
20	1.33	0.1239	1.3010
30	1.52	0.1818	1.4771
40	1.68	0.2253	1.6021
		$\Sigma Y = 0.5563$	$\Sigma X = 5.3802$

Group II

x	y	$Y = \log_{10} y$	$X = \log_{10} x$
50	1.81	0.2577	1.6990
60	1.91	0.2810	1.7782
70	2.01	0.3032	1.8451
80	2.11	0.3243	1.9031
		$\Sigma Y = 1.1662$	$\Sigma X = 7.2254$

Now, using the method of group averages, we determine the constants n and c from Eqs. (5.6) and (5.7) as follows:

$$\frac{1}{4}(Y_1 + Y_2 + Y_3 + Y_4) = \frac{n}{4}(X_1 + X_2 + X_3 + X_4) + c \tag{4}$$

$$\frac{1}{4}(Y_5 + Y_6 + Y_7 + Y_8) = \frac{n}{4}(X_5 + X_6 + X_7 + X_8) + c \tag{5}$$

Substituting the values from the above tables, we get

$$\frac{0.5563}{4} = n\frac{5.3802}{4} + c$$

$$\frac{1.1662}{4} = n\frac{7.2254}{4} + c$$

On simplification, we have

$$4c + 5.3802n - 0.5563 = 0 \tag{6}$$

$$4c + 7.2254n - 1.1662 = 0 \tag{7}$$

Solving Eqs. (6) and (7) by the method of cross-multiplication we obtain

$$c = -0.3055, \qquad n = 0.3305$$

Therefore,

$$m = \text{antilog } c = 0.4949$$

Hence, the required curve fit is

$$y = 0.4949x^{0.3305}$$

Example 5.2 Using the method of group averages, find a curve of the form $y = a + bx^2$, that fits the following data:

x	20	30	35	40	45	50
y	10.0	11.0	11.8	12.4	13.5	14.4

Solution Let the required equation of fit is

$$y = a + bx^2 \tag{1}$$

Setting $x^2 = X$, the above equation becomes

$$y = bX + a \tag{2}$$

Now, the given data can be grouped and tabulated as shown:

Group I

y	x	$X = x^2$
10.0	20	400
11.0	30	900
11.8	35	1225
$\Sigma y = 32.8$		$\Sigma X = 2525$

Group II

y	x	$X = x^2$
12.4	40	1600
13.5	45	2025
14.4	50	2500
$\Sigma y = 40.3$		$\Sigma X = 6125$

To determine the unknown constants a and b, using the method of group averages, Eqs. (5.6) and (5.7) give

$$\frac{32.8}{3} = \frac{2525b}{3} + a$$

$$\frac{40.3}{3} = \frac{6125b}{3} + a$$

On simplification, we have

$$2525b + 3a - 32.8 = 0 \tag{3}$$

$$6125b + 3a - 40.3 = 0 \tag{4}$$

Solving Eqs. (3) and (4), we get

$$b = 2.083 \times 10^{-3}, \qquad a = 9.1799$$

Hence, the required curve fit is

$$y = 9.1799 + 2.083 \times 10^{-3}x^2$$

Example 5.3 Using the method of group averages, fit a curve of the form

$$y = \frac{x}{a + bx}$$

for the following data:

x	8	10	15	20	30	40
y	13	14	15.4	16.3	17.2	17.8

Solution Let the equation of the required fit is

$$y = \frac{x}{a + bx} \tag{1}$$

On rewriting, we get

$$\frac{1}{y} = \frac{a + bx}{x} = \frac{a}{x} + b$$

Setting

$$\frac{1}{y} = Y, \quad \frac{1}{x} = X$$

the above equation becomes

$$Y = aX + b \tag{2}$$

which is linear in X and Y. Now, we shall tabulate the given data into two groups:

Group I

x	$X = 1/x$	y	$Y = 1/y$
8	0.1250	13	0.0769
10	0.1000	14	0.0714
15	0.0667	15.4	0.0649
	$\sum X = 0.2917$		$\sum Y = 0.2132$

Group II

x	$X = 1/x$	y	$Y = 1/y$
20	0.0500	16.3	0.0614
30	0.0333	17.2	0.0581
40	0.0250	17.8	0.0562
	$\sum X = 0.1083$		$\sum Y = 0.1757$

Using the method of group averages, we have

$$\frac{0.2132}{3} = a\frac{0.2917}{3} + b$$

and

$$\frac{0.1757}{3} = a\frac{0.1083}{3} + b$$

On simplification, we get

$$0.2917a + 3b - 0.2132 = 0 \tag{3}$$

$$0.1083a + 3b - 0.1757 = 0 \tag{4}$$

Solving Eqs. (3) and (4), we obain

$$a = 0.2045, \qquad b = 0.0512$$

Hence, the required curve fit is

$$y = \frac{x}{0.2045 + 0.0512x}$$

5.3 THE LEAST SQUARES METHOD

In curve fitting for a given data, the method of least squares is considered to give the best fit, with prior knowledge about the shape of the curve. Suppose $(x_1, y_1), (x_2, y_2), \ldots, (x_n, y_n)$ be n set of observations in an experiment. Let us assume that

$$y = f(x) \tag{5.8}$$

is a relationship between x and y. When $x = x_1$, the observed value of y is y_1 and the expected value of y from Eq. (5.8) is $f(x_1)$. Then the residual is defined by

$$e_1 = y_1 - f(x_1)$$

Similarly, we can define all other residuals $e_2, e_3, \ldots, e_n$ as

$$e_2 = y_2 - f(x_2)$$

$$\vdots$$

$$e_n = y_n - f(x_n)$$

It is clear from the definition of residual that some of them may be positive and some may be negative. In order to give equal importance to both positive and negative residuals, it is better to consider the sum of the squares of residuals, rather than sum of residuals as in the method of group averages. Thus, we consider

$$\begin{aligned} E &= e_1^2 + e_2^2 + \cdots + e_n^2 \\ &= [y_1 - f(x_1)]^2 + [y_2 - f(x_2)]^2 + \cdots + [y_n - f(x_n)]^2 \end{aligned} \tag{5.9}$$

Here, the quantity E is a measure of how well the curve $y = f(x)$ fits the given set of observations as a whole. Therefore, E will be zero if and only if all the points of observations lie on the curve given by Eq. (5.8). The value of E decreases depending on the closeness of the observed data to the curve assumed. Hence, the best representative curve to the given set of observations is one for which E, the sum of the squares of the residuals, is minimum. This concept is known as the *principle of least squares*. In general, we consider, straight line, parabola or an exponential curve for fitting the given data.

5.3.1 Fitting a Straight Line

Suppose $(x_1, y_1), (x_2, y_2), \ldots, (x_n, y_n)$ be a set of n observations in an experiment and we wish to fit a straight line

$$y = ax + b \tag{5.10}$$

to these observations. That is, we have to determine the constants a and b, using the principle of least squares.

For any x_i $(i = 1, 2, \ldots, n)$, the expected value of y is $ax_i + b$, while the observed value of y is y_i. Hence, the residuals are

$$e_i = y_i - (ax_i + b), \qquad \text{for } i = 1, 2, \ldots, n$$

Now, the sum of the squares of residuals is given as

$$E = \sum_{i=1}^{n} [y_i - (ax_i + b)]^2 \tag{5.11}$$

where E is a function of the parameters a and b. The necessary conditions for E to be minimum gives

$$\frac{\partial E}{\partial a} = \frac{\partial E}{\partial b} = 0$$

In other words, the first condition yields

$$2x_1 (y_1 - ax_1 - b) + 2x_2 (y_2 - ax_2 - b) + \cdots + 2x_n (y_n - ax_n - b) = 0$$

That is, in compact form we write it as

$$a\sum_{i=1}^{n} x_i^2 + b\sum_{i=1}^{n} x_i = \sum_{i=1}^{n} x_i y_i \tag{5.12}$$

Similarly, the second condition

$$\frac{\partial E}{\partial b} = 0$$

gives

$$(y_1 - ax_1 - b) + (y_2 - ax_2 - b) + \cdots + (y_n - ax_n - b) = 0$$

That is,

$$a\sum_{i=1}^{n} x_i + nb = \sum_{i=1}^{n} y_i \tag{5.13}$$

Equations (5.12) and (5.13) are called *normal equations*, which when solved give the values of the constants a and b.

To compute E, the sum of the squares of the residuals, Eq. (5.11) can be recast as

$$E = \sum_{i=1}^{n} (y_i - ax_i - b)^2 = \sum_{i=1}^{n} [y_i^2 - 2ax_i y_i - 2by_i + (ax_i + b)^2]$$

Now, using Eq. (5.13), we obtain a convenient formula to compute E as

$$E = \Sigma y_i^2 - a\Sigma x_i y_i - b\Sigma y_i \tag{5.14}$$

Example 5.4 Using the method of least squares, find the straight line $y = ax + b$, that fits the following data:

x	0.5	1.0	1.5	2.0	2.5	3.0
y	15	17	19	14	10	7

Solution The given straight line fit be $y = ax + b$. The normal equations of least square fit are

$$a\Sigma x_i^2 + b\Sigma x_i = \Sigma x_i y_i \tag{1}$$

and

$$a\Sigma x_i + nb = \Sigma y_i \tag{2}$$

From the given data, we have

x	y	xy	x^2
0.5	15	7.5	0.25
1.0	17	17.0	1.00
1.5	19	28.5	2.25
2.0	14	28.0	4.00
2.5	10	25.0	6.25
3.0	7	21.0	9.00
$\Sigma x_i = 10.5$	$\Sigma y_i = 82$	$\Sigma x_i y_i = 127$	$\Sigma x_i^2 = 22.75$

Substituting these summations into Eqs. (1) and (2), we obtain

$$22.75a + 10.5b = 127 \tag{3}$$

$$10.5a + 6b = 82 \tag{4}$$

Solving Eqs. (3) and (4), we get

$$a = -3.7714, \qquad b = 20.2667$$

Hence, the required fit for the given data is

$$y = -3.7714x + 20.2667$$

Example 5.5 Applying the method of least squares find an equation of the form $y = ax + bx^2$ that fits the following data:

x	1	2	3	4	5	6
y	2.6	5.4	8.7	12.1	16.0	20.2

Solution The required curve fit is $y = ax + bx^2$, which can be written as

$$\frac{y}{x} = a + bx$$

Let $y/x = Y$, then the curve to be fitted is $Y = a + bx$. The corresponding data when rewritten takes the form

x	1	2	3	4	5	6
Y	2.6	2.7	2.9	3.025	3.2	3.367

The corresponding normal equations are

$$b\Sigma x_i^2 + a\Sigma x_i = \Sigma x_i Y_i$$

$$b\Sigma x_i + na = \Sigma Y_i$$

From the given data, we have

x	Y	xY	x^2
1	2.6	2.6	1
2	2.7	5.4	4
3	2.9	8.7	9
4	3.025	12.1	16
5	3.2	16.0	25
6	3.367	20.2	36
$\Sigma x_i = 21$	$\Sigma Y_i = 17.792$	$\Sigma x_i Y_i = 65.0$	$\Sigma x_i^2 = 91$

Substituting these summations into normal equations, we obtain

$$91b + 21a = 65 \tag{1}$$

$$21b + 6a = 17.792 \tag{2}$$

Solving Eqs. (1) and (2), we get $b = 0.15589$, $a = 2.41973$. Hence, the required equation of fit for the given data is

$$Y = 0.15589x + 2.41973$$

or

$$y = 0.15589x^2 + 2.41973x$$

5.3.2 Fitting a Parabola

Suppose, we have a set of n obervations (x_i, y_i), $i = 1, 2, \ldots, n$ in an experiment and we wish to fit a parabola to the observed data, using least square method. Let

$$y = ax^2 + bx + c \tag{5.15}$$

be the equation of parabola that fits the given data. Then the problem is to determine the constants a, b and c in the least square sense.

For a given x_i, let the expected value of y is given by $ax_i^2 + bx_i + c$, while the observed value be y_i. Then, the residual e_i is given by

$$e_i = \text{Observed value} - \text{Expected value} = y_i - (ax_i^2 + bx_i + c)$$

For different values of i, we get n residuals. Extending the principle of least squares, we have

$$E = \sum_{i=1}^{n} [y_i - (ax_i^2 + bx_i + c)]^2 \tag{5.16}$$

For E to be minimum, the necessary conditions are

$$\frac{\partial E}{\partial a} = \frac{\partial E}{\partial b} = \frac{\partial E}{\partial c} = 0$$

Now

$$\frac{\partial E}{\partial a} = 0$$

implies

$$2(y_1 - ax_1^2 - bx_1 - c)(-x_1^2) + 2(y_2 - ax_2^2 - bx_2 - c)(-x_2^2) + \cdots + 2(y_n - ax_n^2 - bx_n - c)(-x_n^2) = 0$$

or

$$-(x_1^2 y_1 + x_2^2 y_2 + \cdots + x_n^2 y_n) + a(x_1^4 + x_2^4 + \cdots + x_n^4) + b(x_1^3 + x_2^3 + \cdots + x_n^3) + c(x_1^2 + x_2^2 + \cdots + x_n^2) = 0$$

That is,

$$a\sum_{i=1}^{n} x_i^4 + b\sum_{i=1}^{n} x_i^3 + c\sum_{i=1}^{n} x_i^2 = \sum_{i=1}^{n} x_i^2 y_i \tag{5.17}$$

Similarly

$$\frac{\partial E}{\partial b} = 0$$

gives

$$2(y_1 - ax_1^2 - bx_1 - c)(-x_1) + 2(y_2 - ax_2^2 - bx_2 - c)(-x_2) + \cdots + 2(y_n - ax_n^2 - bx_n - c)(-x_n) = 0$$

That is,

$$a\sum_{i=1}^{n} x_i^3 + b\sum_{i=1}^{n} x_i^2 + c\sum_{i=1}^{n} x_i = \sum_{i=1}^{n} x_i y_i \tag{5.18}$$

Finally, the condition

$$\frac{\partial E}{\partial c} = 0$$

gives

$$a\sum_{i=1}^{n} x_i^2 + b\sum_{i=1}^{n} x_i + nc = \sum_{i=1}^{n} y_i \tag{5.19}$$

Here, Eqs. (5.17)–(5.19) are called normal equations whose solution yields the values of the constants a, b and c. Using normal equations and some algebraic manipulations, Eq. (5.16) will have the following alternate form

$$E = \Sigma y_i^2 - a\Sigma x_i^2 y_i - b\Sigma x_i y_i - c\Sigma y_i \tag{5.20}$$

This form will be more convenient to compute the sum of the squares of the residuals E in the case of parabolic fit.

For illustration of the parabolic fit, we consider an example.

Example 5.6 Fit a parabola to the following data using the method of least squares.

x	1.0	1.2	1.4	1.6	1.8	2.0
y	0.98	1.40	1.86	2.55	2.28	3.20

Solution Let the equation of the curve of parabolic fit is

$$y = ax^2 + bx + c \tag{1}$$

The corresponding normal equations are

$$\left.\begin{aligned} a\Sigma x_i^4 + b\Sigma x_i^3 + c\Sigma x_i^2 &= \Sigma x_i^2 y_i \\ a\Sigma x_i^3 + b\Sigma x_i^2 + c\Sigma x_i &= \Sigma x_i y_i \\ a\Sigma x_i^2 + b\Sigma x_i + nc &= \Sigma y_i \end{aligned}\right\} \tag{2}$$

From the given data, we have

x	y	x^2	x^3	x^4	xy	x^2y
1.0	0.98	1.00	1.0	1.0000	0.980	0.9800
1.2	1.40	1.44	1.728	2.0736	1.680	2.0160
1.4	1.86	1.96	2.744	3.8416	2.604	3.6456
1.6	2.55	2.56	4.096	6.5536	4.080	6.5280
1.8	2.28	3.24	5.832	10.4976	4.104	7.3872
2.0	3.20	4.00	8.000	16.0000	6.400	12.8000
$\Sigma x_i = 9.0$	$\Sigma y_i = 12.27$	$\Sigma x_i^2 = 14.20$	$\Sigma x_i^3 = 23.4$	$\Sigma x_i^4 = 39.9664$	$\Sigma x_i y_i = 19.848$	$\Sigma x_i^2 y_i = 33.3568$

Substituting these summations into normal equations (2), we get

$$39.9664a + 23.4b + 14.2c = 33.3568 \tag{3}$$

$$23.4a + 14.2b + 9.0c = 19.848 \tag{4}$$

$$14.2a + 9.0b + 6c = 12.27 \tag{5}$$

Solving Eqs. (3)–(5), we obtain

$$a = -0.1875, \qquad b = 2.6239, \qquad c = -1.4471$$

Hence, from Eq. (1) the required parabolic fit to the given data is

$$y = -0.1875x^2 + 2.6239x - 1.4471$$

5.3.3 Fitting a Curve of the Form $y = ax^b$

Data from experimental observations need not be always linear. Sometimes, we may be interested to fit a curve of the form,

$$y = ax^b \tag{5.21}$$

for a given data. This form can be linearized by taking logarithms on both sides of Eq. (5.21), thus

$$\log_{10} y = \log_{10} a + b \log_{10} x$$

Let

$$\log_{10} y = Y, \qquad \log_{10} a = A, \qquad \log_{10} x = X$$

then the above equation reduces to

$$Y = A + bX$$

which is linear in Y and X. To illustrate the method, we consider an example.

Example 5.7 By using the method of least squares, find a relation of the form $y = ax^b$, that fits the data

x	2	3	4	5
y	27.8	62.1	110	161

Solution Let the equation of fit be,

$$y = ax^b \tag{1}$$

Taking logarithms on both sides, we get

$$\log_{10} y = \log_{10} a + b \log_{10} x$$

which is of the form

$$Y = A + bX = bX + A$$

where,

$$\log_{10} y = Y, \qquad \log_{10} a = A, \qquad \log_{10} x = X$$

Now, the data in modified variables X and Y is

X	0.3010	0.4771	0.6021	0.6990
Y	1.4440	1.7931	2.0414	2.2068

The corresponding normal equations are

$$b\Sigma X_i^2 + A\Sigma X_i = \Sigma X_i Y_i \tag{2}$$

$$b\Sigma X_i + nA = \Sigma Y_i \tag{3}$$

From the above data, we have

X	Y	XY	X^2
0.3010	1.4440	0.4346	0.0906
0.4771	1.7931	0.8555	0.2276
0.6021	2.0414	1.2291	0.3625
0.6990	2.2068	1.5426	0.4886
$\Sigma X_i = 2.0792$	$\Sigma Y_i = 7.4853$	$\Sigma X_i Y_i = 4.0618$	$\Sigma X_i^2 = 1.1693$

Substituting these summations into normal equations (2) and (3) we get

$$1.1693b + 2.0792A = 4.0618$$

$$2.0792b + 4A = 7.4853$$

Solving these equations, we obtain $b = 1.9311$, $A = 0.8678$. Therefore, $a = \text{antilog } 0.8678 = 7.375$. Hence, the required fit for the given data is

$$y = 7.375x^{1.9311}$$

5.3.4 Fitting an Exponential Curve

Suppose we have n set of observations (x_i, y_i), $i = 1, 2, \ldots, n$ in an experiment and we wish to fit an exponential curve of the form,

$$y = ae^{bx} \tag{5.22}$$

to these observations. At first, we take logarithms on both sides to get

$$\log_{10} y = \log_{10} a + bx \log_{10} e$$

Let

$$\log_{10} y = Y, \qquad \log_{10} a = A, \qquad b \log_{10} e = B$$

then the above equation becomes

$$Y = Bx + A \tag{5.23}$$

Thus, it is equivalent to fitting a straight line. The procedure is illustrated through the following example.

Example 5.8 Using the principle of least squares, fit an equation of the form $y = ae^{bx}$ to the data

x	1	2	3	4
y	1.65	2.70	4.50	7.35

Solution Let

$$y = ae^{bx} \tag{1}$$

is the desired fit to the given data. Taking logarithms on both sides, we have

$$\log_{10} y = \log_{10} a + bx \log_{10} e$$

Suppose

$$\log_{10} y = Y, \qquad \log_{10} a = A, \qquad b \log_{10} e = B$$

then the above equation assumes the form

$$Y = Bx + A \tag{2}$$

The least square normal equations are

$$B\Sigma x_i^2 + A\Sigma x_i = \Sigma x_i Y_i \tag{3}$$

$$B\Sigma x_i + 4A = \Sigma Y_i \tag{4}$$

From the given data, we have

x	y	$Y = \log_{10} y$	x^2	xY
1	1.65	0.2175	1	0.2175
2	2.70	0.4314	4	0.8628
3	4.50	0.6532	9	1.9596
4	7.35	0.8663	16	3.4652
$\Sigma x_i = 10$	$\Sigma y_i = 16.2$	$\Sigma Y_i = 2.1684$	$\Sigma x_i^2 = 30$	$\Sigma x_i Y_i = 6.5051$

Substituting these summations into normal equations (3) and (4), we get

$$30B + 10A = 6.5051$$

$$10B + 4A = 2.1684$$

Solving these equations, we obtain $B = 0.2168$, $A = 0$. Now

$$B = b \log_{10} e = 0.4343b = 0.2168$$

Therefore,

$$b = 0.4992, \qquad a = 1$$

Hence, the required equation of fit is

$$y = e^{0.4992x}$$

5.4 METHOD OF MOMENTS

Consider n set of observations (x_i, y_i), $i = 1, 2, \ldots, n$ in an experiment, such that x_i's are equally spaced. That is,

$$x_2 - x_1 = x_3 - x_2 = \cdots = x_n - x_{n-1} = \Delta x$$

For such set of observation, we shall define the observed moments as follows.

$$\left.\begin{aligned} \mu_1 &= \text{The first moment} = \Sigma\, y_i\, \Delta x = \Delta x\, \Sigma\, y_i \\ \mu_2 &= \text{The second moment} = \Sigma\, x_i y_i\, \Delta x = \Delta x\, \Sigma\, x_i y_i \\ \mu_3 &= \text{The third moment} = \Sigma\, x_i^2\, y_i\, \Delta x = \Delta x\, \Sigma\, x_i^2\, y_i \end{aligned}\right\} \tag{5.24}$$

and so on. These moments are known as *moments of the observed values* of y. We shall also define the *expected moments of the computed values* of y, assuming $y = f(x)$ be an equation of the curve fitting the given data, as

$$\left.\begin{aligned} \gamma_1 &= \text{The first moment} = \int y\, dx = \int f(x)\, dx \\ \gamma_2 &= \text{The second moment} = \int xy\, dx = \int xf(x)\, dx \\ \gamma_3 &= \text{The third moment} = \int x^2 y\, dx = \int x^2 f(x)\, dx \end{aligned}\right\} \tag{5.25}$$

The method of moments states that the moments of the observed y's are respectively equal to the expected moments. That is,

$$\mu_1 = \gamma_1, \qquad \mu_2 = \gamma_2, \qquad \mu_3 = \gamma_3 \tag{5.26}$$

and so on. The observed moments can be easily calculated from the given data, while the moments of the expected y's can be computed as follows.

$$\left.\begin{aligned}
\gamma_1 &= \int y\,dx = \int_{x_1-(\Delta x/2)}^{x_n+(\Delta x/2)} f(x)\,dx \\
\gamma_2 &= \int xy\,dx = \int_{x_1-(\Delta x/2)}^{x_n+(\Delta x/2)} xf(x)\,dx \\
\gamma_3 &= \int x^2y\,dx = \int_{x_1-(\Delta x/2)}^{x_n+(\Delta x/2)} x^2f(x)\,dx
\end{aligned}\right\} \tag{5.27}$$

Thus, from the principle of method of moments, we have the following set of equations:

$$\left.\begin{aligned}
\Delta x\Sigma y_i &= \int_{x_1-(\Delta x/2)}^{x_n+(\Delta x/2)} f(x)\,dx \\
\Delta x\Sigma x_i y_i &= \int_{x_1-(\Delta x/2)}^{x_n+(\Delta x/2)} xf(x)\,dx \\
\Delta x\Sigma x_i^2 y_i &= \int_{x_1-(\Delta x/2)}^{x_n+(\Delta x/2)} x^2f(x)\,dx
\end{aligned}\right\} \tag{5.28}$$

Given the form of $f(x)$, we can use Eqs. (5.28) to determine the unknown constants in $f(x)$. This method is illustrated through the following examples.

Example 5.9 Fit a straight line of the form $y = ax + b$, to the following data by the method of moments

x	2	3	4	5
y	27	40	55	68

Solution Since we have to determine two constants for the required fit, it is enough to compute the first-two moments. In this example, we note that $\Delta x = 1$. Thus, the observed moments are

$$\mu_1 = \Delta x \sum_{i=1}^{4} y_i = 1\,(27 + 40 + 55 + 68) = 190$$

$$\mu_2 = \Delta x \sum_{i=1}^{4} x_i y_i = 1(54 + 120 + 220 + 340) = 734$$

Now, we shall compute the expected moments for the given y. Therefore,

$$\gamma_1 = \int_{1.5}^{5.5} (ax + b)\,dx = \left[a\frac{x^2}{2} + bx\right]_{1.5}^{5.5} = 14a + 4b$$

The second moment of the expected value of y is

$$\gamma_2 = \int xy\,dx = \int_{1.5}^{5.5} x(ax+b)\,dx = \left[a\frac{x^3}{3} + b\frac{x^2}{2}\right]_{1.5}^{5.5} = \frac{163}{3}a + 14b$$

using the method of moments, we have

$$14a + 4b = 190, \qquad \frac{163}{3}a + 14b = 734$$

Solving these equations, we get $a = 12.9375$, $b = 2.2188$. Hence, the required straight line fit is

$$y = 12.9375x + 2.2188$$

Example 5.10 Fit a parabola of the form $y = a + bx + cx^2$ to the data given below, using the method of moments.

x	3	4	5	6	7
y	31.9	34.6	33.8	27.0	31.6

Solution Since we have to fit a parabola of the form $y = a + bx + cx^2$ to the given data, we have to determine three constants. Therefore, it is necessary to compute the first-three moments. In this problem x's are equally spaced and we also note that $\Delta x = 1$.

Now, we compute the first-three moments as

$$\begin{aligned}\mu_1 &= \text{The first moment of the observed } y\text{'s}\\ &= \Delta x\,\Sigma y_i\\ &= 1\,[31.9 + 34.6 + 33.8 + 27.0 + 31.6] = 158.9\end{aligned}$$

$$\begin{aligned}\mu_2 &= \text{The second moment of the observed } y\text{'s}\\ &= \Delta x\,\Sigma x_i\,y_i\\ &= [95.7 + 138.4 + 169.0 + 162 + 221.2] = 786.3\end{aligned}$$

$$\begin{aligned}\mu_3 &= \text{The third moment of the observed } y\text{'s}\\ &= \Delta x\,\Sigma x_i^2\,y_i\\ &= [9 \times 31.9 + 16 \times 34.6 + 25 \times 33.8 + 36 \times 27 + 49 \times 31.6] = 4206.1\end{aligned}$$

We also need to compute the moments of the expected values of y's, given by $y = a + bx + cx^2$. Thus, we have

$$\begin{aligned}\gamma_1 &= \text{The first moment of the expected value of } y\\ &= \int y\,dx = \int_{2.5}^{7.5} (a + bx + cx^2)\,dx\\ &= \left[ax + b\frac{x^2}{2} + c\frac{x^3}{3}\right]_{2.5}^{7.5}\\ &= 5a + 25b + 135.4167c\end{aligned}$$

Similarly,

$$\gamma_2 = \int xy\,dx = \int_{2.5}^{7.5} x(a + bx + cx^2)\,dx$$

$$= \left[a\frac{x^2}{2} + b\frac{x^3}{3} + c\frac{x^4}{4} \right]_{2.5}^{7.5}$$

$$= 25a + 135.42b + 781.25c$$

and

$$\gamma_3 = \int x^2 y\,dy = \int_{2.5}^{7.5} x^2(a + bx + cx^2)\,dx$$

$$= \left[a\frac{x^3}{3} + b\frac{x^4}{4} + c\frac{x^5}{5} \right]_{2.5}^{7.5}$$

$$= 135.42a + 781.25b + 4726.562c$$

From the method of moments, we have at once

$$5a + 25b + 135.42c = 158.9$$

$$25a + 135.42b + 781.25c = 786.3$$

$$135.42a + 781.25b + 4726.562c = 4206.1$$

Solving these equations, we obtain

$$a = 15.7217, \qquad b = 7.9427, \qquad c = -0.8734$$

Hence, the required parabolic fit to the given data is

$$y = 15.7217 + 7.9427x - 0.8734x^2$$

EXERCISES

5.1 Applying the method of group averages, find a curve of the form $y = ax^2 + b$ that fits the following data:

x	1	2	3	4
y	0.43	0.83	1.40	2.33

5.2 Using the method of least squares, find an equation of the form $y = ax + b$, that fits the following data.
Also, calculate the sum of the squares of the residuals E.

x	0	1	2	3	4
y	1	5	10	22	38

5.3 It has been observed that the rate of flow of water through a fire engine hose is a quadratic in pressure p, at the nozzle end. The observed data is

Q	9.4	11.8	14.7	18.0	23.0
p	1.0	1.6	2.5	4.0	6.0

Fit a parabola in the form $Q = ap^2 + bp + c$ by the least squares method.

5.4 By the method of least squares, find a relation of the form $y = ax^b$, that fits the data:

x	1	2	3	4	5
y	0.5	2.0	4.5	8.0	12.5

5.5 Using the principle of least squares, find an equation of the form $y = ae^{bx}$, that fits the following data:

x	77	100	185	239	285
y	2.4	3.4	7.0	11.1	19.6

5.6 Using the method of moments, fit a straight line $y = ax + b$ to the following data:

x	1	2	3	4
y	16	19	23	26

5.7 By the method of least squares, fit a curve of the form $y = a + bx + cx^2$ to the data given below:

x	0.0	0.5	1.0
y	1.0000	1.6487	2.7183

5.8 Using the least squares method, fit a polynomial of the type $y = a + bx + cx^2$ to the data given below:

x	36.9	46.7	63.7	77.8	84.0	87.5
y	181	197	235	270	283	292

5.9 Using the method of least squares, find an equation of the form $y = ax + b$, that fits the following data:

x	−2	−1	0	1	2
y	1	2	3	3	4

5.10 Find the least squares parabolic fit of the form $y = ax^2 + bx + c$ to the following data:

x	−3	−1	1	3
y	15	5	1	5

5.11 Using the principle of least squares, find an equation of the form $y = ae^{bx}$, that fits the following data:

x	1	2	3	4	5
y	0.6	1.9	4.3	7.6	12.6

Chapter 6

Interpolation

6.1 INTRODUCTION

Finite differences play an important role in numerical techniques, where tabulated values of the functions are available. For instance, consider a function $y = f(x)$. As x takes values $x_0, x_1, x_2, \ldots, x_n$, let the corresponding values of y be $y_0, y_1, y_2, \ldots, y_n$. That is, for a given table of values, (x_k, y_k), $k = 0, 1, 2, \ldots, n$; the process of estimating the value of y, for any intermediate value of x, is called *interpolation*. However, the method of computing the value of y, for a given value of x, lying outside the table of values of x is known as *extrapolation*. It may be noted that if the function $f(x)$ is known, the value of y corresponding to any x can be readily computed to the desired accuracy. But, in practice, it may be difficult or sometimes impossible to know the function $y = f(x)$ in its exact form.

To look at a practical example, let us consider the computation of trajectory of a rocket flight, where we solve the Euler's dynamical equations of motion to compute its position and velocity vectors at specified times during the flight. Under the same conditions, suppose, we require the position and velocity vector, at some other intermediate times; we need not compute the trajectory again by solving the dynamical equations. Instead, we can use the best kncwn interpolation technique to get the desired values.

In general, for interpolation of a tabulated function, the concept of finite differences is important. The knowledge about various finite difference operators and their symbolic relations are very much needed to establish various interpolation formulae.

6.2 FINITE DIFFERENCE OPERATORS

6.2.1 Forward Differences

For a given table of values (x_k, y_k), $k = 0, 1, 2, \ldots, n$ with equally-spaced abscissas of a function $y = f(x)$, we define the *forward difference* operator Δ as follows: The first forward difference is usually expressed as

$$\Delta y_i = y_{i+1} - y_i, \qquad i = 0, 1, \ldots, (n-1) \tag{6.1}$$

To be explicit, we write

$$\Delta y_0 = y_1 - y_0$$
$$\Delta y_1 = y_2 - y_1$$
$$\vdots \quad \vdots \quad \vdots$$
$$\Delta y_{n-1} = y_n - y_{n-1}$$

These differences are called *first differences of the function y* and are denoted by the symbol Δy_i. Here, Δ is called *forward difference operator.*

Similarly, the differences of the first differences are called *second differences*, defined by

$$\Delta^2 y_0 = \Delta y_1 - \Delta y_0, \qquad \Delta^2 y_1 = \Delta y_2 - \Delta y_1$$

Thus, in general

$$\Delta^2 y_i = \Delta y_{i+1} - \Delta y_i \tag{6.2}$$

Here Δ^2 is called the *second difference operator*. Thus, continuing, we can define, r th difference of y, as

$$\Delta^r y_i = \Delta^{r-1}\, y_{i+1} - \Delta^{r-1}\, y_i \tag{6.3}$$

By defining a difference table as a convenient device for displaying various differences, the above defined differences can be written down systematically by constructing a difference table for values (x_k, y_k), $k = 0, 1, \ldots, 6$ as shown below:

Table 6.1 Forward Difference Table

x	y	Δy	$\Delta^2 y$	$\Delta^3 y$	$\Delta^4 y$	$\Delta^5 y$	$\Delta^6 y$
x_0	y_0						
		Δy_0					
x_1	y_1		$\Delta^2 y_0$				
		Δy_1		$\Delta^3 y_0$			
x_2	y_2		$\Delta^2 y_1$		$\Delta^4 y_0$		
		Δy_2		$\Delta^3 y_1$		$\Delta^5 y_0$	
x_3	y_3		$\Delta^2 y_2$		$\Delta^4 y_1$		$\Delta^6 y_0$
		Δy_3		$\Delta^3 y_2$		$\Delta^5 y_1$	
x_4	y_4		$\Delta^2 y_3$		$\Delta^4 y_2$		
		Δy_4		$\Delta^3 y_3$			
x_5	y_5		$\Delta^2 y_4$				
		Δy_5					
x_6	y_6						

This difference table is called *forward difference table* or *diagonal difference table*. Here, each difference is located in its appropriate column, mid-way between the elements of the previous column. It can be noted that the subscript remains constant along each diagonal of the table. The first term in the table, that is y_0 is called the *leading term*, while the differences Δy_0, $\Delta^2 y_0$, $\Delta^3 y_0$, ... are called leading differences.

Example 6.1 Construct a forward difference table for the following values of x and y:

x	0.1	0.3	0.5	0.7	0.9	1.1	1.3
y	0.003	0.067	0.148	0.248	0.370	0.518	0.697

Solution

x	y	Δy	$\Delta^2 y$	$\Delta^3 y$	$\Delta^4 y$	$\Delta^5 y$
0.1	0.003					
		0.064				
0.3	0.067		0.017			
		0.081		0.002		
0.5	0.148		0.019		0.001	
		0.100		0.003		0.000
0.7	0.248		0.022		0.001	
		0.122		0.004		0.000
0.9	0.370		0.026		0.001	
		0.148		0.005		
1.1	0.518		0.031			
		0.179				
1.3	0.697					

Example 6.2 Express $\Delta^2 y_0$ and $\Delta^3 y_0$ in terms of the values of the function y.

Solution Noting that each higher order difference is defined in terms of the lower order difference, we have

$$\Delta^2 y_0 = \Delta y_1 - \Delta y_0 = (y_2 - y_1) - (y_1 - y_0) = y_2 - 2y_1 + y_0$$

and

$$\begin{aligned}\Delta^3 y_0 &= \Delta^2 y_1 - \Delta^2 y_0 = (\Delta y_2 - \Delta y_1) - (\Delta y_1 - \Delta y_0)\\ &= (y_3 - y_2) - (y_2 - y_1) - (y_2 - y_1) + (y_1 - y_0)\\ &= y_3 - 3y_2 + 3y_1 - y_0\end{aligned}$$

Hence, we observe that the coefficients of the values of y, in the expansion of $\Delta^2 y_0$, $\Delta^3 y_0$ are binomial coefficients. Thus, in general, we arrive at the following result.

$$\Delta^n y_0 = y_n - {}^nC_1 y_{n-1} + {}^nC_2 y_{n-2} - {}^nC_3 y_{n-3} + \cdots + (-1)^n y_0 \tag{6.4}$$

Example 6.3 Show that the value of y_n can be expressed in terms of the leading value y_0 and the leading differences $\Delta y_0, \Delta^2 y_0, \ldots, \Delta^n y_0$.

Solution We have from the forward difference table

$$\left.\begin{aligned} y_1 - y_0 = \Delta y_0 \quad &\text{or} \quad y_1 = y_0 + \Delta y_0\\ y_2 - y_1 = \Delta y_1 \quad &\text{or} \quad y_2 = y_1 + \Delta y_1\\ y_3 - y_2 = \Delta y_2 \quad &\text{or} \quad y_3 = y_2 + \Delta y_2 \end{aligned}\right\} \tag{6.5}$$

and so on. Similarly

$$\left.\begin{aligned} \Delta y_1 - \Delta y_0 = \Delta^2 y_0 \quad &\text{or} \quad \Delta y_1 = \Delta y_0 + \Delta^2 y_0\\ \Delta y_2 - \Delta y_1 = \Delta^2 y_1 \quad &\text{or} \quad \Delta y_2 = \Delta y_1 + \Delta^2 y_1 \end{aligned}\right\} \tag{6.6}$$

and so on. Similarly, we can write

$$\left.\begin{aligned} \Delta^2 y_1 - \Delta^2 y_0 = \Delta^3 y_0 \quad &\text{or} \quad \Delta^2 y_1 = \Delta^2 y_0 + \Delta^3 y_0\\ \Delta^2 y_2 - \Delta^2 y_1 = \Delta^3 y_1 \quad &\text{or} \quad \Delta^2 y_2 = \Delta^2 y_1 + \Delta^3 y_1 \end{aligned}\right\} \tag{6.7}$$

and so on. Also, from Eqs. (6.6) and (6.7), we can rewrite Δy_2 as

$$\Delta y_2 = (\Delta y_0 + \Delta^2 y_0) + (\Delta^2 y_0 + \Delta^3 y_0)$$
$$= \Delta y_0 + 2\Delta^2 y_0 + \Delta^3 y_0 \tag{6.8}$$

From Eqs. (6.5)–(6.8), y_3 can be rewritten

$$\begin{aligned} y_3 &= y_2 + \Delta y_2 \\ &= (y_1 + \Delta y_1) + (\Delta y_1 + \Delta^2 y_1) \\ &= (y_0 + \Delta y_0) + 2(\Delta y_0 + \Delta^2 y_0) + (\Delta^2 y_0 + \Delta^3 y_0) \\ &= y_0 + 3\Delta y_0 + 3\Delta^2 y_0 + \Delta^3 y_0 \\ &= (1 + \Delta)^3 y_0 \end{aligned} \tag{6.9}$$

Similarly, we can symbolically write

$$y_1 = (1 + \Delta)y_0, \qquad y_2 = (1 + \Delta)^2 y_0, \qquad y_3 = (1 + \Delta)^3 y_0$$

Continuing this procedure, we can show, in general

$$y_n = (1 + \Delta)^n y_0$$

Hence, we obtain

$$y_n = y_0 + {}^nC_1 \Delta y_0 + {}^nC_2 \Delta^2 y_0 + \cdots + \Delta^n y_0 \tag{6.10}$$

Equivalently, we can also write the result as

$$y_n = \sum_{i=0}^{n} {}^nC_i \, \Delta^i y_0 \tag{6.11}$$

6.2.2 Backward Differences

For a given table of values (x_k, y_k), $k = 0, 1, 2, \ldots, n$ of a function $y = f(x)$ with equally spaced abscissas, the first backward differences are usually expressed in terms of the backward difference operator ∇ as

$$\nabla y_i = y_i - y_{i-1}, \qquad i = n, (n - 1), \ldots, 1 \tag{6.12}$$

Thus,

$$\begin{aligned} \nabla y_1 &= y_1 - y_0 \\ \nabla y_2 &= y_2 - y_1 \\ &\vdots \\ \nabla y_n &= y_n - y_{n-1} \end{aligned}$$

The differences of these differences are called *second differences* and they are denoted by $\nabla^2 y_2, \nabla^2 y_3, \ldots, \nabla^2 y_n$. That is,

$$\begin{aligned} \nabla^2 y_2 &= \nabla y_2 - \nabla y_1 \\ \nabla^2 y_3 &= \nabla y_3 - \nabla y_2 \\ &\vdots \\ \nabla^2 y_n &= \nabla y_n - \nabla y_{n-1} \end{aligned}$$

Thus, in general, the second backward differences are

$$\nabla^2 y_i = \nabla y_i - \nabla y_{i-1}, \qquad i = n, (n-1), \ldots, 2 \tag{6.13}$$

while the kth backward differences are given as

$$\nabla^k y_i = \nabla^{k-1} y_i - \nabla^{k-1} y_{i-1}, \qquad i = n, (n-1), \ldots, k \tag{6.14}$$

These backward differences can be systematically arranged for a table of values (x_k, y_k), $k = 0, 1, \ldots, 6$ as indicated in Table 6.2.

Table 6.2 Backward Difference Table

x	y	∇y	$\nabla^2 y$	$\nabla^3 y$	$\nabla^4 y$	$\nabla^5 y$	$\nabla^6 y$
x_0	y_0						
		∇y_1					
x_1	y_1		$\nabla^2 y_2$				
		∇y_2		$\nabla^3 y_3$			
x_2	y_2		$\nabla^2 y_3$		$\nabla^4 y_4$		
		∇y_3		$\nabla^3 y_4$		$\nabla^5 y_5$	
x_3	y_3		$\nabla^2 y_4$		$\nabla^4 y_5$		$\nabla^6 y_6$
		∇y_4		$\nabla^3 y_5$		$\nabla^5 y_6$	
x_4	y_4		$\nabla^2 y_5$		$\nabla^4 y_6$		
		∇y_5		$\nabla^3 y_6$			
x_5	y_5		$\nabla^2 y_6$				
		∇y_6					
x_6	y_6						

From this table, it can be observed that the subscript remains constant along every backward diagonal.

Example 6.4 Show that any value of y can be expressed in terms of y_n and its backward differences.

Solution From Eq. (6.12) we have

$$y_{n-1} = y_n - \nabla y_n \quad \text{and} \quad y_{n-2} = y_{n-1} - \nabla y_{n-1}$$

Also from the definition as given in Eq. (6.13), we get

$$\nabla y_{n-1} = \nabla y_n - \nabla^2 y_n$$

From these equations, we obtain

$$y_{n-2} = y_n - 2\nabla y_n + \nabla^2 y_n$$

Similarly, we can show that

$$y_{n-3} = y_n - 3\nabla y_n + 3\nabla^2 y_n - \nabla^3 y_n$$

Symbolically, these results can be rewritten as follows:

$$y_{n-1} = (1 - \nabla) y_n, \qquad y_{n-2} = (1 - \nabla)^2 y_n, \qquad y_{n-3} = (1 - \nabla)^3 y_n$$

Thus is general, we can write

$$y_{n-r} = (1 - \nabla)^r y_n$$

That is,

$$y_{n-r} = y_n - {}^rC_1 \nabla y_n + {}^rC_2 \nabla^2 y_n - \cdots + (-1)^r \nabla^r y_n \tag{6.15}$$

6.2.3 Central Differences

In some applications, central difference notation is found to be more convenient to represent the successive differences of a function. Here, we use the symbol δ to represent central difference operator and the subscript of δy for any difference as the average of the subscripts of the two members of the difference. Thus, we write

$$\delta y_{1/2} = y_1 - y_0, \qquad \delta y_{3/2} = y_2 - y_1, \text{ etc.}$$

In general

$$\delta y_i = y_{i+(1/2)} - y_{i-(1/2)} \tag{6.16}$$

Higher order differences are defined as follows:

$$\delta^2 y_i = \delta y_{i+(1/2)} - \delta y_{i-(1/2)} \tag{6.17}$$

$$\delta^n y_i = \delta^{n-1} y_{i+(1/2)} - \delta^{n-1} y_{i-(1/2)} \tag{6.18}$$

These central differences can be systematically arranged as indicated in Table 6.3:

Table 6.3 Central Difference Table

x	y	δy	$\delta^2 y$	$\delta^3 y$	$\delta^4 y$	$\delta^5 y$	$\delta^6 y$
x_0	y_0						
		$\delta y_{1/2}$					
x_1	y_1		$\delta^2 y_1$				
		$\delta y_{3/2}$		$\delta^3 y_{3/2}$			
x_2	y_2		$\delta^2 y_2$		$\delta^4 y_2$		
		$\delta y_{5/2}$		$\delta^3 y_{5/2}$		$\delta^5 y_{5/2}$	
x_3	y_3		$\delta^2 y_3$		$\delta^4 y_3$		$\delta^6 y_3$
		$\delta y_{7/2}$		$\delta^3 y_{7/2}$		$\delta^5 y_{7/2}$	
x_4	y_4		$\delta^2 y_4$		$\delta^4 y_4$		
		$\delta y_{9/2}$		$\delta^3 y_{9/2}$			
x_5	y_5		$\delta^2 y_5$				
		$\delta y_{11/2}$					
x_6	y_6						

Thus, we observe that all the odd differences have a fractional suffix and all the even differences with the same subscript lie horizontally.

The following alternative notation may also be adopted to introduce finite difference operators. Let $y = f(x)$ be a functional relation between x and y, which is also denoted by y_x. Suppose, we are given consecutive values of x differing by h say x, $x + h$, $x + 2h$, $x + 3h$, etc. The corresponding values of y are y_x, y_{x+h}, y_{x+2h}, y_{x+3h}, etc. As before, we can form the differences of these values. Thus

$$\Delta y_x = y_{x+h} - y_x = f(x + h) - f(x) \tag{6.19}$$

$$\Delta^2 y_x = \Delta y_{x+h} - \Delta y_x$$

Similarly

$$\nabla y_x = y_x - y_{x-h} = f(x) - f(x - h) \tag{6.20}$$

and

$$\delta y_x = y_{x+(h/2)} - y_{x-(h/2)} = f\left(x + \frac{h}{2}\right) - f\left(x - \frac{h}{2}\right) \tag{6.21}$$

Shift operator, E

Let $y = f(x)$ be a function of x, and let x takes the consecutive values x, $x + h$, $x + 2h$, etc. We then define an operator E having the property

$$E\,f(x) = f(x + h) \tag{6.22}$$

Thus, when E operates on $f(x)$, the result is the next value of the function. Here, E is called the *shift operator*. If we apply the operator E twice on $f(x)$, we get

$$E^2 f(x) = E\,[E\,f(x)] = E\,[f(x + h)] = f(x + 2h)$$

Thus, in general, if we apply the operator E n times on $f(x)$, we arrive at

$$E^n f(x) = f(x + nh)$$

In terms of new notation, we can write

$$E^n y_x = y_{x+nh}$$

or

$$E^n f(x) = f(x + nh) \tag{6.23}$$

for all real values of n. Also, if $y_0, y_1, y_2, y_3, \ldots$ are the consecutive values of the function y_x, then we can also write

$$Ey_0 = y_1, \qquad E^2 y_0 = y_2, \qquad E^4 y_0 = y_4, \qquad \ldots, \qquad E^2 y_2 = y_4$$

and so on. The inverse operator E^{-1} is defined as

$$E^{-1} f(x) = f(x - h)$$

and similarly

$$E^{-n} f(x) = f(x - nh) \tag{6.24}$$

Average operator, μ

The average operator μ is defined as

$$\mu\,f(x) = \frac{1}{2}\left[f\left(x + \frac{h}{2}\right) + f\left(x - \frac{h}{2}\right)\right] = \frac{1}{2}[y_{x+(h/2)} + y_{x-(h/2)}] \tag{6.25}$$

Differential operator, D

It is known that D represents a differential operator having a property

$$\left.\begin{aligned} Df(x) &= \frac{d}{dx} f(x) = f'(x) \\ D^2 f(x) &= \frac{d^2}{dx^2} f(x) = f''(x) \end{aligned}\right\} \tag{6.26}$$

Having defined various difference operators Δ, ∇, δ, E, μ and D, we can obtain the following relations easily:

From the definition of operators Δ and E, we have

$$\Delta y_x = y_{x+h} - y_x = Ey_x - y_x = (E - 1)y_x$$

Therefore,

$$\Delta = E - 1 \tag{6.27}$$

Following the definition of operators ∇ and E^{-1}, we have

$$\nabla y_x = y_x - y_{x-h} = y_x - E^{-1} y_x = (1 - E^{-1})\, y_x$$

Therefore,

$$\nabla = 1 - E^{-1} = \frac{E - 1}{E} \tag{6.28}$$

The definition of operators δ and E gives

$$\delta y_x = y_{x+(h/2)} - y_{x-(h/2)} = E^{1/2} y_x - E^{-1/2} y_x = (E^{1/2} - E^{-1/2})\, y_x$$

Hence,

$$\delta = E^{1/2} - E^{-1/2} \tag{6.29}$$

The definition of μ and E similarly yields

$$\mu y_x = \frac{1}{2}[y_{x+(h/2)} + y_{x-(h/2)}] = \frac{1}{2}(E^{1/2} + E^{-1/2}) y_x$$

Therefore,

$$\mu = \frac{1}{2}\left(E^{1/2} + E^{-1/2}\right) \tag{6.30}$$

It is known that

$$E y_x = y_{x+h} = f(x + h)$$

using Taylor series expansion, we have

$$E y_x = f(x) + hf'(x) + \frac{h^2}{2!} f''(x) + \cdots$$

$$= f(x) + hDf(x) + \frac{h^2}{2!} D^2 f(x) + \cdots$$

$$= \left(1 + \frac{hD}{1!} + \frac{h^2 D^2}{2!} + \cdots\right) f(x) = e^{hD} y_x$$

Thus,

$$hD = \log E \tag{6.31}$$

Hence, all the operators are expressed in terms of E.

Example 6.5 Prove that

$$hD = \log (1 + \Delta) = - \log (1 - \nabla) = \sinh^{-1} (\mu\delta)$$

Solution Using the standard relations (6.27)–(6.31), we have

$$hD = \log E = \log (1 + \Delta) = - \log E^{-1} = -\log (1 - \nabla) \tag{1}$$

Also,

$$\mu\delta = \frac{1}{2}(E^{1/2} + E^{-1/2})(E^{1/2} - E^{-1/2}) = \frac{1}{2}(E - E^{-1}) = \frac{1}{2}(e^{hD} - e^{-hD}) = \sinh (hD)$$

Therefore,

$$hD = \sinh^{-1}(\mu\delta) \tag{2}$$

Equations (1) and (2) constitute the required result.

Example 6.6 If Δ, ∇, δ denote forward, backward and central difference operators, E and μ are respectively the shift and average operators, in the analysis of data with equal spacing h, show that

(i) $1 + \delta^2\mu^2 = \left(1 + \dfrac{\delta^2}{2}\right)^2$ (ii) $E^{1/2} = \mu + \dfrac{\delta}{2}$

(iii) $\Delta = \dfrac{\delta^2}{2} + \delta\sqrt{1 + (\delta^2/4)}$ (iv) $\mu\delta = \dfrac{\Delta E^{-1}}{2} + \dfrac{\Delta}{2}$

(v) $\mu\delta = \dfrac{\Delta + \nabla}{2}$

Solutions (i) From the definition of operators, we have

$$\mu\delta = \frac{1}{2}(E^{1/2} + E^{-1/2})(E^{1/2} - E^{-1/2}) = \frac{1}{2}(E - E^{-1})$$

Therefore,

$$1 + \mu^2\delta^2 = 1 + \frac{1}{4}(E^2 - 2 + E^{-2}) = \frac{1}{4}(E + E^{-1})^2 \tag{1}$$

Also,

$$1 + \frac{\delta^2}{2} = 1 + \frac{1}{2}(E^{1/2} - E^{-1/2})^2 = \frac{1}{2}(E + E^{-1}) \tag{2}$$

From Eqs. (1) and (2), the first result follows.

(ii) Now

$$\mu + \frac{\delta}{2} = \frac{1}{2}(E^{1/2} + E^{-1/2} + E^{1/2} - E^{-1/2}) = E^{1/2}$$

Thus, the second result is proved.

(iii) We can write

$$\frac{\delta^2}{2} + \delta\sqrt{1 + (\delta^2/4)} = \frac{\left(E^{1/2} - E^{-1/2}\right)^2}{2} + \frac{\left(E^{1/2} - E^{-1/2}\right)\sqrt{1 + \frac{1}{4}\left(E^{1/2} - E^{-1/2}\right)^2}}{1}$$

$$= \frac{E - 2 + E^{-1}}{2} + \frac{1}{2}\left(E^{1/2} - E^{-1/2}\right)\left(E^{1/2} + E^{-1/2}\right)$$

$$= \frac{E - 2 + E^{-1}}{2} + \frac{E - E^{-1}}{2}$$

$$= E - 1$$

Using Eq. (6.27), we get

$$E - 1 = \Delta$$

(iv) We have

$$\mu\delta = \frac{1}{2}\left(E^{1/2} + E^{-1/2}\right)\left(E^{1/2} - E^{-1/2}\right) = \frac{1}{2}\left(E - E^{-1}\right)$$

Now, using Eq. (6.27), we get

$$= \frac{1}{2}\left(1 + \Delta - E^{-1}\right) = \frac{\Delta}{2} + \frac{1}{2}(1 - E^{-1})$$

$$= \frac{\Delta}{2} + \frac{1}{2}\left(\frac{E-1}{E}\right) = \frac{\Delta}{2} + \frac{\Delta}{2E}$$

(v) We can write

$$\mu\delta = \frac{1}{2}\left(E^{1/2} + E^{-1/2}\right)\left(E^{1/2} - E^{-1/2}\right) = \frac{1}{2}\left(E - E^{-1}\right)$$

Now using Eqs. (6.27) and (6.28), we have

$$\mu\delta = \frac{1}{2}(1 + \Delta - 1 + \nabla) = \frac{1}{2}(\Delta + \nabla)$$

Example 6.7 Show that the operations μ and E commute.

Solution From the definition of operators μ and E, we have

$$\mu E y_0 = \mu y_1 = \frac{1}{2}(y_{3/2} + y_{1/2}) \tag{1}$$

While

$$E\mu y_0 = \frac{1}{2}E\,(y_{1/2} + y_{-1/2}) = \frac{1}{2}(y_{3/2} + y_{1/2}) \tag{2}$$

Equating (1) and (2), we have

$$\mu E = E\mu$$

Therefore, the operators μ and E commute.

Theorem 6.1 (Differences of a polynomial). The *n*th differences of a polynomial of degree n is constant, when the values of the independent variable are given at equal intervals.

Proof Let us consider a polynomial of degree n in the form

$$y_x = a_0x^n + a_1x^{n-1} + a_2x^{n-2} + \cdots + a_{n-1}x + a_n,$$

where $a_0 \neq 0$ and $a_0, a_1, a_2, \ldots, a_n$ are constants. Let h be the interval of differencing. Then

$$y_{x+h} = a_0(x + h)^n + a_1(x + h)^{n-1} + a_2(x + h)^{n-2} + \cdots + a_{n-1}(x + h) + a_n$$

We now examine the differences of the polynomial:

$$\Delta y_x = y_{x+h} - y_x = a_0[(x + h)^n - x^n] + a_1[(x + h)^{n-1} - x^{n-1}]$$
$$+ a_2[(x + h)^{n-2} - x^{n-2}] + \cdots + a_{n-1}(x + h - x)$$

Binomial expansion yields

$$\begin{aligned}\Delta y_x &= a_0(x^n + {}^nC_1x^{n-1}h + {}^nC_2x^{n-2}h^2 + \cdots + h^n - x^n)\\ &\quad + a_1[x^{n-1} + {}^{(n-1)}C_1x^{n-2}h + {}^{(n-1)}C_2x^{n-3}h^2 + \cdots + h^{n-1} - x^{n-1}] + \cdots\\ &\quad + a_{n-1}h\\ &= a_0nhx^{n-1} + [a_0{}^nC_2h^2 + a_1{}^{(n-1)}C_1h]x^{n-2} + \cdots + a_{n-1}h\end{aligned}$$

Therefore,

$$\Delta y_x = a_0nhx^{n-1} + b'x^{n-2} + c'x^{n-3} + \cdots + k'x + l'$$

where b', c', ..., k', l' are constants involving h but not x. Thus, the first difference of a polynomial of degree n is another polynomial of degree $(n - 1)$. Similarly

$$\begin{aligned}\Delta^2 y_x &= \Delta(\Delta y_x) = \Delta y_{x+h} - \Delta y_x\\ &= a_0nh[(x + h)^{n-1} - x^{n-1}] + b'[(x + h)^{n-2} - x^{n-2}] + \cdots\\ &\quad + k'(x + h - x)\end{aligned}$$

On simplification, it reduces to the form

$$\Delta^2 y_x = a_0n(n - 1)\, h^2x^{n-2} + b''x^{n-3} + c''x^{n-4} + \cdots + q''$$

Therefore, $\Delta^2 y_x$ is a polynomial of degree $(n - 2)$ in x. Similarly, we can form the higher order differences, and every time we observe that the degree of the polynomial is reduced by one. After differencing n times, we are left with only the first term in the form

$$\Delta^n y_x = a_0n(n - 1)\,(n - 2)\,\ldots\,(2)(1)\,h^n = a_0\,(n!)h^n = \text{Constant}$$

This constant is independent of x. Since $\Delta^n y_x$ is a constant, $\Delta^{n+1} y_x = 0$. Hence the $(n + 1)$-th and higher order differences of a polynomial of degree n are zero.

6.3 NEWTON'S FORWARD DIFFERENCE INTERPOLATION FORMULA

Let $y = f(x)$ be a function which takes values $f(x_0)$, $f(x_0 + h)$, $f(x_0 + 2h)$, ..., corresponding to various equispaced values of x with spacing h, say x_0, $x_0 + h$, $x_0 + 2h$, ... Suppose, we wish to evaluate the function $f(x)$ for a value $x_0 + ph$, where p is any real number, then for any real number p, we have the operator E such that $E^pf(x) = f(x + ph)$. Therefore, using Eq. (6.27) we have

$$f(x_0 + ph) = E^pf(x_0) = (1 + \Delta)^pf(x_0)$$

$$= \left[1 + p\Delta + \frac{p(p-1)}{2!}\Delta^2 + \frac{p(p-1)(p-2)}{3!}\Delta^3 + \cdots\right]f(x_0)$$

That is,

$$f(x_0 + ph) = f(x_0) + p\Delta f(x_0) + \frac{p(p-1)}{2!}\Delta^2 f(x_0) + \frac{p(p-1)(p-2)}{3!}\Delta^3 f(x_0) + \cdots$$

$$+ \frac{p(p-1)\cdots(p-n+1)}{n!}\Delta^n f(x_0) + \text{Error} \qquad (6.32)$$

This is known as *Newton's forward difference formula for interpolation*, which gives the value of $f(x_0 + ph)$ in terms of $f(x_0)$ and its leading differences. This formula is also known as *Newton–Gregory forward difference interpolation formula*. Here, $p = (x - x_0)/h$. Equation (6.32) can also be written in another alternate form as

$$y_x = y_0 + p\Delta y_0 + \frac{p(p-1)}{2!}\Delta^2 y_0 + \frac{p(p-1)(p-2)}{3!}\Delta^3 y_0 + \cdots$$

$$+ \frac{p(p-1)(p-n+1)}{n!}\Delta^n y_0 + \text{Error} \tag{6.33}$$

If we retain $(r + 1)$ terms in Eq. (6.33), we obtain a polynomial of degree r agreeing with y_x at $x_0, x_1, \ldots, x_r$.

This formula is mainly used for interpolating the values of y near the beginning of a set of tabular values and for extrapolating values of y, a short distance backward from y_0. We shall illustrate these formulae by considering the following simple examples.

Example 6.8 Evaluate $f(15)$, given the following table of values:

x	10	20	30	40	50
$y = f(x)$	46	66	81	93	101

Solution We may note that $x = 15$ is very near to the beginning of the table. Hence, we use Newton's forward difference interpolation formula. The forward differences are calculated and tabulated as given below:

x	$y = f(x)$	Δy	$\Delta^2 y$	$\Delta^3 y$	$\Delta^4 y$
10	46				
		20			
20	66		–5		
		15		2	
30	81		–3		–3
		12		–1	
40	93		–4		
		8			
50	101				

We have Newton's forward difference interpolation formula as

$$y = y_0 + p\Delta y_0 + \frac{p(p-1)}{2!}\Delta^2 y_0 + \frac{p(p-1)(p-2)}{3!}\Delta^3 y_0$$

$$+ \frac{p(p-1)(p-2)(p-3)}{4!}\Delta^4 y_0 \tag{1}$$

In this example, from the above table, we have

$$x_0 = 10, \quad y_0 = 46, \quad \Delta y_0 = 20, \quad \Delta^2 y_0 = -5, \quad \Delta^3 y_0 = 2, \quad \Delta^4 y_0 = -3$$

Let y_{15} be the value of y when $x = 15$, then

$$p = \frac{x - x_0}{h} = \frac{15 - 10}{10} = 0.5$$

Substituting these values in Eq. (1), we get

$$f(15) = y_{15} = 46 + (0.5)(20) + \frac{(0.5)(0.5 - 1)}{2}(-5)$$

$$+ \frac{(0.5)(0.5 - 1)(0.5 - 2)}{6}(2) + \frac{(0.5)(0.5 - 1)(0.5 - 2)(0.5 - 3)}{24}(-3)$$

$$= 46 + 10 + 0.625 + 0.125 + 0.1172$$

Therefore, $f(15) = 56.8672$ correct to four decimal places.

Example 6.9 Find Newtons forward difference interpolating polynomial for the following data:

x	0.1	0.2	0.3	0.4	0.5
$y = f(x)$	1.40	1.56	1.76	2.00	2.28

Solution We shall first construct the forward difference table to the given data as indicated below:

x	$y = f(x)$	Δy	$\Delta^2 y$	$\Delta^3 y$	$\Delta^4 y$
0.1	1.40				
		0.16			
0.2	1.56		0.04		
		0.20		0.00	
0.3	1.76		0.04		0.00
		0.24		0.00	
0.4	2.00		0.04		
		0.28			
0.5	2.28				

Since, third and fourth leading differences are zero, we have Newton's forward difference interpolating formula as

$$y = y_0 + p\Delta y_0 + \frac{p(p - 1)}{2}\Delta^2 y_0 \tag{1}$$

In this problem, $x_0 = 0.1$, $y_0 = 1.40$, $\Delta y_0 = 0.16$, $\Delta^2 y_0 = 0.04$, and

$$p = \frac{x - 0.1}{0.1} = 10x - 1$$

Substituting these values in Eq. (1), we obtain

$$y = f(x) = 1.40 + (10x - 1)(0.16) + \frac{(10x - 1)(10x - 2)}{2}(0.04)$$

That is, $y = 2x^2 + x + 1.28$. This is the required Newton's interpolating polynomial.

Example 6.10 Estimate the missing figure in the following table:

x	1	2	3	4	5
$y = f(x)$	2	5	7	–	32

Solution Since we are given four entries in the table, the function $y = f(x)$ can be represented by a polynomial of degree three. Using Theorem 6.1, we have

$$\Delta^3 f(x) = \text{Constant} \quad \text{and} \quad \Delta^4 f(x) = 0$$

for all x. In particular, $\Delta^4 f(x_0) = 0$. Equivalently, $(E-1)^4 f(x_0) = 0$. Expanding, we have

$$(E^4 - 4E^3 + 6E^2 - 4E + 1) f(x_0) = 0$$

That is,

$$f(x_4) - 4f(x_3) + 6f(x_2) - 4f(x_1) + f(x_0) = 0$$

Using the values given in the table, we obtain

$$32 - 4f(x_3) + 6 \times 7 - 4 \times 5 + 2 = 0$$

which gives $f(x_3)$, the missing value equal to 14.

Example 6.11 Find a cubic polynomial in x which takes on the values –3, 3, 11, 27, 57 and 107, when $x = 0, 1, 2, 3, 4$ and 5 respectively.

Solution Here, the observations are given at equal intervals of unit width. To determine the required polynomial, we first construct the difference table as follows:

x	$f(x)$	$\Delta f(x)$	$\Delta^2 f(x)$	$\Delta^3 f(x)$
0	–3			
		6		
1	3		2	
		8		6
2	11		8	
		16		6
3	27		14	
		30		6
4	57		20	
		50		
5	107			

Since the fourth and higher order differences are zero, we have the required Newton's interpolation formula in the form

$$f(x_0 + ph) = f(x_0) + p\Delta f(x_0) + \frac{p(p-1)}{2}\Delta^2 f(x_0) + \frac{p(p-1)(p-2)}{6}\Delta^3 f(x_0) \tag{1}$$

Here,

$$p = \frac{x - x_0}{h} = \frac{x - 0}{1} = x, \quad \Delta f(x_0) = 6, \quad \Delta^2 f(x_0) = 2, \quad \Delta^3 f(x_0) = 6$$

Substituting these values into Eq. (1), we have

$$f(x) = -3 + 6x + \frac{x(x-1)}{2}(2) + \frac{x(x-1)(x-2)}{6}(6)$$

That is, $f(x) = x^3 - 2x^2 + 7x - 3$, is the required cubic polynomial.

6.4 NEWTON'S BACKWARD DIFFERENCE INTERPOLATION FORMULA

If one wishes to interpolate the value of the function $y = f(x)$ near the end of table of values, and to extrapolate value of the function a short distance forward from y_n, Newton's backward interpolation formula is used, which can be derived as follows:

Let $y = f(x)$ be a function which takes on values $f(x_n), f(x_n - h), f(x_n - 2h), \ldots, f(x_0)$ corresponding to equispaced values $x_n, x_n - h, x_n - 2h, \ldots, x_0$. Suppose, we wish to evaluate the function $f(x)$ at $(x_n + ph)$, where p is any real number, then we have the shift operator E, such that

$$f(x_n + ph) = E^p f(x_n) = (E^{-1})^{-p} f(x_n) = (1 - \nabla)^{-p} f(x_n)$$

Binomial expansion yields,

$$f(x_n + ph) = \left[1 + p\nabla + \frac{p(p+1)}{2!}\nabla^2 + \frac{p(p+1)(p+2)}{3!}\nabla^3 + \cdots \right.$$
$$\left. + \frac{p(p+1)(p+2)\cdots(p+n-1)}{n!}\nabla^n + \text{Error}\right] f(x_n)$$

That is,

$$f(x_n + ph) = f(x_n) + p\nabla f(x_n) + \frac{p(p+1)}{2!}\nabla^2 f(x_n)$$
$$+ \frac{p(p+1)(p+2)}{3!}\nabla^3 f(x_n) + \cdots$$
$$+ \frac{p(p+1)(p+2)\cdots(p+n-1)}{n!}\nabla^n f(x_n) + \text{Error} \quad (6.34)$$

This formula is known as *Newton's backward interpolation formula*. This formula is also known as *Newtons–Gregory backward difference interpolation formula*. If we retain $(r + 1)$ terms in Eq. (6.34), we obtain a polynomial of degree r agreeing with $f(x)$ at $x_n, x_{n-1}, \ldots, x_{n-r}$. Alternatively, this formula can also be written as

$$y_x = y_n + p\nabla y_n + \frac{p(p+1)}{2!}\nabla^2 y_n + \frac{p(p+1)(p+2)}{3!}\nabla^3 y_n + \cdots$$
$$+ \frac{p(p+1)(p+2)\cdots(p+n-1)}{n!}\nabla^n y_n + \text{Error} \quad (6.35)$$

where

$$p = \frac{x - x_n}{h}$$

Here follows a couple of examples for illustration.

Example 6.12 For the following table of values, estimate $f(7.5)$.

x	1	2	3	4	5	6	7	8
$y = f(x)$	1	8	27	64	125	216	343	512

Solution The value to be interpolated is at the end of the table. Hence, it is appropriate to use Newton's backward interpolation formula. We shall first construct the backward difference table for the given data:

x	$y = f(x)$	∇y	$\nabla^2 y$	$\nabla^3 y$	$\nabla^4 y$
1	1				
		7			
2	8		12		
		19		6	
3	27		18		0
		37		6	
4	64		24		0
		61		6	
5	125		30		0
		91		6	
6	216		36		0
		127		6	
7	343		42		
		169			
8	512				

Since the fourth and higher order differences are zero, the required Newton's backward interpolation formula is

$$y_x = y_n + p\nabla y_n + \frac{p(p+1)}{2!}\nabla^2 y_n + \frac{p(p+1)(p+2)}{3!}\nabla^3 y_n$$

In this problem,

$$p = \frac{x - x_n}{h} = \frac{7.5 - 8.0}{1} = -0.5$$

and

$$\nabla y_n = 169, \quad \nabla^2 y_n = 42, \quad \nabla^3 y_n = 6$$

Therefore,

$$y_{7.5} = 512 + (-0.5)(169) + \frac{(-0.5)(0.5)}{2}(42) + \frac{(-0.5)(0.5)(1.5)}{6}(6)$$

$$= 512 - 84.5 - 5.25 - 0.375$$

$$= 421.875$$

Example 6.13 The sales in a particular department store for the last five years is given in the following table:

Year	1974	1976	1978	1980	1982
Sales (in lakhs)	40	43	48	52	57

Estimate the sales for the year 1979.

Solution At the outset, we shall construct Newton's backward difference table for the given data as

x	y	∇y	$\nabla^2 y$	$\nabla^3 y$	$\nabla^4 y$
1974	40				
		3			
1976	43		2		
		5		−3	
1978	48		−1		5
		4		2	
1980	52		1		
		5			
1982	57				

In this example,

$$p = \frac{1979 - 1982}{2} = -1.5$$

and

$$\nabla y_n = 5, \qquad \nabla^2 y_n = 1, \qquad \nabla^3 y_n = 2, \qquad \nabla^4 y_n = 5$$

Newton's interpolation formula gives

$$y_{1979} = 57 + (-1.5)5 + \frac{(-1.5)(-0.5)}{2}(1)$$

$$+ \frac{(-1.5)(-0.5)(0.5)}{6}(2) + \frac{(-1.5)(-0.5)(0.5)(1.5)}{24}(5)$$

$$= 57 - 7.5 + 0.375 + 0.125 + 0.1172$$

Therefore,

$$y_{1979} = 50.1172$$

6.5 LAGRANGE'S INTERPOLATION FORMULA

Newton's interpolation formulae developed in the earlier sections can be used only when the values of the independent variable x are equally spaced. Also the differences of y must ultimately become small. If the values of the independent variable are not given at equidistant intervals, then we have the basic formula associated with the name of *Lagrange* which is derived as follows:

Let $y = f(x)$ be a function which takes the values $y_0, y_1, y_2, \ldots, y_n$ corresponding to $x_0, x_1, x_2, \ldots, x_n$. Since there are $(n + 1)$ values of y corresponding to $(n + 1)$ values of x, we can represent the function $f(x)$ by a polynomial of degree n. Suppose we write this polynomial in the form

$$f(x) = A_0 x^n + A_1 x^{n-1} + \cdots + A_n$$

or, more conveniently, in the form

$$y = f(x) = a_0 (x - x_1)(x - x_2) \cdots (x - x_n) + a_1 (x - x_0)(x - x_2) \cdots (x - x_n)$$
$$+ a_2 (x - x_0)(x - x_1) \cdots (x - x_n) + \cdots + a_n (x - x_0)(x - x_1) \cdots (x - x_{n-1}) \tag{6.36}$$

Here, the coefficients a_k are so chosen as to satisfy Eq. (6.36) by the $(n + 1)$ pairs

(x_i, y_i). Thus, Eq. (6.36) yields

$$y_0 = f(x_0) = a_0 (x_0 - x_1)(x_0 - x_2) \ldots (x_0 - x_n)$$

Therefore,

$$a_0 = \frac{y_0}{(x_0 - x_1)(x_0 - x_2) \cdots (x_0 - x_n)}$$

Similarly, we obtain

$$a_1 = \frac{y_1}{(x_1 - x_0)(x_1 - x_2) \cdots (x_1 - x_n)}$$

$$a_i = \frac{y_i}{(x_i - x_0)(x_i - x_1) \cdots (x_i - x_{i-1})(x_i - x_{i+1}) \cdots (x_i - x_n)}$$

and

$$a_n = \frac{y_n}{(x_n - x_0)(x_n - x_1) \cdots (x_n - x_{n-1})}$$

Now, substituting the values of $a_0, a_1, \ldots, a_n$ into Eq. (6.36), we get

$$\begin{aligned} y = f(x) &= \frac{(x - x_1)(x - x_2) \cdots (x - x_n)}{(x_0 - x_1)(x_0 - x_2) \cdots (x_0 - x_n)} y_0 \\ &+ \frac{(x - x_0)(x - x_2) \cdots (x - x_n)}{(x_1 - x_0)(x_1 - x_2) \cdots (x_1 - x_n)} y_1 + \cdots \\ &+ \frac{(x - x_0)(x - x_1) \cdots (x - x_{i-1})(x - x_{i+1}) \cdots (x - x_n)}{(x_i - x_0)(x_i - x_1) \cdots (x_i - x_{i-1})(x_i - x_{i+1}) \cdots (x_i - x_n)} y_i + \cdots \\ &+ \frac{(x - x_0)(x - x_1)(x - x_2) \cdots (x - x_{n-1})}{(x_n - x_0)(x_n - x_1)(x_n - x_2) \cdots (x_n - x_{n-1})} y_n \end{aligned} \tag{6.37}$$

Equation (6.37) is Lagrange's formula for interpolation. This formula can be used whether the values $x_0, x_1, x_2, \ldots, x_n$ are equally spaced or not. Alternatively, Eq. (6.37) can also be written in compact form as

$$\begin{aligned} y = f(x) &= L_0(x) y_0 + L_1(x) y_1 + \cdots + L_i(x) y_i + \cdots + L_n(x) y_n \\ &= \sum_{k=0}^{n} L_k(x) y_k \\ &= \sum_{k=0}^{n} L_k(x) f(x_k) \end{aligned} \tag{6.38}$$

where,

$$L_i(x) = \frac{(x - x_0)(x - x_1) \cdots (x - x_{i-1})(x - x_{i+1}) \cdots (x - x_n)}{(x_i - x_0)(x_i - x_1) \cdots (x_i - x_{i-1})(x_i - x_{i+1}) \cdots (x_i - x_n)} \tag{6.39}$$

we can easily observe that, $L_i(x_i) = 1$ and $L_i(x_j) = 0$, $i \neq j$. Thus introducing

Kronecker delta notation

$$L_i(x_j) = \delta_{ij} = \begin{cases} 1, & \text{if } i = j \\ 0, & \text{if } i \neq j \end{cases}$$

Further, if we introduce the notation

$$\Pi(x) = \prod_{i=0}^{n} (x - x_i) = (x - x_0)(x - x_1) \cdots (x - x_n) \tag{6.40}$$

that is, $\Pi(x)$ is a product of $(n + 1)$ factors. Clearly, its derivative $\Pi'(x)$ contains a sum of $(n + 1)$ terms in each of which one of the factors of $\Pi(x)$ will be absent. We also define,

$$P_k(x) = \prod_{i \neq k} (x - x_i) \tag{6.41}$$

which is same as $\Pi(x)$ except that the factor $(x - x_k)$ is absent. Then

$$\Pi'(x) = P_0(x) + P_1(x) + \cdots + P_n(x) \tag{6.42}$$

But, when $x = x_k$, all terms in the above sum vanishes except $P_k(x_k)$. Hence,

$$\Pi'(x_k) = P_k(x_k) = (x_k - x_0) \cdots (x_k - x_{k-1})(x_k - x_{k+1}) \cdots (x_k - x_n) \tag{6.43}$$

Therefore, using Eqs. (6.40)–(6.43), Eq. (6.39) can be rewritten as

$$L_k(x) = \frac{P_k(x)}{P_k(x_k)} = \frac{P_k(x)}{\Pi'(x_k)} = \frac{\Pi(x)}{(x - x_k)\Pi'(x_k)} \tag{6.44}$$

Finally, the Lagrange's interpolation polynomial of degree n can be written as

$$y(x) = f(x) = \sum_{k=0}^{n} \frac{\Pi(x)}{(x - x_k)\Pi'(x_k)} f(x_k) = \sum_{k=0}^{n} L_k(x) f(x_k) = \sum_{k=0}^{n} L_k(x) y_k \tag{6.45}$$

Lagrange's interpolation is illustrated through the following examples.

Example 6.14 Find Lagrange's interpolation polynomial fitting the points $y(1) = -3$, $y(3) = 0$, $y(4) = 30$, $y(6) = 132$. Hence find $y(5)$.

Solution The given data can be arranged as follows:

x	1	3	4	6
$y = f(x)$	–3	0	30	132

using Lagrange's interpolation formula (6.37), we have

$$y(x) = f(x) = \frac{(x-3)(x-4)(x-6)}{(1-3)(1-4)(1-6)}(-3) + \frac{(x-1)(x-4)(x-6)}{(3-1)(3-4)(3-6)}(0)$$

$$+ \frac{(x-1)(x-3)(x-6)}{(4-1)(4-3)(4-6)}(30) + \frac{(x-1)(x-3)(x-4)}{(6-1)(6-3)(6-4)}(132)$$

$$= \frac{x^3 - 13x^2 + 54x - 72}{-30}(-3) + \frac{x^3 - 11x^2 + 34x - 24}{6}(0)$$

$$+ \frac{x^3 - 10x^2 + 27x - 18}{-6}(30) + \frac{x^3 - 8x^2 + 19x - 12}{30}(132)$$

On simplification, we get

$$y(x) = \frac{1}{10}\left(-5x^3 + 135x^2 - 460x + 300\right) = \frac{1}{2}(-x^3 + 27x^2 - 92x + 60)$$

which is the required Lagrange's interpolation polynomial. Now, $y(5) = 75$.

Example 6.15 Given the following data, evaluate $f(3)$ using Lagrange's interpolating polynomial.

x	1	2	5
$f(x)$	1	4	10

Solution Using Lagrange's interpolation formula given by Eq. (6.37), we have

$$f(x) = \frac{(x - x_1)(x - x_2)}{(x_0 - x_1)(x_0 - x_2)}f(x_0) + \frac{(x - x_0)(x - x_2)}{(x_1 - x_0)(x_1 - x_2)}f(x_1)$$

$$+ \frac{(x - x_0)(x - x_1)}{(x_2 - x_0)(x_2 - x_1)}f(x_2)$$

Therefore,

$$f(3) = \frac{(3-2)(3-5)}{(1-2)(1-5)}(1) + \frac{(3-1)(3-5)}{(2-1)(2-5)}(4) + \frac{(3-1)(3-2)}{(5-1)(5-2)}(10) = 6.4999$$

6.6 DIVIDED DIFFERENCES

When the function values are given at non-equispaced points, we have already developed the Lagrange's interpolation formula for interpolation in Section 6.5. Now, we shall introduce the concept of divided differences and then develop Newton's divided difference interpolation formula, whose accuracy is same as that of Lagrange's formula, but has the advantage of being computationally economical in the sense that it involves less number of arithmetic operations.

Let us assume that the function $y = f(x)$ is known for several values of x, (x_i, y_i), for $i = 0\ (1)\ n$. The divided differences of orders 0, 1, 2, …, n are defined recursively as follows:

$$y\,[x_0] = y\,(x_0) = y_0$$

is the 0th order divided difference. The first order divided difference is defined as

$$y\,[x_0, x_1] = \frac{y_1 - y_0}{x_1 - x_0}$$

Similarly, the higher order divided differences are defined in terms of lower order divided differences by the relations (Hildebrand, 1982) of the form

$$y[x_0, x_1, x_2] = \frac{y[x_1, x_2] - y[x_0, x_1]}{x_2 - x_0}$$

while

$$y[x_0, x_1, \cdots, x_n] = \frac{y[x_1, x_2, \cdots, x_n] - y[x_0, x_1, \cdots, x_{n-1}]}{x_n - x_0} \tag{6.46}$$

The standard format of the divided differences are displayed in Table 6.4.

Table 6.4 Divided Differences

x	$y(x)$	1st order	2nd order	3rd order	4th order
x_0	y_0				
		$y[x_1, x_0]$			
x_1	y_1		$y[x_0, x_1, x_2]$		
		$y[x_2, x_1]$		$y[x_0, x_1, x_2, x_3]$	
x_2	y_2		$y[x_1, x_2, x_3]$		$y[x_0, x_1, x_2, x_3, x_4]$
		$y[x_3, x_2]$		$y[x_1, x_2, x_3, x_4]$	
x_3	y_3		$y[x_2, x_3, x_4]$		
		$y[x_4, x_3]$			
x_4	y_4				

We can easily verify that the divided difference is a symmetric function of its arguments. That is,

$$y[x_1, x_0] = y[x_0, x_1] = \frac{y_0}{x_0 - x_1} + \frac{y_1}{x_1 - x_0}$$

Now,

$$y[x_0, x_1, x_2] = \frac{y[x_1, x_2] - y[x_0, x_1]}{x_2 - x_0} = \frac{1}{x_2 - x_0}\left(\frac{y_2 - y_1}{x_2 - x_1} - \frac{y_1 - y_0}{x_1 - x_0}\right)$$

Therefore,

$$y[x_0, x_1, x_2] = \frac{y_0}{(x_0 - x_1)(x_0 - x_2)} + \frac{y_1}{(x_1 - x_0)(x_1 - x_2)} + \frac{y_2}{(x_2 - x_0)(x_2 - x_1)}$$

which is a symmetric form, hence suggests the general result as

$$y[x_0, \cdots, x_k] = \frac{y_0}{(x_0 - x_1)\cdots(x_0 - x_k)} + \frac{y_1}{(x_1 - x_0)\cdots(x_1 - x_k)} + \cdots$$

$$+ \frac{y_k}{(x_k - x_0)\cdots(x_k - x_{k-1})}$$

$$= \sum_{i=0}^{k} \frac{y_i}{\prod\limits_{\substack{i=0 \\ i \neq j}}^{k} (x_i - x_j)} \tag{6.47}$$

In Eq. (6.47), it can be noted that zero factor $(x_i - x_i)$ is omitted in the denominator of each term of the sum.

6.6.1 Newton's Divided Difference Interpolation Formula

Let $y = f(x)$ be a function which takes values $y_0, y_1, \ldots, y_n$ corresponding to $x = x_i$, $i = 0, 1, \ldots, n$. We choose an interpolating polynomial, interpolating at $x = x_i$, $i = 0, 1, \ldots, n$ in the following convenient form

$$y = f(x) = a_0 + a_1 (x - x_0) + a_2(x - x_0)(x - x_1) + \cdots + a_n (x - x_0)(x - x_1) \cdots (x - x_{n-1}) \quad (6.48)$$

Here, the coefficients a_k are so chosen as to satisfy Eq. (6.48) by the $(n + 1)$ pairs (x_i, y_i). Thus, we have

$$\left.\begin{aligned}
&y(x_0) = f(x_0) = y_0 = a_0 \\
&y(x_1) = f(x_1) = y_1 = a_0 + a_1(x_1 - x_0) \\
&y(x_2) = f(x_2) = y_2 = a_0 + a_1(x_2 - x_0) + a_2(x_2 - x_0)(x_2 - x_1) \\
&\vdots \\
&\quad y_n = a_0 + a_1(x_n - x_0) + a_2(x_n - x_0)(x_n - x_1) + \cdots \\
&\qquad + a_n(x_n - x_0) \cdots (x_n - x_{n-1})
\end{aligned}\right\} \quad (6.49)$$

The coefficients $a_0, a_1, \ldots, a_n$ can be easily obtained from the system of Eqs. (6.49), as they form a lower triangular matrix. The first equation of (6.49) gives

$$a_0 = y(x_0) = y_0 \quad (6.50)$$

The second equation of (6.49) and Eq. (6.50) gives

$$a_1 = \frac{y_1 - y_0}{x_1 - x_0} = y[x_0, x_1] \quad (6.51)$$

The third equation of (6.49) after using a_0 and a_1 as given in Eqs. (6.50) and (6.51) yields

$$a_2 = \frac{y_2 - y_0 - (x_2 - x_0)y[x_0, x_1]}{(x_2 - x_0)(x_2 - x_1)}$$

which can be rewritten as

$$a_2 = \frac{\left[y_2 - y_1 + \dfrac{y_1 - y_0}{x_1 - x_0}(x_1 - x_0)\right] - (x_2 - x_0)y[x_0, x_1]}{(x_2 - x_0)(x_2 - x_1)}$$

That is,

$$a_2 = \frac{y_2 - y_1 + y[x_0, x_1](x_1 - x_2)}{(x_2 - x_0)(x_2 - x_1)} = \frac{y[x_1, x_2] - y[x_0, x_1]}{x_2 - x_0}$$

Thus, in terms of second order divided differences, we have

$$a_2 = y[x_0, x_1, x_2] \quad (6.52)$$

Similarly, we can show that

$$a_n = y[x_0, x_1, \ldots, x_n] \quad (6.53)$$

Hence, Newton's divided difference interpolation formula can be written as

$$y = f(x) = y_0 + (x - x_0)\, y[x_0, x_1] + (x - x_0)(x - x_1)\, y\,[x_0, x_1, x_2] + \cdots$$
$$+ (x - x_0)(x - x_1) \ldots (x - x_{n-1})\, y\,[x_0, x_1, \ldots, x_n] \tag{6.54}$$

Newton's divided differences can also be expressed in terms of forward, backward and central differences. They can be easily derived as follows: Assuming equispaced values of abscissa, we have

$$y[x_0, x_1] = \frac{y_1 - y_0}{x_1 - x_0} = \frac{\Delta y_0}{h}$$

$$y[x_0, x_1, x_2] = \frac{y[x_1, x_2] - y[x_0, x_1]}{x_2 - x_0} = \frac{\dfrac{\Delta y_1}{h} - \dfrac{\Delta y_0}{h}}{2h} = \frac{\Delta^2 y_0}{2!h^2}$$

By induction, we can in general arrive at the result

$$y[x_0, x_1, \cdots, x_n] = \frac{\Delta^n y_0}{n!h^n} \tag{6.55}$$

Similarly

$$y[x_0, x_1] = \frac{y_1 - y_0}{x_1 - x_0} = \frac{\nabla y_1}{h}$$

$$y[x_0, x_1, x_2] = \frac{y[x_1, x_2] - y[x_0, x_1]}{x_2 - x_0} = \frac{\dfrac{\nabla y_2}{h} - \dfrac{\nabla y_1}{h}}{2h} = \frac{\nabla^2 y_2}{2!\,h^2}$$

In general, we have

$$y[x_0, x_1, \cdots, x_n] = \frac{\nabla^n y_n}{n!\,h^n} \tag{6.56}$$

Also, in terms of central differences, we have

$$y[x_0, x_1] = \frac{y_1 - y_0}{x_1 - x_0} = \frac{\delta y_{1/2}}{h}$$

$$y[x_0, x_1, x_2] = \frac{y[x_1, x_2] - y[x_0, x_1]}{x_2 - x_0} = \frac{\dfrac{\delta y_{3/2}}{h} - \dfrac{\delta y_{1/2}}{h}}{2h} = \frac{\delta^2 y_1}{2!\,h^2}$$

In general, the following pattern is arrived:

$$y\,[x_0, x_1, \ldots, x_{2m}] = \frac{\delta^{2m} y_m}{(2m)!\,h^{2m}}$$

or

$$y\,[x_0, x_1, \ldots, x_{2m+1}] = \frac{\delta^{2m+1} y_{m+(1/2)}}{(2m+1)!\,h^{2m+1}} \tag{6.57}$$

We present below few examples for illustration.

Example 6.16 Find the interpolating polynomial by (i) Lagrange's formula, and (ii) Newton's divided difference formula for the following data, and hence show that they represent the same interpolating polynomial.

x	0	1	2	4
y	1	1	2	5

Solution The divided difference table for the given data is constructed as follows:

x	y	1st divided difference	2nd divided difference	3rd divided difference
0	1			
		0		
1	1		1/2	
		1		−1/12
2	2		1/6	
		3/2		
4	5			

(i) Lagrange's interpolation formula (6.37) gives

$$y = f(x) = \frac{(x-1)(x-2)(x-4)}{(-1)(-2)(-4)}(1) + \frac{(x-0)(x-2)(x-4)}{(1-0)(1-2)(1-4)}(1)$$

$$+ \frac{(x-0)(x-1)(x-4)}{(2)(2-1)(2-4)}(2) + \frac{(x-0)(x-1)(x-2)}{4(4-1)(4-2)}(5)$$

$$= \frac{-(x^3 - 7x^2 + 14x - 8)}{8} + \frac{x^3 - 6x^2 + 8x}{3} - \frac{x^3 - 5x^2 + 4x}{2}$$

$$+ \frac{5(x^3 - 3x^2 + 2x)}{24}$$

$$= -\frac{x^3}{12} + \frac{3x^2}{4} - \frac{2}{3}x + 1 \tag{1}$$

(ii) Newton's divided difference formula gives

$$y = f(x) = 1 + (x-0)(0) + (x-0)(x-1)\left(\frac{1}{2}\right) + (x-0)(x-1)(x-2)\left(-\frac{1}{12}\right)$$

$$= -\frac{x^3}{12} + \frac{3x^2}{4} - \frac{2}{3}x + 1 \tag{2}$$

From Eqs. (1) and (2) we observe that the interpolating polynomial by both Lagrange's and Newton's divided difference formulae is one and the same. Also Newton's formula involves less number of arithmetic operations than that of Lagrange's.

Example 6.17 Using Newton's divided difference formula, find the quadratic equation for the following data. Hence find $y(2)$.

x	0	1	4
y	2	1	4

Solution The divided difference table for the given data is constructed as follows:

x	y	1st divided difference	2nd divided difference
0	2		
		–1	
1	1		1/2
		1	
4	4		

Now, using Newton's divided difference formula, we have

$$y = 2 + (x - 0)(-1) + (x - 0)(x - 1)\left(\frac{1}{2}\right) = \frac{1}{2}(x^2 - 3x + 4)$$

Hence, $y(2) = 1$.

Example 6.18 A function $y = f(x)$ is given at the sample points $x = x_0$, x_1 and x_2. Show that the Newton's divided difference interpolation formula and the corresponding Lagrange's interpolation formula are identical.

Solution For the function $y = f(x)$, we have the data (x_i, y_i), $i = 0, 1, 2$. The interpolation polynomial using Newton's divided difference formula is given as

$$y = f(x) = y_0 + (x - x_0)\, y[x_0, x_1] + (x - x_0)(x - x_1)\, y[x_0, x_1, x_2] \qquad (1)$$

Using the definition of divided differences and Eq. (6.47), we can rewrite Eq. (1) in the form

$$y = y_0 + (x - x_0)\frac{(y_1 - y_0)}{(x_1 - x_0)} + (x - x_0)(x - x_1)\left[\frac{y_0}{(x_0 - x_1)(x_0 - x_2)}\right.$$

$$\left. + \frac{y_1}{(x_1 - x_0)(x_1 - x_2)} + \frac{y_2}{(x_2 - x_0)(x_2 - x_1)}\right]$$

$$= \left[1 - \frac{(x_0 - x)}{(x_0 - x_1)} + \frac{(x - x_0)(x - x_1)}{(x_0 - x_1)(x_0 - x_2)}\right] y_0$$

$$+ \left[\frac{(x - x_0)}{(x_1 - x_0)} + \frac{(x - x_0)(x - x_1)}{(x_1 - x_0)(x_1 - x_2)}\right] y_1 + \frac{(x - x_0)(x - x_1)}{(x_2 - x_0)(x_2 - x_1)}\, y_2$$

On simplification, it reduces to

$$y = \frac{(x - x_1)(x - x_2)}{(x_0 - x_1)(x_0 - x_2)} y_0 + \frac{(x - x_0)(x - x_2)}{(x_1 - x_0)(x_1 - x_2)} y_1 + \frac{(x - x_0)(x - x_1)}{(x_2 - x_0)(x_2 - x_1)} y_2 \quad (2)$$

which is the Lagrange's form of interpolation polynomial. Hence Eqs. (1) and (2) are identical.

6.6.2 Newton's Divided Difference Formula with Error Term

Following the basic definition (6.46) of divided differences, we have for any x

$$\left.\begin{aligned} y(x) &= y_0 + (x - x_0)\, y[x, x_0] \\ y[x, x_0] &= y[x_0, x_1] + (x - x_1)\, y[x, x_0, x_1] \\ y[x, x_0, x_1] &= y[x_0, x_1, x_2] + (x - x_2)\, y[x, x_0, x_1, x_2] \\ &\vdots \\ y[x, x_0, \ldots, x_{n-1}] &= y[x_0, x_1, \ldots, x_n] + (x - x_n)\, y[x, x_0, \ldots, x_n] \end{aligned}\right\} \quad (6.58)$$

Multiplying the second of Eqs. (6.58) by $(x - x_0)$, third by $(x - x_0)(x - x_1)$ and so on, while the last one by $(x - x_0)(x - x_1) \ldots (x - x_{n-1})$ and adding the resulting equations, we finally obtain

$$\begin{aligned} y(x) = y_0 &+ (x - x_0)\, y[x_0, x_1] + (x - x_0)(x - x_1)\, y[x_0, x_1, x_2] + \cdots \\ &+ (x - x_0)(x - x_1) \cdots (x - x_{n-1})\, y[x_0, x_1, \ldots, x_n] + \varepsilon(x) \end{aligned} \quad (6.59)$$

where

$$\varepsilon(x) = (x - x_0)(x - x_1) \cdots (x - x_n)\, y[x, x_0, \ldots, x_n] \quad (6.60)$$

It may be noted that for $x = x_0, x_1, \ldots, x_n$, the error term $\varepsilon(x)$ vanishes.

6.6.3 Error Term in Interpolation Formulae

We have seen in Section 6.6.2 that if $y(x)$ is approximated by a polynomial $P_n(x)$ of degree n then the error is given by

$$\varepsilon(x) = y(x) - P_n(x),$$

where,

$$\varepsilon(x) = (x - x_0)(x - x_1) \ldots (x - x_n) y\, [x, x_0, \ldots, x_n]$$

Alternatively it is also expressed as

$$\varepsilon(x) = \Pi(x)\, y\, [x, x_0, \ldots, x_n] = K\Pi(x) \quad (6.61)$$

Now, consider a function $F(x)$, such that

$$F(x) = y(x) - P_n(x) - K\Pi(x) \quad (6.62)$$

and determine the constant K in such a way that $F(x)$ vanishes for $x = x_0, x_1, \ldots, x_n$ and also for an arbitrarily chosen point $\bar{x}$, which is different from the given $(n + 1)$ points. Let I denotes the closed interval spanned by the values $x_0, \ldots, x_n, \bar{x}$. Then $F(x)$ vanishes $(n + 2)$ times in the interval I. By Rolle's theorem $F'(x)$ vanishes at least $(n + 1)$ times in the interval I, $F''(x)$ vanishes at least n times, and so on. Eventually, we can show that $F^{(n+1)}(x)$ vanishes at least once in the interval I, say at $x = \xi$. Thus, we obtain

$$0 = y^{(n+1)}(\xi) - P_n^{(n+1)}(\xi) - K\Pi^{(n+1)}(\xi) \tag{6.63}$$

Since $P_n(x)$ is a polynomial of degree n, its $(n + 1)$th derivative is zero. Also, from the definition of $\Pi(x)$, $\Pi^{(n+1)}(x) = (n + 1)!$. Therefore, Eq. (6.63) gives

$$K = \frac{y^{(n+1)}(\xi)}{(n + 1)!}$$

Substituting the value of constant K into Eq. (6.61) gives

$$\varepsilon(x) = y(x) - P_n(x) = \frac{y^{(n+1)}(\xi)}{(n + 1)!}\Pi(x) \tag{6.64}$$

for some $\xi = \xi(x)$ in the interval I. Incidentally, by equating Eqs. (6.61) and (6.64), we observe that

$$y[x, x_0, \ldots, x_n] = \frac{y^{(n+1)}(\xi)}{(n + 1)!} \tag{6.65}$$

Thus, the error committed in replacing $y(x)$ by either Newton's divided difference formula or by an identical Lagrange's formula is given by

$$\varepsilon(x) = \Pi(x)y[x, x_0, \cdots, x_n] = \Pi(x)\frac{y^{(n+1)}(\xi)}{(n + 1)!} \tag{6.66}$$

6.7 INTERPOLATION IN TWO DIMENSIONS

Let u be a polynomial function in two variables, say x and y, in particular quadratic in x and cubic in y, which in general can be written as

$$\begin{aligned} u = f(x, y) &= a_0 + a_1x + a_2y + a_3x^2 + a_4xy + a_5y^2 + a_6y^3 \\ &+ a_7y^2x + a_8yx^2 + a_9y^3x + a_{10}y^2x^2 + a_{11}y^3x^2 \end{aligned} \tag{6.67}$$

This relation involves many terms. If we have to write a relation involving three or more variables, even low degree polynomials give rise to prohibitatively long expressions. If necessary, we can certainly write, but such complications can be avoided by handling each variable separately.

If we let x, a constant, say $x = c$, Eq. (6.67) immediately simplifies to the form

$$u\,|_{x=c} = b_0 + b_1y + b_2y^2 + b_3y^3 \tag{6.68}$$

Now, we adopt the following procedure to interpolate at a point (l, m) in a table of two variables, by treating one variable a constant say $x = x_1$. The problem reduces to that of a single variable interpolation. Any one of the methods discussed in preceding sections can then be applied to get $f(x_1, m)$. Then we repeat this procedure for various values of x say $x = x_2, x_3, \ldots, x_n$ keeping y constant. Thus, we get a new table with y constant at the value $y = m$ and with x varying. We can then interpolate at $x = l$. We shall illustrate this procedure by considering the following example.

Example 6.19 Tabulate the values of the function

$$f(x, y) = x^2 + y^2 - y$$

for $x = 0, 1, 2, 3, 4$ and $y = 0, 1, 2, 3, 4$. Using this table of values, compute $f(2.5, 3.5)$ by numerical double interpolation.

Solution The values of the function for the given values of x and y are given in the following table:

x	y = 0	1	2	3	4
0	0	0	2	6	12
1	1	1	3	7	13
2	4	4	6	10	16
3	9	9	11	15	21
4	16	16	18	22	28

Using quadratic interpolation in both x and y directions we need to consider three points in x and y directions. To start with, we have to treat one variable constant, say x. Keeping $x = 2.5$, $y = 3.5$ as the near centre of the set, we choose the table of values corresponding to $x = 1, 2, 3$ and $y = 2, 3, 4$. The region of fit for the construction of our interpolation polynomial is shown with dots in the above table.

Thus, using Newton's forward difference formula, we have

At $x = 1$

y	f	Δf	$\Delta^2 f$
2	3		
		4	
3	7		2
		6	
4	13		

At $x = 2$

y	f	Δf	$\Delta^2 f$
2	6		
		4	
3	10		2
		6	
4	16		

	At $x = 3$		
y	f	Δf	$\Delta^2 f$
2	11		
		4	
3	15		2
		6	
4	21		

with

$$p = \frac{y - y_0}{h} = \frac{3.5 - 2}{1} = 1.5$$

We obtain,

$$f(1, 3.5) = f_0 + p\Delta f_0 + \frac{p(p-1)}{2!}\Delta^2 f_0 = 3 + (1.5)(4) + \frac{(1.5)(0.5)}{2}(2) = 9.75$$

$$f(2, 3.5) = 6 + (1.5)(4) + \frac{(1.5)(0.5)}{2}(2) = 12.75$$

$$f(3, 3.5) = 11 + (1.5)(4) + \frac{(1.5)(0.5)}{2}(2) = 17.75$$

Therefore, we arrive at the following result

	At $y = 3.5$		
x	f	Δf	$\Delta^2 f$
1	9.75		
		3	
2	12.75		2
		5	
3	17.75		

Now, defining

$$p = \frac{2.5 - 1}{1} = 1.5$$

and using Newton's formula, we obtain

$$f(2.5, 3.5) = 9.75 + (1.5)(3) + \frac{(1.5)(0.5)}{2}(2) = 15$$

From the functional relation, we also find that

$$f(2.5, 3.5) = (2.5)^2 + (3.5)^2 - 3.5 = 15$$

and hence no error in interpolation.

6.8 CUBIC SPLINE INTERPOLATION

The name *spline function* is derived from a device known as mechanical spline used by draftsmen for drawing smooth curves. It consists of a flexible steel strip to which weights may be attached in such a way as to constrain it to pass through a given set of points. Mathematically, a *spline function* is one whose graph is a composite curve made up of a number of polynomial arcs of a given degree fitted together in such a way that the junctions of the successive arcs are as smooth as they could be made without going to a single polynomial over the entire range. Fitting to an empirical data by a spline function offers a numerical method for obtaining a curve similar to the one produced by a French curve. This technique has been used effectively in the areas of computer graphics, flow simulations and for smoothing of satellite data which is received at a tracking station with noise.

Definition 6.1 Suppose, we have $(n + 1)$ data points (x_i, y_i), $i = 0, 1, 2, \ldots, n$; where x_i may not be equally spaced and $x_0 = a$, $x_n = b$ and we wish to determine a cubic spline function $S(x)$, such that it has the following properties:

(i) The cubic spline function has the form

$$S(x) = a_i(x - x_i)^3 + b_i(x - x_i)^2 + c_i(x - x_i) + d_i$$

in each interval (x_i, x_{i+1}), $i = 0, 1, 2, \ldots, (n - 1)$.

(ii) $S(x_i) = y_i$, $(i = 0, 1, 2, \ldots, n)$.

(iii) The cubics are so joined that the function $S(x)$ and both its slope $S'(x)$ and curvature $S''(x)$ are continuous in (x_0, x_n). It means that the spline curve $S(x)$ will not have sharp corners and the radius of curvature is defined at each point.

Thus, the cubic spline function will have the form $S(x) = S_i(x)$ in the interval (x_i, x_{i+1}). To get $S(x)$, we have to put together the cubics $S_i(x)$ as shown in Fig. 6.1. A detailed account of the basic properties of the cubic spline can be found in Ahlberg et al. (1967).

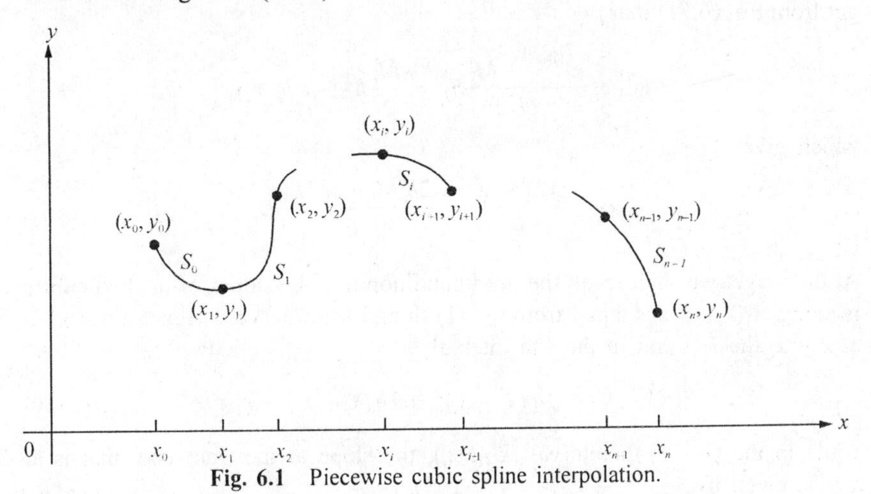

Fig. 6.1 Piecewise cubic spline interpolation.

6.8.1 Construction of Cubic Spline

Spath (1969) suggested a cubic spline $S(x)$ by the piecewise cubic polynomials of the form

$$S(x) = a_i(x - x_i)^3 + b_i(x - x_i)^2 + c_i(x - x_i) + d_i \tag{6.69}$$

in each interval (x_i, x_{i+1}), $i = 0, 1, 2, \ldots, (n - 1)$.

Since condition (ii) in Definition 6.1 implies that the cubic spline fits exactly at the two end points x_i and x_{i+1} of the i-th interval, we have

$$S(x_i) = y_i = a_i(x_i - x_i)^3 + b_i(x_i - x_i)^2 + c_i(x_i - x_i) + d_i = d_i \quad (6.70)$$

$$S(x_{i+1}) = y_{i+1} = a_i(x_{i+1} - x_i)^3 + b_i(x_{i+1} - x_i)^2 + c_i(x_{i+1} - x_i) + d_i \quad (6.71)$$

Now, introducing the notation $h_i = x_{i+1} - x_i$, Eq. (6.71) becomes

$$y_{i+1} = a_i h_i^3 + b_i h_i^2 + c_i h_i + d_i \quad (6.72)$$

To satisfy the third condition relating to the slope and curvature of the joining cubics, we obtain from Eq. (6.69) that

$$S'(x) = y'(x) = 3a_i(x - x_i)^2 + 2b_i(x - x_i) + c_i \quad (6.73)$$

$$S''(x) = y''(x) = 6a_i(x - x_i) + 2b_i \quad (6.74)$$

Using the notation $S_i'' = y''(x_i) = M_i$, we can determine a_i, b_i and c_i in terms of M_i. From Eq. (6.74), we have

$$M_i = 6a_i(x_i - x_i) + 2b_i = 2b_i \quad (6.75)$$

$$M_{i+1} = 6a_i(x_{i+1} - x_i) + 2b_i = 6a_i h_i + 2b_i \quad (6.76)$$

which gives us

$$b_i = \frac{M_i}{2}, \quad a_i = \frac{M_{i+1} - M_i}{6h_i} \quad (6.77, 6.78)$$

Now, using the values of d_i, b_i and a_i given by Eqs. (6.70), (6,77) and (6.78), we get from Eq. (6.71) that

$$y_{i+1} = \frac{M_{i+1} - M_i}{6h_i} h_i^3 + \frac{M_i}{2} h_i^2 + c_i h_i + y_i$$

which gives

$$c_i = \frac{y_{i+1} - y_i}{h_i} - \frac{2h_i M_i + h_i M_{i+1}}{6} \quad (6.79)$$

At this stage, we shall recall the third condition that the slopes of the two cubics meeting at (x_i, y_i) are equal from $(i - 1)$-th and i-th intervals. Hence Eq. (6.73) at $x = x_i$ the left end in the i-th interval is

$$y'(x_i) = y'_i = 3a_i(x_i - x_i)^2 + 2b_i(x_i - x_i) + c_i = c_i \quad (6.80)$$

while in the $(i - 1)$-th interval (x_{i-1}, x_i), the slope at the right end, that is at $x = x_i$ given by

$$y'(x_i) = y'_i = 3a_{i-1}(x_i - x_{i-1})^2 + 2b_{i-1}(x_i - x_{i-1}) + c_{i-1}$$

$$= 3a_{i-1}h_{i-1}^2 + 2b_{i-1}h_{i-1} + c_{i-1} \quad (6.81)$$

Now, equating Eqs. (6.80) and (6.81) and using Eqs. (6.77)–(6.79) for b_{i-1}, a_{i-1} and c_{i-1} respectively, we get

$$\frac{y_{i+1}-y_i}{h_i}-\frac{2h_iM_i+h_iM_{i+1}}{6}=3\,\frac{M_i-M_{i-1}}{6h_{i-1}}\,h_{i-1}^2+M_{i-1}h_{i-1}$$
$$+\frac{y_i-y_{i-1}}{h_{i-1}}-\frac{2h_{i-1}M_{i-1}+h_{i-1}M_i}{6} \tag{6.82}$$

On simplification, we obtain

$$h_{i-1}M_{i-1}+(2h_{i-1}+2h_i)\,M_i+h_iM_{i+1}=6\left(\frac{y_{i+1}-y_i}{h_i}-\frac{y_i-y_{i-1}}{h_{i-1}}\right) \tag{6.83}$$

for $i = 1, 2, \ldots, (n-1)$. From Eq. (6.83), it may be observed that only M_i's are unknowns, while all other terms can be computed from the given data points. In fact, it represents a system of $(n-1)$ linear equations in $(n+1)$ unknowns M_0, $M_1, \ldots, M_n$. Hence, two more additional conditions are required, relating to the end points of the complete spline curve, to generate two more additional equations. Many types of end conditions are specified and discussed in the literature. However, we shall consider only two types of end conditions.

6.8.2 End Conditions

Type I: We specify $S_0 = S_n = 0$, which means $M_0 = 0$, $M_n = 0$. In this case, the end cubics linearly approach to their extremities. This is called *natural spline.* This form of specification of end conditions is very popular. In this case, Eq. (6.83) readily gives us the system

$$\left.\begin{aligned}
2(h_0+h_1)M_1+h_1M_2&=6\left(\frac{y_2-y_1}{h_1}-\frac{y_1-y_0}{h_0}\right)\\
h_{i-1}M_{i-1}+(2h_{i-1}+2h_i)M_i+h_iM_{i+1}&=6\left(\frac{y_{i+1}-y_i}{h_i}-\frac{y_i-y_{i-1}}{h_{i-1}}\right)\\
\text{where } i=2,3,\ldots,(n-2)\qquad&\\
h_{n-2}M_{n-2}+2(h_{n-2}+h_{n-1})M_{n-1}&=6\left(\frac{y_n-y_{n-1}}{h_{n-1}}-\frac{y_{n-1}-y_{n-2}}{h_{n-2}}\right)
\end{aligned}\right\} \tag{6.84}$$

In compact matrix notation, we present it as

$$\begin{bmatrix}
2(h_0+h_1) & h_1 & & & \\
h_1 & 2(h_1+h_2) & h_2 & & \\
 & h_2 & 2(h_2+h_3) & h_4 & \\
 & & & \vdots & \\
 & & & h_{n-2} & 2(h_{n-2}+h_{n-1})
\end{bmatrix}
\begin{pmatrix} M_1\\ M_2\\ M_3\\ \vdots\\ M_{n-1}\end{pmatrix}$$

$$= 6 \begin{pmatrix} \dfrac{y_2 - y_1}{h_1} - \dfrac{y_1 - y_0}{h_0} \\ \dfrac{y_3 - y_2}{h_2} - \dfrac{y_2 - y_1}{h_1} \\ \dfrac{y_4 - y_3}{h_3} - \dfrac{y_3 - y_2}{h_2} \\ \vdots \\ \dfrac{y_n - y_{n-1}}{h_{n-1}} - \dfrac{y_{n-1} - y_{n-2}}{h_{n-2}} \end{pmatrix} \tag{6.85}$$

This being an $(n - 1 \times n - 1)$ tridiagonal system can be solved economically using Crout's reduction method as explained in Section 3.4 or Thomas Algorithm as explained in Appendix.

Type II: In this case, we specify the slopes at the end of entire spline curve; that is, we are given $y'(x_0) = A$ and $y'(x_n) = B$. This is called *clamped cubic spline.* From Eqs. (6.79) and (6.80), we have

$$y_i' = -\frac{h_i}{3} M_i - \frac{h_i}{6} M_{i+1} + \frac{y_{i+1} - y_i}{h_i} \tag{6.86}$$

using the left-end condition for $i = 0$, we obtain

$$-\frac{h_0}{3} M_0 - \frac{h_0}{6} M_1 + \frac{y_1 - y_0}{h_0} = A$$

That is,

$$2M_0 + M_1 = \frac{6}{h_0}\left(\frac{y_1 - y_0}{h_0} - A\right) \tag{6.87}$$

Similarly from Eq. (6.82), we have

$$y_i' = \frac{M_i h_{i-1}}{3} + \frac{M_{i-1} h_{i-1}}{6} + \frac{y_i - y_{i-1}}{h_{i-1}}$$

Now, using the right-end condition for $i = n$, the above equation becomes

$$B = \frac{M_n h_{n-1}}{3} + \frac{M_{n-1} h_{n-1}}{6} + \frac{y_n - y_{n-1}}{h_{n-1}}$$

Further simplification yields

$$M_{n-1} + 2M_n = \frac{6}{h_{n-1}}\left(B - \frac{y_n - y_{n-1}}{h_{n-1}}\right) \tag{6.88}$$

For $i = 1$, Eq. (6.83) gives

$$h_0 M_0 + (2h_0 + 2h_1) M_1 + h_1 M_2 = 6\left(\frac{y_2 - y_1}{h_1} - \frac{y_1 - y_0}{h_0}\right) \tag{6.89}$$

Eliminating M_0 from Eqs. (6.87) and (6.89), we get

$$\left(\frac{3}{2}h_0 + 2h_1\right)M_1 + h_1 M_2 = 6\,\frac{y_2 - y_1}{h_1} - 9\,\frac{y_1 - y_0}{h_0} + 3A \qquad (6.90)$$

Also, for $i = n - 1$, Eq. (6.83) gives

$$h_{n-2}M_{n-2} + (2h_{n-2} + 2h_{n-1})M_{n-1} + h_{n-1}M_n = 6\left(\frac{y_n - y_{n-1}}{h_{n-1}} - \frac{y_{n-1} - y_{n-2}}{h_{n-2}}\right) \qquad (6.91)$$

Eliminating M_n from Eqs. (6.88) and (6.91), we get

$$h_{n-2}M_{n-2} + \left(2h_{n-2} + \frac{3}{2}h_{n-1}\right)M_{n-1} = 9\frac{y_n - y_{n-1}}{h_{n-1}} - 6\frac{y_{n-1} - y_{n-2}}{h_{n-2}} - 3B \qquad (6.92)$$

Hence, in type II, we have to solve Eqs. (6.83), (6.90) and (6.92). These equations together constitute an $(n - 1 \times n - 1)$ tridiagonal system in unknowns M_1, M_2, ..., M_{n-1}, which can be solved using Crout's reduction technique or Thomas Algorithm. In order to see the sequence of steps involved to construct a cubic spline $S(x)$ for a given data set, using cubic spline interpolation, we shall consider below a couple of simple examples.

Example 6.20 Fit a cubic spline curve that passes through (0, 1), (1, 4), (2, 0), (3, –2) with the natural end boundary conditions $S''(0) = S''(3) = 0.0$.

Solution From the given data, we observe that there are three intervals, in each of which we can construct a cubic spline function. These piecewise cubic spline polynomials when put together determine the cubic spline curve $S(x)$ in the entire interval (0, 3).

At the outset, we observe that $h_0 = h_1 = h_2 = 1$. For natural spline, we obtain Eqs. (6.84) as

$$\begin{pmatrix} 4 & 1 \\ 1 & 4 \end{pmatrix}\begin{pmatrix} M_1 \\ M_2 \end{pmatrix} = 6\begin{pmatrix} -4 & -3 \\ -2 & +4 \end{pmatrix} = \begin{pmatrix} -42 \\ 12 \end{pmatrix}$$

That is,

$$4M_1 + M_2 = -42$$
$$M_1 + 4\,M_2 = 12$$

Its solution is $M_1 = -12$, $M_2 = 6$. Natural-end conditions imply $M_0 = M_3 = 0.0$. Let the natural cubic spline is given by

$$S(x) = a_i(x - x_i)^3 + b_i(x - x_i)^2 + c_i(x - x_i) + d_i$$

where the coefficients are given by the relations

$$a_i = \frac{M_{i+1} - M_i}{6h_i}, \qquad b_i = \frac{M_i}{2}$$

$$c_i = \frac{y_{i+1} - y_i}{h_i} - \frac{2h_i M_i + h_i M_{i+1}}{6}, \qquad d_i = y_i$$

for $i = 0, 1, 2$. Using the data and the values of M_0 and M_3, we compute the coefficients as

$$a_0 = -2.0, \quad a_1 = 3, \quad a_2 = -1$$
$$b_0 = 0.0, \quad b_1 = -6, \quad b_2 = 3$$
$$c_0 = 3 + \frac{12}{6} = 5, \quad c_1 = -1, \quad c_2 = -4$$
$$d_0 = 1.0, \quad d_1 = 4.0, \quad d_2 = 0.0$$

Hence the required piecewise cubic splines in each interval is given by

$$S_0(x) = -2.0x^3 + 5x + 1 \qquad \text{for } 0 \le x \le 1$$
$$S_1(x) = 3(x-1)^3 - 6(x-1)^2 - (x-1) + 4 \qquad \text{for } 1 \le x \le 2$$
$$S_2(x) = -(x-2)^3 + 3(x-2)^2 - 4(x-2) \qquad \text{for } 2 \le x \le 3$$

Example 6.21 Fit a cubic spline curve that passes through points (0, 1), (1, 4), (2, 0) and (3, –2) with the given derivative boundary conditions

$$S'(0) = 2, \qquad S'(3) = 2$$

Solution In this example, we have three intervals, in each of which, we can construct a cubic spline functions denoted by S_0, S_1 and S_2. At the outset, we observe that $h_0 = h_1 = h_2 = 1$. For derivative boundary conditions, we use Eqs. (6.90) and (6.92) and get

$$\frac{7}{2}M_1 + M_2 = 6(-4) - 9(3) + 3 \times 2 = -45$$

$$M_1 + \frac{7}{2}M_2 = 9(-2) - 6(-4) - 3 \times 2 = 0$$

Its solution is $M_1 = -14$, $M_2 = 4$. Now from Eq. (6.87), we obtain $2M_0 - 14 = 6$, which gives $M_0 = 10$. Also, Eq. (6.88) gives $M_2 + 2M_3 = 6(2 + 2) = 24$. Using the value of M_2, we get $M_3 = 10$.

Let the cubic spline in each interval is given by

$$S(x) = a_i(x - x_i)^3 + b_i(x - x_i)^2 + c_i(x - x_i) + d_i$$

The coefficients are computed as

$$a_0 = -4, \quad a_1 = 3, \quad a_2 = 1$$
$$b_0 = 5, \quad b_1 = -7, \quad b_2 = 2$$
$$c_0 = 2, \quad c_1 = 0, \quad c_2 = -5$$
$$d_0 = 1, \quad d_1 = 4, \quad d_2 = 0.0$$

Hence the required piecewise cubic spline polynomials in each interval is given by

$$S_0(x) = -4x^3 + 5x^2 + 2x + 1, \qquad \text{for } 0 \le x \le 1$$
$$S_1(x) = 3(x-1)^3 - 7(x-1)^2 + 4, \qquad \text{for } 1 \le x \le 2$$
$$S_2(x) = (x-2)^3 + 2(x-2)^2 - 5(x-2), \qquad \text{for } 2 \le x \le 3$$

6.9 MAXIMA AND MINIMA OF A TABULATED FUNCTION

The idea of finding the maxima and minima of a tabulated function is useful in many practical problems. Recalling the Newton's forward interpolation formula (6.33) as

$$y = y_0 + p\Delta y_0 + \frac{p(p-1)}{2!}\Delta^2 y_0 + \frac{p(p-1)(p-2)}{3!}\Delta^3 y_0 + \cdots$$

and noting that $p = (x - x_0)/h$; its differentiation with respect to x yields

$$\frac{dy}{dx} = \frac{1}{h}\left(\Delta y_0 + \frac{2p-1}{2}\Delta^2 y_0 + \frac{3p^2 - 6p + 2}{6}\Delta^3 y_0 + \cdots\right) \quad (6.93)$$

From elementary calculus, it is known that the maxima and minimum values of a function $y = f(x)$ are obtained by equating its first derivative with respect to x to zero and solving it for x. Same idea holds in the case of tabulated function, too. Thus, we have

$$\Delta y_0 + \frac{2p-1}{2}\Delta^2 y_o + \frac{3p^2 - 6p + 2}{6}\Delta^3 y_0 + \cdots = 0 \quad (6.94)$$

By retaining terms up to third difference only, we arrive at the expression

$$\Delta y_0 + \frac{2p-1}{2}\Delta^2 y_0 + \frac{3p^2 - 6p + 2}{6}\Delta^3 y_0 = 0$$

Here, the first, second and third differences can be obtained from the difference table. Thus, Eq. (6.94) gives a polynomial in x, whose solution gives us those values of x, at which the given tabulated function may be maximum or minimum. This idea is illustrated in the following couple of examples.

Example 6.22 Find, for what value of x, y is minimum using the data given below:

x	3	4	5	6	7	8
y	0.205	0.240	0.259	0.262	0.250	0.224

Solution From the given data, it can be seen that the arguments are equally spaced and therefore, we construct the forward difference table as

x	y	Δy	$\Delta^2 y$	$\Delta^3 y$	$\Delta^4 y$
3	0.205 →				
		0.035 →			
4	0.240		–0.016 →		
		0.019		0.0 →	
5	0.259		–0.016		–0.001
		0.003		–0.001	
6	0.262		–0.015		0.002
		–0.012		0.001	
7	0.250		–0.014		
		–0.026			
8	0.224				

In this example, $p = (x - x_0)/h = (x - 3)$, $\Delta y_0 = 0.035$, $\Delta^2 y_0 = -0.016$, $\Delta^3 y_0 = 0$, $\Delta^4 y_0 = -0.001$. Recalling Newton's forward difference formula given by Eq. (6.33) and retaining up to third differences only, we have

$$y = y_0 + p\Delta y_0 + \frac{p(p-1)}{2!}\Delta^2 y_0 + \frac{p(p-1)(p-2)}{3!}\Delta^3 y_0$$

Substituting the values of p and the differences, the above equation becomes

$$y = 0.205 + (x-3)(0.035) + \frac{(x-3)(x-4)(-0.016)}{2}$$

That is,

$$y = -0.008x^2 + 0.091x + 0.004 \qquad (1)$$

For maxima or minima, we require

$$\frac{dy}{dx} = -0.016x + 0.091 = 0$$

which gives $x = 5.6875$. Thus, the minimum value of y at $x = 5.6875$ is found from Eq. (1) as $y = -0.008\ (33.414) + 0.091(5.6875) + 0.004 = 0.25425$. Hence the minimum value at $x = 5.6875$ of the given tabulated function is $= 0.25425$.

Example 6.23 Find, for what value of x, y is maximum, from the following data:

x	–1	1	2	3
y	–21	15	12	3

Solution From the given data, it can be seen that the arguments are not equally-spaced, and therefore, we can use either Lagrange's interpolation or Newton's divided difference interpolation formula. We choose the former and hence the Lagrange's interpolation formula for the given data given as

$$y = \frac{(x-1)(x-2)(x-3)}{(-1-1)(-1-2)(-1-3)}(-21) + \frac{(x+1)(x-2)(x-3)}{(1+1)(1-2)(1-3)} \quad (15)$$

$$+ \frac{(x+1)(x-1)(x-3)}{(2+1)(2-1)(2-3)}(12) + \frac{(x+1)(x-1)(x-2)}{(3+1)(3-1)(3-2)} \quad (3)$$

$$= \frac{21}{24}(x^3 - 6x^2 + 11x - 6) + \frac{15}{4}(x^3 - 4x^2 + x + 6)$$

$$-\frac{12}{3}(x^3 - 3x^2 - x + 3) + \frac{3}{8}(x^3 - 2x^2 - x + 2)$$

which simplifies to

$$y = x^3 - 9x^2 + 17x + 6 \qquad (1)$$

For maxima or minima, it is required that

$$\frac{dy}{dx} = 3x^2 - 18x + 17 = 0$$

which is a quadratic equation, whose solution is given by

$$x = \frac{18 \pm \sqrt{18^2 - 12 \times 17}}{6} = 4.8257 \quad \text{or} \quad 1.1743$$

Here, we note that $x = 4.8257$ is outside the considered range. However,

$$\left.\frac{d^2y}{dx^2}\right|_{x=1.1743} = (6x - 18)\Big|_{x=1.1743} = \text{negative}$$

Hence, the maximum value of y at $x = 1.1743$ is given as

$$y = (1.1743)^3 - 9(1.1743)^2 + 17(1.1743) + 6 = 15.1716$$

Example 6.24 Given $\Sigma_1^{10} f(x) = 500426$, $\Sigma_4^{10} f(x) = 329240$, $\Sigma_7^{10} f(x) = 175212$ and $f(10) = 40365$, find $f(2)$.

Solution In this example, we are given the cumulative values of the function and therefore, we adopt the following notation:

$$F(1) = \sum_{1}^{10} f(x) = 500426$$

$$F(4) = \sum_{4}^{10} f(x) = 329240$$

$$F(7) = \sum_{7}^{10} f(x) = 175212$$

$$F(10) = f(10) = 40365$$

and construct the forward difference table as

x	$F(x)$	$\Delta F(x)$	$\Delta^2 F(x)$	$\Delta^3 F(x)$
1	500426			
		–171186		
4	329240		17158	
		–154028		2023
7	175212		19181	
		–134847		
10	40365			

Now, to find $F(2) = \Sigma_2^{10} f(x)$, we may recall Newton's forward difference formula given by Eq. (6.33) as

$$F(2) = F(1) + p\Delta F(1) + \frac{p(p-1)}{2!}\Delta^2 F(1) + \frac{p(p-1)(p-2)}{3!}\Delta^3 F(1)$$

where

$$p = \frac{x - x_0}{h} = \frac{2 - 1}{3} = \frac{1}{3}$$

Also, from the table we note that $\Delta F(1) = -171186$, $\Delta^2 F(1) = 17158$ and $\Delta^3 F(1) = 2023$. Substituting these values, the above equation gives.

$$F(2) = 500426 - \frac{171186}{3} - \frac{17158}{9} + \frac{5(2023)}{81} = 441582.4325$$

Therefore, we finally have

$$f(2) = F(1) - F(2) = 500426 - 441582.4325 = 58843.5675$$

6.10 HERMITE INTERPOLATION

In Hermite interpolation, we use the expansion involving not only the function values but also its first derivative. To state the problem: given a set of data points (x_i, y_i, y'_i), $i = 0, 1, 2, ..., n$, we have to determine a polynomial $P(x)$ of degree $(2n + 1)$. Thus, keeping in mind the Lagrange interpolation formula, we seek $P(x)$ in the form

$$P(x) = \sum_{i=0}^{n} U_i(x) y_i + \sum_{i=0}^{n} V_i(x) y'_i \tag{6.95}$$

where $U_i(x)$ and $V_i(x)$ are polynomials of degree $(2n + 1)$ that satisfy the relations

$$\left.\begin{aligned} & U_i(x_j) = \delta_{ij} \\ & \left.\frac{\partial U_i}{\partial x}\right|_{x = x_j} = 0 \\ \text{and} \quad & V_i(x_j) = 0 \\ & \left.\frac{\partial V_i}{\partial x}\right|_{x = x_j} = \delta_{ij} \end{aligned}\right\} \tag{6.96}$$

Here, δ_{ij} is a kronecker delta, whose value is unity if $i = j$, otherwise zero. Polynomials satisfying the above conditions are called Hermite polynomials. Now, we define

$$U_i = \left\{1 - 2(x - x_i)\left.\frac{dL_i}{dx}\right|_{x=x_i}\right\}[L_i(x)]^2$$

and

$$V_i = (x - x_i)[L_i(x)]^2 \tag{6.97}$$

which of course meets the requirements as defined in Eq. (6.96), where $L_i(x)$ is a Lagrange polynomial satisfying

$$L_i(x_j) = \delta_{ij}$$

Substituting $x = x_i$ in Eq. (6.97), we find that

$$U_i(x_i) = [L_i(x_i)]^2 = 1$$

and

$$V_i(x_i) = 0$$

Now, differentiating Eqs. (6.97), we have

$$U_i'(x) = [1 - 2L_i'(x_i)(x - x_i)]2\,L_i(x)\,L_i'(x)$$

$$-2L_i'(x_i)[L_i(x)]^2$$

and

$V'_i(x) = (x - x_i)\,2L_i(x)\,L'_i(x) + L_i(x)^2$ observe that $U'_i(x_j) = 0$, $V'_i(x_j) = 0$ for $i \neq j$. Since $L_i(x_i) = 1$, we get

$$U_i'(x_i) = 2L_i'(x_i) - 2L_i'(x_i) = 0$$

and

$$V_i'(x_i) = [L_i(x_i)]^2 = 1$$

Hence, the Hermite Interpolation formula is given as

$$P(x) = \sum_{i=0}^{n} [1 - 2L_i'(x_i)(x - x_i)][L_i(x)]^2\, y_i$$

$$+ (x - x_i)[L_i(x)]^2\, y_i' \tag{6.98}$$

For illustration, we consider the following example.

Example 6.25 Estimate the value of $y(1.05)$ using Hermite interpolation formula from the following data:

x	y	y'
1.00	1.00000	0.5000
1.10	1.04881	0.47673

Solution: At first we compute

$$L_0(x) = \frac{x - x_1}{x_0 - x_1} = \frac{1.05 - 1.10}{1.00 - 1.10} = 0.5$$

$$L_1(x) = \frac{x - x_0}{x_1 - x_0} = \frac{1.05 - 1.00}{1.10 - 1.00} = 0.5$$

$$L_0'(x) = \frac{1}{x_0 - x_1} = -\frac{1}{0.10}$$

$$L_1'(x) = \frac{1}{x_1 - x_0} = \frac{1}{0.1}$$

Substituting these expressions in Hermite formula,

$$P(x) = \sum_{i=0}^{n} [1 - 2L_i'(x_i)(x - x_i)][L_i(x)]^2 y_i$$

$$+ (x - x_i)[L_i(x)]^2 y_i'$$

we find

$$y(1.05) = \left[1 - 2\left(-\frac{1}{0.1}\right)(0.05)\right]\left(\frac{1}{2}\right)^2 (1) + (0.05)\left(\frac{1}{2}\right)^2 (0.5)$$

$$+ \left[1 - 2\left(\frac{1}{0.1}\right)(-0.05)\right]\left(\frac{1}{2}\right)^2 (1.04881)$$

$$+ (-0.05)\left(\frac{1}{2}\right)^2 (0.47673)$$

$$= 1.0247$$

EXERCISES

6.1 Express $\Delta^2 y_1$ and $\Delta^4 y_0$ in terms of the values of the function y.

6.2 Compute the missing values of y_n and Δy_n in the following table

y_n	Δy_n	$\Delta^2 y_n$
–		
	–	
–		1
	–	
–		4
	5	
6		13
	–	
–		18
	–	
–		24
	–	
–		

6.3 Show that $E\nabla = \Delta = \delta E^{1/2}$.

6.4 Prove that (i) $\delta = 2 \sinh (hD/2)$ and, (ii) $\mu = 2 \cosh (hD/2)$.

6.5 Show that the operators δ, μ, E, Δ and ∇ commute with one another.

6.6 Explain the concept of linear interpolation. Using linear interpolation, find $f(3)$ for $f(x) = 5^x$. Compare with the actual value. Comment on the result obtained.

6.7 The following table gives pressure of a steam at a given temperature. Using Newton's formula, compute the pressure for a temperature of 142°C.

Temperature, °C	140	150	160	170	180
Pressure, kgf/cm^2	3.685	4.854	6.302	8.076	10.225

6.8 Find Newton's backward interpolating polynomial for the following data:

x	1	2	3	4	5
y	1	–1	1	–1	1

6.9 A second degree polynomial passes through (0, 1) (1, 3), (2, 7), and (3, 13). Find the polynomial, using Newton's forward difference formula.

6.10 The following data gives the melting point of an alloy of lead and zinc; where T is the temperature in °C and P is the percentage of lead in the alloy. Find the melting point of the alloy containing 84% of lead using Newton's interpolation method.

P	60	70	80	90
T	226	250	276	304

6.11 Find the interpolating polynomial for the function $f(x)$ given by

x	0	1	2	5
$y = f(x)$	2	3	12	147

6.12 Find the interpolating polynomial for the following data using Lagrange's formula

x	1	2	–4
$y = f(x)$	3	–5	4

6.13 Starting from Newton's divided difference interpolation formula (6.54) and making use of Eq. (6.47) and recalling the definitions of $\Pi(x)$ from Eq. (6.40), show that it can be reduced to Lagrange's form given by Eq. (6.38).

6.14 Find the interpolating polynomial by (i) Newton's divided difference formula (ii) Lagrange's formula, for the following data and hence show that both the methods give raise to the same polynomial.

x	1	2	3	5
y	0	7	26	124

6.15 Show that the nth differences of a polynomial of degree n is constant.

6.16 Tabulate the values of the function

$$u(x, y) = e^x \sin y + y - 0.1$$

for

$$x = 0.5, 1.0, 1.5, 2.0, 2.5, 3.0, 3.5$$

and

$$y = 0.1, 0.2, 0.3, 0.4, 0.5, 0.6$$

Hence, using this generated table and quadratic interpolation in x-direction and cubic in y-direction, compute $u(1.6, 0.33)$ by numerical two-dimensional interpolation.

6.17 Find the equation of a cubic curve which passes through the points (4, –43), (7, 83), (9, 327) and (12, 1053) using divided difference formula.

6.18 Using a polynomial of third degree, complete the record of the export of a certain commodity during five years, as given below.

Year, x	1985	1986	1987	1988	1989
Export in tons, y	443	384	–	397	467

6.19 Following is the table of values of x and y

x	3	4	5	6	7	8
y	0.205	0.240	0.259	0.262	0.250	0.224

Find the value of x for which y is minimum using Newton's forward difference formula. Also find the minimum value of y.

6.20 Find the missing values in the following table

x	0	1	2	3	4	5	6
y	–4	–2	–	–	220	546	1148

6.21 Fit a cubic spline curve that passes through (0, 0.0), (1, 0.5), (2, 2.0) and (3, 1.5) with the natural-end boundary conditions, $S''(0) = 0$, $S''(3) = 0$.

6.22 Fit a clamped cubic spline curve that passes through the points (0, 0.0), (1, 0.5), (2, 2.0) and (3, 1.5) with the end conditions $S'(0) = 0.2$, $S'(3) = -1$.

6.23 Fit a natural cubic spline curve that passes through (0.0, 2.0), (1.0, 4.4366), (1.5, 6.7134) and (2.25, 13.9130).

6.24 Using Newton's divided difference formula, evaluate $f(2)$ and $f(15)$ from the following table of values:

x	4	5	7	10	11	13
$f(x)$	48	100	294	900	1210	2028

6.25 Using cubic spline interpolation, find the value of y at $x = 1/2$, for the following data:

(0, 1), (1, 2), (2, 9) and (3, 28).

6.26 Using Hermite interpolation, estimate the value of $y(1.05)$ from the following data.

x	0.9	1.0	1.1	1.2
$y = \sin x$	0.7833	0.8415	0.8912	0.9320
$y' = \cos x$	0.6216	0.5403	0.4536	0.3624

6.27 Using Hermite interpolation, estimate the value of $y(1.3)$ from the following data

x	0.5	1.0	1.5	2.0
y	0.4794	0.8415	0.9975	0.9093
y'	0.8776	0.5403	0.7074	−0.4162

6.28 The natural logarithm and its derivative is given in the following table. Estimate the value of ln (0.6) using Hermite interpolation formula.

x	0.4	0.5	0.7	0.8
$\ln x$	−0.9163	−0.6932	−0.3567	−0.2231
$1/x$	2.5	2.0	1.43	1.25

Chapter 7

Numerical Differentiation and Integration

7.1 INTRODUCTION

Consider a function of single variable $y = f(x)$. If the function is known and simple, we can easily obtain its derivative(s) or can evaluate its definite integral. However, if we do not know the function as such or the function is complicated and is given in a tabular form at a set of points $x_0, x_1, \ldots, x_n$, we use only numerical methods for differentiation or integration of the given function. We shall discuss numerical approximation to derivatives of functions of two or more variables in subsequent chapters to follow under partial differential equations. In the next couple of sections, we shall derive and illustrate various formulae for numerical differentiation of a function of a single variable based on finite difference operators and interpolation. Subsequently, we shall develop Newton–Cotes formulae and related trapezoidal rule and Simpson's rule for numerical integration of a function. Finally, we shall present Gaussian Quadrature formulae for evaluating both simple and multiple integrals.*

7.2 DIFFERENTIATION USING DIFFERENCE OPERATORS

We assume that the function $y = f(x)$ is given for the values of the independent variable $x = x_0 + ph$, for $p = 0, 1, 2, \ldots$ and so on. To find the derivatives of such a tabular function, we proceed as follows.

Case I: Using forward difference operator Δ and combining Eqs. (6.27) and (6.31) we have

$$hD = \log E = \log(1 + \Delta) \tag{7.1}$$

where D is a differential operator, E a shift operator. In terms of Δ, Eq. (7.1) gives

$$D = \frac{1}{h}\left(\Delta - \frac{\Delta^2}{2} + \frac{\Delta^3}{3} - \frac{\Delta^4}{4} + \frac{\Delta^5}{5} - \cdots\right) \tag{7.2}$$

Therefore,

$$Df(x_0) = f'(x_0) = \frac{1}{h}\left[\Delta f(x_0) - \frac{\Delta^2 f(x_0)}{2} + \frac{\Delta^3 f(x_0)}{3} - \frac{\Delta^4 f(x_0)}{4} + \frac{\Delta^5 f(x_0)}{5} - \cdots\right] = \frac{d}{dx} f(x_0)$$

*[For extrapolation methods, the interested reader may consult Hildebrand (1982).]

In other words,

$$Dy_0 = y_0' = \frac{1}{h}\left(\Delta y_0 - \frac{\Delta^2 y_0}{2} + \frac{\Delta^3 y_0}{3} - \frac{\Delta^4 y_0}{4} + \cdots\right) \tag{7.3}$$

Also, we can easily verify

$$D^2 = \frac{1}{h^2}\left(\Delta - \frac{\Delta^2}{2} + \frac{\Delta^3}{3} - \frac{\Delta^4}{4} + \ldots\right)^2 = \frac{1}{h^2}\left(\Delta^2 - \Delta^3 + \frac{11}{12}\Delta^4 - \frac{5}{6}\Delta^5 + \ldots\right) \tag{7.4}$$

Thus,

$$D^2 y_0 = \frac{d^2 y_0}{dx^2} = y_0'' = \frac{1}{h^2}\left(\Delta^2 y_0 - \Delta^3 y_0 + \frac{11}{12}\Delta^4 y_0 - \frac{5}{6}\Delta^5 y_0 + \cdots\right) \tag{7.5}$$

Case II: Using backward difference operator ∇, we have seen in Example 6.5 that $hD = -\log(1 - \nabla)$.
On expansion, we have

$$D = \frac{1}{h}\left(\nabla + \frac{\nabla^2}{2} + \frac{\nabla^3}{3} + \frac{\nabla^4}{4} + \ldots\right) \tag{7.6}$$

we can also verify that

$$D^2 = \frac{1}{h^2}\left(\nabla + \frac{\nabla^2}{2} + \frac{\nabla^3}{3} + \frac{\nabla^4}{4} + \cdots\right)^2 = \frac{1}{h^2}\left(\nabla^2 + \nabla^3 + \frac{11}{12}\nabla^4 + \frac{5}{6}\nabla^5 + \cdots\right) \tag{7.7}$$

Hence,

$$\frac{d}{dx}y_n = Dy_n = y_n' = \frac{1}{h}\left(\nabla y_n + \frac{\nabla^2 y_n}{2} + \frac{\nabla^3 y_n}{3} + \frac{\nabla^4 y_n}{4} + \cdots\right) \tag{7.8}$$

and

$$y_n'' = D^2 y_n = \frac{1}{h^2}\left(\nabla^2 y_n + \nabla^3 y_n + \frac{11}{12}\nabla^4 y_n + \frac{5}{6}\nabla^5 y_n + \cdots\right) \tag{7.9}$$

The formulae (7.3) and (7.5) are useful to calculate the first and second derivatives at the beginning of the table of values in terms of forward differences; while formulae (7.8) and (7.9) are used to compute the first and second derivatives near the end points of the table, in terms of backward differences.

Similar formulae can also be derived for computing higher order derivatives.

To compute the derivatives of a tabular function at points not found in the table, we can proceed as follows:

Recalling Eq. (6.34) in the form

$$y(x_n + ph) = y(x_n) + p\nabla y(x_n) + \frac{p(p+1)}{2!}\nabla^2 y(x_n) + \frac{p(p+1)(p+2)}{3!}\nabla^3 y(x_n)$$

$$+ \frac{p(p+1)(p+2)(p+3)}{4!}\nabla^4 y(x_n) + \cdots \tag{7.9a}$$

Let $x = x_n + ph$, then $p = (x - x_n)/h$. Now, differentiating Eq. (7.9a) with respect to x, we get

$$y' = \frac{dy}{dx} = \frac{dy}{dp}\frac{dp}{dx} = \frac{1}{h}\left[\nabla y_n + \frac{2p+1}{2}\nabla^2 y_n + \frac{3p^2+6p+2}{6}\nabla^3 y_n + \frac{4p^3+18p^2+22p+6}{24}\nabla^4 y_n + \cdots\right] \tag{7.9b}$$

Differentiating this result once again with respect to x, we arrive at the second derivative as

$$y'' = \frac{d^2y}{dx^2} = \frac{d}{dp}(y')\frac{dp}{dx} = \frac{1}{h^2}\left[\nabla^2 y_n + (p+1)\nabla^3 y_n + \frac{6p^2+18p+11}{12}\nabla^4 y_n + \cdots\right] \tag{7.9c}$$

Equations (7.9b) and (7.9c) are *Newton's backward interpolation formulae*, which can be used to compute the first and second derivatives of a tabular function near the end of the table. Similar expressions of Newton's forward interpolation formulae can be derived to compute the first- and higher-order derivatives near the beginning of the table of values.

Case III: Using central difference operator δ and following the definitions of differential operator D, central difference operator δ and the shift operator E, we have

$$\delta = E^{1/2} - E^{-1/2} = e^{hD/2} - e^{-hD/2} = 2\sinh\frac{hD}{2}$$

Therefore, we find

$$\frac{hD}{2} = \sinh^{-1}\frac{\delta}{2} \tag{7.10}$$

But,

$$\sinh^{-1} x = x - \frac{1}{2}\frac{x^3}{3} + \frac{1\times 3}{2\times 4}\frac{x^5}{5} - \frac{1\times 3\times 5}{2\times 4\times 6}\frac{x^7}{7} + \cdots$$

Using this expansion into Eq. (7.10), we get

$$\frac{hD}{2} = \frac{\delta}{2} - \frac{\delta^3}{6\times 8} + \frac{3}{40\times 32}\delta^5 - \cdots$$

That is,

$$D = \frac{1}{h}\left(\delta - \frac{1}{24}\delta^3 + \frac{3}{640}\delta^5 - \cdots\right)$$

Therefore,

$$\frac{d}{dx}y = y' = Dy = \frac{1}{h}\left(\delta y - \frac{1}{24}\delta^3 y + \frac{3}{640}\delta^5 y - \cdots\right) \tag{7.11}$$

Also

$$D^2 = \frac{1}{h^2}\left(\delta^2 - \frac{1}{12}\delta^4 + \frac{1}{90}\delta^6 - \cdots\right)$$

Hence,

$$y'' = D^2 y = \frac{1}{h^2}\left(\delta^2 y - \frac{1}{12}\delta^4 y + \frac{1}{90}\delta^6 y - \cdots\right) \tag{7.12}$$

For calculating the second derivative at an interior tabular point, we use Eq. (7.12), while for computing the first derivative at an interior tabular point, we in general use another convenient form for D, which is derived as follows.

Multiplying the right-hand side of Eq. (7.11) by

$$\frac{\mu}{\sqrt{1 + (\delta^2/4)}}$$

which is unity and noting the Binomial expansion

$$\left(1 + \frac{1}{4}\delta^2\right)^{-1/2} = 1 - \frac{1}{8}\delta^2 + \frac{3}{128}\delta^4 - \frac{15}{48 \times 64}\delta^6 + \cdots$$

We get

$$D = \frac{\mu}{h}\left(1 - \frac{1}{8}\delta^2 + \frac{3}{128}\delta^4 - \cdots\right)\left(\delta - \frac{1}{24}\delta^3 + \frac{3}{640}\delta^5 - \cdots\right)$$

On simplification, we obtain

$$D = \frac{\mu}{h}\left(\delta - \frac{1}{6}\delta^3 + \frac{4}{120}\delta^5 - \cdots\right)$$

Therefore,

$$y' = Dy = \frac{\mu}{h}\left(\delta y - \frac{1}{6}\delta^3 y + \frac{1}{30}\delta^5 y - \cdots\right) \tag{7.13}$$

Equations (7.12) and (7.13) are known as Stirling's formulae for computing the derivatives of a tabular function. Similar formulae can be derived for computing higher order derivatives of a tabular function. Equation (7.13) can also be written in another useful form as

$$y' = \frac{\mu}{h}\left[\delta y - \frac{1^2}{3!}\delta^3 y + \frac{(1^2)(2^2)}{5!}\delta^5 y - \frac{(1^2)(2^2)(3^2)}{7!}\delta^7 y + \cdots\right] \tag{7.14}$$

In order to illustrate the use of formulae derived so far, for computing the derivatives of a tabulated function, we shall consider the following examples.

Example 7.1 Compute $f''(0)$ and $f'(0.2)$ from the following tabular data.

x	0.0	0.2	0.4	0.6	0.8	1.0
$f(x)$	1.00	1.16	3.56	13.96	41.96	101.00

Solution Since $x = 0$ and 0.2 appear at and near beginning of the table, it is appropriate to use formulae based on forward differences to find the derivatives. The difference table for the given data is depicted below:

x	$f(x)$	$\Delta f(x)$	$\Delta^2 f(x)$	$\Delta^3 f(x)$	$\Delta^4 f(x)$	$\Delta^5 f(x)$
0.0	1.00					
		0.16				
0.2	1.16		2.24			
		2.40		5.76		
0.4	3.56		8.00		3.84	
		10.40		9.60		0.00
0.6	13.96		17.60		3.84	
		28.00		13.44		
0.8	41.96		31.04			
		59.04				
1.0	101.00					

Using forward difference formula (7.5) for $D^2 f(x)$, i.e.

$$D^2 f(x) = \frac{1}{h^2}\left[\Delta^2 f(x) - \Delta^3 f(x) + \frac{11}{12}\Delta^4 f(x) - \frac{5}{6}\Delta^5 f(x)\right]$$

We obtain

$$f''(0) = \frac{1}{(0.2)^2}\left[2.24 - 5.76 + \frac{11}{12}(3.84) - \frac{5}{6}(0)\right] = 0.0$$

Also, using the formula (7.3) we have

$$Df(x) = \frac{1}{h}\left[\Delta f(x) - \frac{\Delta^2 f(x)}{2} + \frac{\Delta^3 f(x)}{3} - \frac{\Delta^4 f(x)}{4}\right]$$

Hence,

$$f'(0.2) = \frac{1}{0.2}\left(2.40 - \frac{8.00}{2} + \frac{9.60}{3} - \frac{3.84}{4}\right) = 3.2$$

Example 7.2 Find $y'(2.2)$ and $y''(2.2)$ from the table

x	1.4	1.6	1.8	2.0	2.2
$y(x)$	4.0552	4.9530	6.0496	7.3891	9.0250

Solution Since $x = 2.2$ occurs at the end of the table, it is appropriate to use backward difference formulae for derivatives. The backward difference table for the given data is shown as follows:

x	$y(x)$	∇y	$\nabla^2 y$	$\nabla^3 y$	$\nabla^4 y$
1.4	4.0552				
		0.8978			
1.6	4.9530		0.1988		
		1.0966		0.0441	
1.8	6.0496		0.2429		0.0094
		1.3395		0.0535	
2.0	7.3891		0.2964		
		1.6359			
2.2	9.0250				

Using backward difference formulae (7.8) and (7.9) for $y'(x)$ and $y''(x)$, we have

$$y'_n = \frac{1}{h}\left(\nabla y_n + \frac{\nabla^2 y_n}{2} + \frac{\nabla^3 y_n}{3} + \frac{\nabla^4 y_n}{4}\right)$$

Therefore,

$$y'(2.2) = \frac{1}{0.2}\left(1.6359 + \frac{0.2964}{2} + \frac{0.0535}{3} + \frac{0.0094}{4}\right) = 5(1.8043) = 9.0215$$

Also

$$y''_n = \frac{1}{h^2}\left(\nabla^2 y_n + \nabla^3 y_n + \frac{11}{12}\nabla^4 y_n\right)$$

Therefore,

$$y''(2.2) = \frac{1}{(0.2)^2}\left[0.2964 + 0.0535 + \frac{11}{12}(0.0094)\right] = 25\ (0.3585) = 8.9629$$

Example 7.3 From the following table of values, estimate $y'(2)$ and $y''(2)$ using appropriate central difference formula:

x	0	1	2	3	4
y	6.9897	7.4036	7.7815	8.1281	8.4510

Solution The central difference table for the given data is given below:

x	y	δy	$\delta^2 y$	$\delta^3 y$	$\delta^4 y$
0	6.9897				
		0.4139			
1	7.4036		−0.0360		
		→ 0.3779		0.0047	
2	7.7815		→ −0.0313		→ 0.0029
		→ 0.3466		→ 0.0076	
3	8.1281		−0.0237		
		0.3229			
4	8.4510				

Now, using central difference formula (7.13), we shall compute the first derivative

$$y' = \frac{\mu}{h}\left(\delta y - \frac{1}{6}\delta^3 y + \frac{1}{30}\delta^5 y - \cdots\right)$$

In the present example

$$y'(2) = \frac{1}{1}\left(\frac{0.3779 + 0.3466}{2} - \frac{1}{6}\frac{0.0047 + 0.0076}{2}\right) = 0.3613$$

To compute the second derivative, we shall use formula (7.12). Thus,

$$y'' = \frac{1}{h^2}\left(\delta^2 y - \frac{1}{12}\delta^4 y + \frac{1}{90}\delta^6 y - \cdots\right)$$

In this example,

$$y''(2) = \frac{1}{1}\left(-0.0313 - \frac{0.0029}{12}\right) = -0.0315$$

Case IV (Two- and three-point formulae): Retaining only the first term in Eq. (7.3), we can get another useful form for the first derivative as

$$y_i' = \frac{\Delta y_i}{h} = \frac{y_{i+1} - y_i}{h} = \frac{y(x_i + h) - y(x_i)}{h} \tag{7.15}$$

Similarly, by retaining only the first term in Eq. (7.8), we obtain

$$y_i' = \frac{\nabla y_i}{h} = \frac{y_i - y_{i-1}}{h} = \frac{y(x_i) - y(x_i - h)}{h} \tag{7.16}$$

Adding Eqs. (7.15) and (7.16), we have

$$y_i' = \frac{y(x_i + h) - y(x_i - h)}{2h} \tag{7.17}$$

Equations (7.15) – (7.17) constitute two-point formulae for the first derivative. By retaining only the first term in Eq. (7.5), we get

$$y_i'' = \frac{\Delta^2 y_i}{h^2} = \frac{y_{i+2} - 2y_{i+1} + y_i}{h^2} = \frac{y(x_i + 2h) - 2y(x_i + h) + y(x_i)}{h^2} \tag{7.18}$$

Similarly, Eq. (7.9) gives

$$y_i'' = \frac{\nabla^2 y_i}{h^2} = \frac{y(x_i) - 2y(x_i - h) + y(x_i - 2h)}{h^2} \tag{7.19}$$

While retaining only the first term in Eq. (7.12), we obtain

$$y_i'' = \frac{\delta^2 y_i}{h^2} = \frac{\delta y_{i+(1/2)} - \delta y_{i-(1/2)}}{h^2} = \frac{y_{i+1} - 2y_i + y_{i-1}}{h^2}$$

$$= \frac{y(x_i - h) - 2y(x_i) + y(x_i + h)}{h^2} \tag{7.20}$$

Equations (7.18)–(7.20) constitute three-point formulae for computing the second derivative. We shall see later that these two- and three-point formulae become handy for developing extrapolation methods to numerical differentiation and integration.

7.3 DIFFERENTIATION USING INTERPOLATION

If the given tabular function $y(x)$ is reasonably well approximated by a polynomial $P_n(x)$ of degree n, it is hoped that the result of $P_n'(x)$ will also satisfactorily approximate the corresponding derivative of $y(x)$. However, even if $P_n(x)$ and $y(x)$ coincide at the tabular points, their derivatives or slopes may substantially differ at these points as is illustrated in Fig. 7.1.

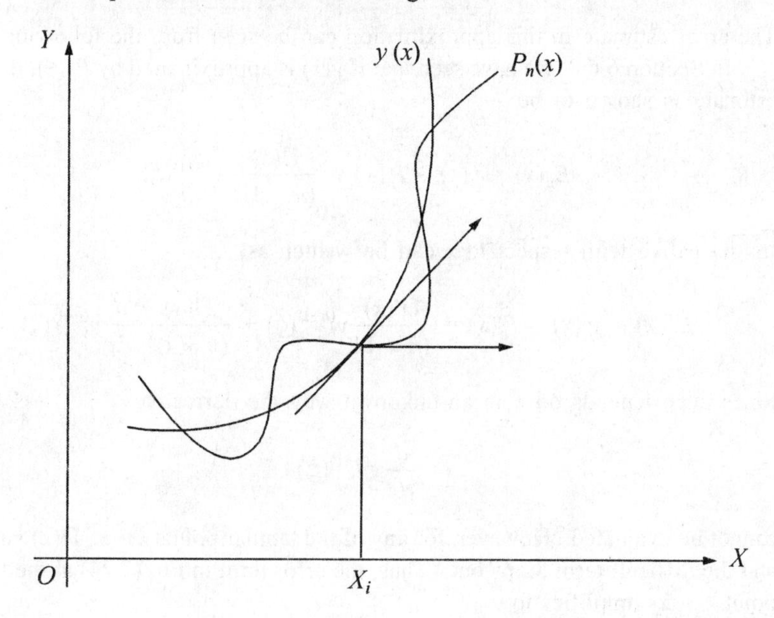

Fig. 7.1 Deviation of derivatives.

For higher order derivatives, the deviations may be still worst. However, we can estimate the error involved in such an approximation.

For non-equidistant tabular pairs (x_i, y_i), $i = 0, \ldots, n$ we can fit the data by using either Lagrange's interpolating polynomial or by using Newton's divided difference interpolating polynomial. In view of economy of computation, we prefer the use of the latter polynomial. Thus, recalling the Newton's divided difference interpolating polynomial for fitting this data as

$$P_n(x) = y[x_0] + (x - x_0)y[x_0, x_1] + (x - x_0)(x - x_1)y[x_0, x_1, x_2] + \cdots$$

$$+ \prod_{i=0}^{n-1} (x - x_i)y[x_0, x_1, \ldots, x_n] \tag{7.21}$$

Assuming that $P_n(x)$ is a good approximation to $y(x)$, the polynomial approximation to $y'(x)$ can be obtained by differentiating $P_n(x)$. Using product rule of differentiation, the derivative of the products in $P_n(x)$ can be seen as follows:

$$\frac{d}{dx}\prod_{i=0}^{n-1} (x - x_i) = \sum_{i=0}^{n-1} \frac{(x - x_0)(x - x_1)\cdots(x - x_{n-1})}{x - x_i}$$

Thus, $y'(x)$ is approximated by $P_n'(x)$ which is given by

$$P_n'(x) = y[x_0, x_1] + [(x - x_1) + (x - x_0)]\, y[x_0, x_1, x_2] + \cdots$$
$$+ \sum_{i=0}^{n-1} \frac{(x - x_0)(x - x_1)\cdots(x - x_{n-1})}{x - x_i}\, y[x_0, x_1, \cdots, x_n] \quad (7.22)$$

The error estimate in this approximation can be seen from the following.

In Section 6.6.3, we have seen that if $y(x)$ is approximated by $P_n(x)$, the error estimate is shown to be

$$E_n(x) = y(x) - P_n(x) = \frac{\Pi(x)}{(n+1)!}\, y^{(n+1)}(\xi) \quad (7.23)$$

Its derivative with respect to x can be written as

$$E_n'(x) = y'(x) - P_n'(x) = \frac{\Pi'(x)}{(n+1)!}\, y^{(n+1)}(\xi) + \frac{\Pi(x)}{(n+1)!}\, \frac{d}{dx} y^{(n+1)}(\xi) \quad (7.24)$$

Since $\xi(x)$ depends on x in an unknown way the derivative

$$\frac{d}{dx} y^{(n+1)}(\xi)$$

cannot be evaluated. However, for any of the tabular points $x = x_i$, $\Pi(x)$ vanishes and the difficult term drops out. Thus, the error term in Eq. (7.24) at the tabular point $x = x_i$ simplifies to

$$E_n'(x_i) = \text{Error} = \Pi'(x_i)\frac{y^{(n+1)}(\xi)}{(n+1)!} \quad (7.25)$$

for some ξ in the interval I defined by the smallest and largest of $x, x_0, x_1, \ldots, x_n$ and

$$\Pi'(x_i) = (x_i - x_0)\cdots(x_i - x_n) = \prod_{\substack{j=0 \\ j \neq i}}^{n} (x_i - x_j) \quad (7.26)$$

The error in the rth derivative at the tabular points can indeed be expressed analogously.

To understand this method better, we consider the following example.

Example 7.4 Find $y'(0.25)$ and $f''(0.25)$ from the following data using the method based on divided differences:

x	0.15	0.21	0.23	0.27	0.32	0.35
$y = f(x)$	0.1761	0.3222	0.3617	0.4314	0.5051	0.5441

Solution We first construct divided difference table for the given data as follows:

x	y	1st divided difference	2nd divided difference	3rd divided difference	4th divided difference	5th divided difference
$x_0 = 0.15$	0.1761					
		2.4350				
$x_1 = 0.21$	0.3222		−5.7500			
		1.9750		15.6250		
$x_2 = 0.23$	0.3617		−3.8750		−44.23	
		1.7425		8.1064		172.2
$x_3 = 0.27$	0.4314		−2.9833		−9.79	
		1.4740		6.7358		
$x_4 = 0.32$	0.5051		−2.1750			
		1.3000				
$x_5 = 0.35$	0.5441					

Using Newton's divided difference formula (7.21), we have

$$y(x) = p_5(x) = y[x_0] + (x - x_0)\, y[x_0, x_1] + (x - x_0)(x - x_1)\, y[x_0, x_1, x_2]$$
$$+ (x - x_0)(x - x_1)(x - x_2)\, y[x_0, x_1, x_2, x_3]$$
$$+ (x - x_0)(x - x_1)(x - x_2)(x - x_3) y[x_0, x_1, x_2, x_3, x_4]$$

Now, using values from the above table of divided differences, we obtain

$$y(x) = 0.1761 + (x - 0.15)\, 2.4350 + (x - 0.15)\,(x - 0.21)\,(-5.75)$$
$$+ (x - 0.15)\,(x - 0.21)\,(x - 0.23)15.625$$
$$+ (x - 0.15)\,(x - 0.21)\,(x - 0.23)\,(x - 0.27)\,(-44.23)$$
$$+ (x - 0.15)\,(x - 0.21)\,(x - 0.23)\,(x - 0.27)\,(x - 0.32)\, 172.2 \tag{1}$$

Differentiating Eq. (1) with respect to x, we get

$$y'(x) = 2.4350 - (2x - 0.36)5.75 + 15.625(3x^2 - 1.18x + 0.1143)$$
$$- 44.23(4x^3 - 2.58x^2 + 0.5472x - 38.105 \times 10^{-3})$$
$$+ 172.2(5x^4 - 4.72x^3 + 1.6464x^2 - 0.2515x + 14.15 \times 10^{-3}) \tag{2}$$

Which immediately gives

$$y'(0.25) = 2.4350 - 0.805 + 0.10625 + 2.432 \times 10^{-3} - 7.5338 \times 10^{-3} = 1.7312$$

Now, differentiating Eq. (2) once again with respect to x, we obtain

$$y''(x) = 3444x^3 - 2969.112x^2 + 888.99696x - 91.700456$$

which gives at once

$$y''(0.25) = 53.8125 - 185.5695 + 222.24924 - 91.700456 = -1.208216$$

7.4 RICHARDSON'S EXTRAPOLATION METHOD

To improve the accuracy of the derivative of a function, which is computed by starting with an arbitrarily selected value of h, Richardson's extrapolation method is often employed in practice, as explained below:

Suppose we use two-point formula (7.17) to compute the derivative of a function, then we have

$$y'(x) = \frac{y(x+h) - y(x-h)}{2h} + E_T = F(h) + E_T$$

where E_T is the *truncation error*. Using Taylor's series expansion, we can see that

$$E_T = c_1 h^2 + c_2 h^4 + c_3 h^6 + \cdots$$

The idea of *Richardson's extrapolation* is to combine two computed values of $y'(x)$ using the same method but with two different step sizes usually h and $h/2$ to yield a higher order method. Thus, we have

$$y'(x) = F(h) + c_1 h^2 + c_2 h^4 + \cdots$$

and

$$y'(x) = F\left(\frac{h}{2}\right) + c_1 \frac{h^2}{4} + c_2 \frac{h^4}{16} + \cdots$$

Here, c_i are constants, independent of h, and $F(h)$ and $F(h/2)$ represent approximate values of derivatives. Eliminating c_1 from the above pair of equations, we get

$$y'(x) = \frac{4F\left(\frac{h}{2}\right) - F(h)}{3} + d_1 h^4 + O(h^6) \tag{7.27}$$

Now, assuming

$$F_1\left(\frac{h}{2}\right) = \frac{4F\left(\frac{h}{2}\right) - F(h)}{3} \tag{7.28}$$

Equation (7.27) reduces to

$$y'(x) = F_1\left(\frac{h}{2}\right) + d_1 h^4 + O(h^6)$$

Thus, we have obtained a fourth-order accurate differentiation formula by combining two results which are of second-order accurate. Now, repeating the above argument, we have

$$y'(x) = F_1\left(\frac{h}{2}\right) + d_1 h^4 + O(h^6)$$

$$y'(x) = F_1\left(\frac{h}{4}\right) + \frac{d_1 h^4}{16} + O(h^6)$$

Eliminating d_1 from the above pair of equations, we get a better approximation as

$$y'(x) = F_2\left(\frac{h}{4}\right) + O(h^6)$$

which is of sixth-order accurate, where

$$F_2\left(\frac{h}{4}\right) = \frac{4^2 F_1\left(\frac{h}{2^2}\right) - F_1\left(\frac{h}{2}\right)}{4^2 - 1} \tag{7.29}$$

This extrapolation process can be repeated further until the required accuracy is achieved, which is called an *extrapolation to the limit*. Equation (7.29) can be generalized as

$$F_m\left(\frac{h}{2^m}\right) = \frac{4^m F_{m-1}\left(\frac{h}{2^m}\right) - F_{m-1}\left(\frac{h}{2^{m-1}}\right)}{4^m - 1}, \qquad m = 1, 2, 3, \ldots \tag{7.30}$$

where $F_0(h) = F(h)$.

To illustrate this procedure, we consider the following example.

Example 7.5 Using the Richardson's extrapolation limit, find $y'(0.05)$ to the function $y = -1/x$, with $h = 0.0128, 0.0064, 0.0032$.

Solution To start with, we take, $h = 0.0128$, then compute $F(h)$ as

$$F(h) = \frac{y(x+h) - y(x-h)}{2h} = \frac{-\dfrac{1}{0.05 + 0.0128} + \dfrac{1}{0.05 - 0.0128}}{2(0.0128)}$$

$$= \frac{-15.923566 + 26.88172}{0.0256} = 428.05289 \tag{1}$$

Similarly, $F(h/2) = 406.66273$. Therefore, using Eq. (7.30), we get

$$F_1\left(\frac{h}{2}\right) = \frac{4F\left(\frac{h}{2}\right) - F(h)}{4 - 1} = 399.5327 \tag{2}$$

which is accurate to $O(h^4)$. Halving the step size further, we compute

$$F\left(\frac{h}{2^2}\right) = \frac{-\dfrac{1}{0.05 + 0.0032} + \dfrac{1}{0.05 - 0.0032}}{2(0.0032)} = 401.64515 \tag{3}$$

and

$$F_1\left(\frac{h}{2^2}\right) = \frac{4F\left(\frac{h}{2^2}\right) - F\left(\frac{h}{2}\right)}{4 - 1} = 399.97263 \tag{4}$$

Again, using Eq. (7.30), we obtain

$$F_2\left(\frac{h}{2^2}\right) = \frac{4^2 F_1\left(\frac{h}{2^2}\right) - F_1\left(\frac{h}{2}\right)}{4^2 - 1} = 400.00195 \tag{5}$$

The above computation can be summarized in the following table:

h	F	F_1	F_2
0.0128	428.0529		
		399.5327	
0.0064	406.6627		400.00195
		399.9726	
0.0032	401.6452		

Thus, after two steps, it is found that $y'(0.05) = 400.00195$ while the exact value is

$$y'(0.05) = \left(\frac{1}{x^2}\right)_{x=0.05} = \frac{1}{0.0025} = 400$$

7.5 NUMERICAL INTEGRATION

Consider the definite integral

$$I = \int_{x=a}^{b} f(x)\, dx \tag{7.31}$$

where $f(x)$ is known either explicitly or is given as a table of values corresponding to some values of x, whether equispaced or not. Integration of such functions can be carried out using numerical techniques. Of course, we assume that the function to be integrated is smooth and Riemann integrable in the interval of integration. In the following section, we shall develop Newton–Cotes formulae based on interpolation which form the basis for trapezoidal rule and Simpson's rule of numerical integration.

7.6 NEWTON–COTES INTEGRATION FORMULAE

In this method, as in the case of numerical differentiation, we shall approximate the given tabulated function, by a polynomial $P_n(x)$ and then integrate this polynomial. Suppose, we are given the data (x_i, y_i), $i = 0\ (1)\ n$, at equispaced points with spacing $h = x_{i+1} - x_i$, we can represent the polynomial by any standard interpolation polynomial. Suppose, we use Lagrangian approximation given by Eq. (6.45), then we have

$$f(x) \approx \sum L_k(x) y(x_k) \tag{7.32}$$

with associated error given by

$$E(x) = \frac{\Pi(x)}{(n+1)!} y^{(n+1)}(\xi) \tag{7.33}$$

where

$$L_k(x) = \frac{\Pi(x)}{(x - x_k)\Pi'(x_k)} \tag{7.34}$$

and

$$\Pi(x) = (x - x_0)(x - x_1) \ldots (x - x_n) \tag{7.35}$$

Then, we obtain an equivalent integration formula to the definite integral (7.31) in the form

$$\int_a^b f(x)\, dx \approx \sum_{k=1}^{n} c_k y(x_k) \tag{7.36}$$

where c_k are the weighting coefficients given by

$$c_k = \int_a^b L_k(x)\, dx \tag{7.37}$$

which are also called Cotes numbers. Let the equispaced nodes are defined by

$$x_0 = a, \quad x_n = b, \quad h = \frac{b-a}{n}, \quad x_k = x_0 + kh$$

so that $x_k - x_1 = (k-1)h$ etc. Now, we shall change the variable x to p such that, $x = x_0 + ph$, then we can rewrite Eqs. (7.35) and (7.34) respectively as

$$\Pi(x) = h^{n+1} p(p-1) \ldots (p-n) \tag{7.38}$$

and

$$L_k(x) = \frac{(x - x_0)(x - x_1)\cdots(x - x_{k-1})(x - x_{k+1})\cdots(x - x_n)}{(x_k - x_0)(x_k - x_1)\cdots(x_k - x_{k-1})(x_k - x_{k+1})\cdots(x_k - x_n)}$$

$$= \frac{(ph)(p-1)h\cdots(p-k+1)h(p-k-1)h\cdots(p-n)h}{(kh)(k-1)h\cdots(1)h(-1)h\cdots(k-n)h}$$

or

$$L_k(x) = (-1)^{(n-k)} \frac{p(p-1)\cdots(p-k+1)(p-k-1)\cdots(p-n)}{k!(n-k)!} \tag{7.39}$$

Also, noting that $dx = h\, dp$. The limits of the integral in Eq. (7.37) change from 0 to n and Eq. (7.37) reduces to

$$c_k = \frac{(-1)^{n-k} h}{k!(n-k)!} \int_0^n p(p-1) \cdots (p-k+1)(p-k-1) \cdots (p-n) dp \tag{7.40}$$

The error in approximating the integral (7.36) can be obtained by substituting (7.38) into Eq. (7.33) in the form

$$E_n = \frac{h^{n+2}}{(n+1)!} \int_0^n p(p-1) \cdots (p-n) y^{(n+1)}(\xi)\, dp \tag{7.41}$$

where $x_0 < \xi < x_n$. For illustration, let us consider the cases for $n = 1, 2$. From Eq. (7.40), we get

$$c_0 = -h \int_0^1 (p-1)\, dp = \frac{h}{2}, \qquad c_1 = h \int_0^1 p\, dp = \frac{h}{2}$$

and Eq. (7.41) gives

$$E_1 = \frac{h^3}{2} y''(\xi) \int_0^1 p(p-1)\, dp = -\frac{h^3}{12} y''(\xi)$$

Thus, the integration formula corresponding to integral (7.36) is found to be

$$\int_{x_0}^{x_1} f(x)\, dx = c_0 y_0 + c_1 y_1 + \text{Error} = \frac{h}{2}(y_0 + y_1) - \frac{h^3}{12} y''(\xi) \tag{7.42}$$

This equation represents the Trapezoidal rule in the interval $[x_0, x_1]$ with error term. Geometrically, it represents an area between the curve $y = f(x)$, the x-axis and the ordinates erected at $x = x_0$ $(=a)$ and $x = x_1$ as shown in Fig. 7.2. This area is approximated by the trapezium formed by replacing the curve with its secant line drawn between the end points (x_0, y_0) and (x_1, y_1).

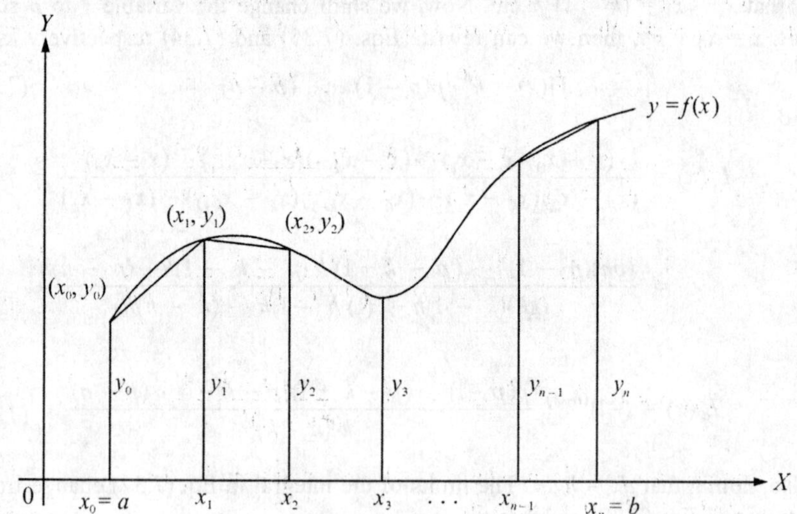

Fig. 7.2 Trapezoidal rule.

For $n = 2$, Eq. (7.40) gives

$$c_0 = \frac{h}{2} \int_0^2 (p-1)(p-2)\, dp = \frac{h}{3}$$

$$c_1 = -h\int_0^2 p(p-2)\,dp = \frac{4}{3}h$$

$$c_2 = \frac{h}{2}\int_0^2 p(p-1)\,dp = \frac{h}{3}$$

and the error term is given by

$$E_2 = -\frac{h^5}{90}y^{(iv)}(\xi)$$

Thus, for $n = 2$, the integration (7.36) takes the form

$$\int_{x_0}^{x_2} f(x)\,dx = c_0 y_0 + c_1 y_1 + c_2 y_2 + \text{Error}$$

$$= \frac{h}{3}(y_0 + 4y_1 + y_2) - \frac{h^5}{90}y^{(iv)}(\xi) \tag{7.43}$$

This is known as *Simpson's 1/3 rule*. Geometrically, this equation represents the area between the curve $y = f(x)$, the x-axis and the ordinates at $x = x_0$ and x_2 after replacing the arc of the curve between (x_0, y_0) and (x_2, y_2) by an arc of a quadratic polynomial as shown in Fig. 7.3. Thus Simpson's 1/3 rule is based on fitting three points with a quadratic.

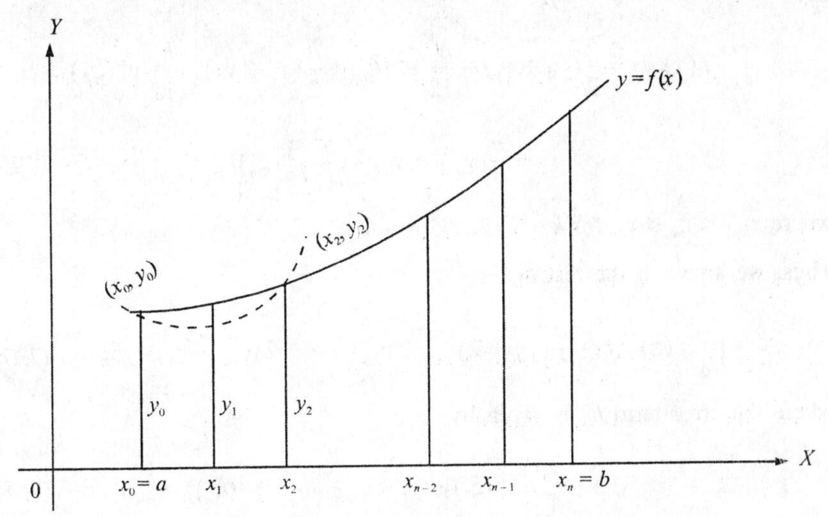

Fig. 7.3 Simpson's rule.

Similarly, for $n = 3$, the integration (7.36) is found to be

$$\int_{x_0}^{x_3} f(x)\,dx = \frac{3}{8}h(y_0 + 3y_1 + 3y_2 + y_3) - \frac{3}{80}h^5 y^{(iv)}(\xi) \tag{7.44}$$

This is known as *Simpson's 3/8 rule*, which is based on fitting four points by a cubic. Still higher order Newton–Cotes integration formulae can be derived for large values of n. But for all practical purposes, Simpson's 1/3 rule is found to be sufficiently accurate.

7.6.1 The Trapezoidal Rule (Composite Form)

The Newton–Cotes formula (7.42) is based on approximating $y = f(x)$ between (x_0, y_0) and (x_1, y_1) by a straight line, thus forming a trapezium, is called *trapezoidal rule*. In order to evaluate the definite integral

$$I = \int_a^b f(x)\, dx$$

we divide the interval $[a, b]$ into n sub-intervals, each of size $h = (b - a)/n$ and denote the sub-intervals by $[x_0, x_1], [x_1, x_2], \ldots, [x_{n-1}, x_n]$, such that $x_0 = a$ and $x_n = b$ and $x_k = x_0 + kh$, $k = 1, 2, \ldots, n - 1$. Thus, we can write the above definite integral as a sum. Therefore,

$$I = \int_{x_0}^{x_n} f(x)\, dx = \int_{x_0}^{x_1} f(x)\, dx + \int_{x_1}^{x_2} f(x)\, dx + \cdots + \int_{x_{n-1}}^{x_n} f(x)\, dx \qquad (7.45)$$

As shown in Fig. 7.2, the area under the curve in each sub-interval is approximated by a trapezium. The integral I, which represents an area between the curve $y = f(x)$, the x-axis and the ordinates at $x = x_0$ and $x = x_n$ is obtained by adding all the trapezoidal areas in each sub-interval.

Now, using the trapezoidal rule as expressed in Eq. (7.42) into Eq. (7.45), we get

$$\int_{x_0}^{x_n} f(x)\, dx = \frac{h}{2}(y_0 + y_1) - \frac{h^3}{12} y''(\xi_1) + \frac{h}{2}(y_1 + y_2) - \frac{h^3}{12} y''(\xi_2)$$

$$+ \cdots + \frac{h}{2}(y_{n-1} + y_n) - \frac{h^3}{12} y''(\xi_n) \qquad (7.46)$$

where $x_{k-1} < \xi_k < x_k$, for $k = 1, 2, \ldots, n - 1$.

Thus, we arrive at the result

$$\int_{x_0}^{x_n} f(x)\, dx = \frac{h}{2}(y_0 + 2y_1 + 2y_2 + \cdots + 2y_{n-1} + y_n) + E_n \qquad (7.47)$$

where the error term E_n is given by

$$E_n = -\frac{h^3}{12}[y''(\xi_1) + y''(\xi_2) + \cdots + y''(\xi_n)] \qquad (7.48)$$

Equation (7.47) represents the trapezoidal rule over $[x_0, x_n]$, which is also called the *composite form of the trapezoidal rule.*

The error term given by Eq. (7.48) is called the *global error*. However, if we assume that $y''(x)$ is continuous over $[x_0, x_n]$ then there exists some ξ in $[x_0, x_n]$ such that $x_n = x_0 + nh$ and

$$E_n = -\frac{h^3}{12}[ny''(\xi)] = -\frac{x_n - x_0}{12} h^2 y''(\xi) \qquad (7.49)$$

Then the global error can be conveniently written as $O(h^2)$.

7.6.2 Simpson's Rules (Composite Forms)

In deriving Eq. (7.43), the Simpson's 1/3 rule, we have used two sub-intervals of equal width. In order to get a composite formula, we shall divide the interval of integration $[a, b]$ into an even number of sub-intervals say $2N$, each of width $(b - a)/2N$, thereby we have $x_0 = a, x_1, \ldots, x_{2N} = b$ and $x_k = x_0 + kh$, $k = 1, 2, \ldots, (2N - 1)$. Thus, the definite integral I can be written as

$$I = \int_a^b f(x)\,dx = \int_{x_0}^{x_2} f(x)\,dx + \int_{x_2}^{x_4} f(x)\,dx + \cdots + \int_{x_{2N-2}}^{x_{2N}} f(x)\,dx \quad (7.50)$$

Applying Simpson's 1/3 rule as in Eq. (7.43) to each of the integrals on the right-hand side of Eq. (7.50), we obtain

$$I = \frac{h}{3}[(y_0 + 4y_1 + y_2) + (y_2 + 4y_3 + y_4) + \cdots$$
$$+ (y_{2N-2} + 4y_{2N-1} + y_{2N})] - \frac{N}{90}h^5 y^{(iv)}(\xi)$$

That is,

$$\int_{x_0}^{x_{2N}} f(x)\,dx = \frac{h}{3}[y_0 + 4(y_1 + y_3 + \cdots + y_{2N-1})$$
$$+ 2(y_2 + y_4 + \cdots + y_{2N-2}) + y_{2N}] + \text{Error term} \quad (7.51)$$

This formula is called *composite Simpson's 1/3 rule*. The error term E, which is also called *global error*, is given by

$$E = -\frac{N}{90}h^5 y^{(iv)}(\xi) = -\frac{x_{2N} - x_0}{180}h^4 y^{(iv)}(\xi) \quad (7.52)$$

for some ξ in $[x_0, x_{2N}]$. Thus, in Simpson's 1/3 rule, the global error is of $O(h^4)$.

Similarly in deriving composite Simpson's 3/8 rule, we divide the interval of integration into n sub-intervals, where n is divisible by 3, and applying the integration formula (7.44) to each of the integral given below

$$\int_{x_0}^{x_n} f(x)\,dx = \int_{x_0}^{x_3} f(x)\,dx + \int_{x_3}^{x_6} f(x)\,dx + \cdots + \int_{x_{n-3}}^{x_n} f(x)\,dx$$

we obtain the composite form of Simpson's 3/8 rule as

$$\int_a^b f(x)\,dx = \frac{3}{8}h[y(a) + 3y_1 + 3y_2 + 2y_3 + 3y_4 + 3y_5 + 2y_6 + \cdots$$
$$+ 2y_{n-3} + 3y_{n-2} + 3y_{n-1} + y(b)] \quad (7.53)$$

with the global error E given by

$$E = -\frac{x_n - x_0}{80}h^4 y^{(iv)}(\xi) \quad (7.54)$$

It may be noted from Eqs. (7.52) and (7.54), the global error in Simpson's 1/3 and 3/8 rules are of the same order. However, if we consider the magnitudes of the error terms, we notice that Simpson's 1/3 rule is superior to Simpson's 3/8 rule. For illustration, we consider few examples.

Example 7.6 Find the approximate value of

$$y = \int_0^{\pi} \sin x \, dx$$

using (i) trapezoidal rule, (ii) Simpson's 1/3 rule by dividing the range of integration into six equal parts. Calculate the percentage error from its true value in both the cases.

Solution We shall at first divide the range of integration (0, π) into six equal parts so that each part is of width $\pi/6$ and write down the table of values:

x	0	$\pi/6$	$\pi/3$	$\pi/2$	$2\pi/3$	$5\pi/6$	π
$y = \sin x$	0.0	0.5	0.8660	1.0	0.8660	0.5	0.0

Applying trapezoidal rule, we have

$$\int_0^{\pi} \sin x \, dx = \frac{h}{2}[y_0 + y_6 + 2(y_1 + y_2 + y_3 + y_4 + y_5)]$$

Here, h, the width of the interval is $\pi/6$. Therefore,

$$y = \int_0^{\pi} \sin x \, dx = \frac{\pi}{12}[0 + 0 + 2(3.732)] = \frac{3.1415}{6} \times 3.732 = 1.9540$$

Applying Simpson's 1/3 rule (7.41), we have

$$\int_0^{\pi} \sin x \, dx = \frac{h}{3}[y_0 + y_6 + 4(y_1 + y_3 + y_5) + 2(y_2 + y_4)]$$

$$= \frac{\pi}{18}[0 + 0 + (4 \times 2) + (2)(1.732)] = \frac{3.1415}{18} \times 11.464 = 2.0008$$

But the actual value of the integral is

$$\int_0^{\pi} \sin x \, dx = [-\cos x]_0^{\pi} = 2$$

Hence, in the case of trapezoidal rule

$$\text{The percentage of error} = \frac{2 - 1.954}{2} \times 100 = 2.3$$

While in the case of Simpson's rule the percentage error is

$$\frac{2 - 2.0008}{2} \times 100 = 0.04 \qquad \text{(sign ignored)}$$

Example 7.7 From the following data, estimate the value of

$$\int_1^5 \log x \, dx$$

using Simpson's 1/3 rule. Also, obtain the value of h, so that the value of the integral will be accurate up to five decimal places.

x	1.0	1.5	2.0	2.5	3.0	3.5	4.0	4.5	5.0
$y = \log x$	0.0000	0.4055	0.6931	0.9163	1.0986	1.2528	1.3863	1.5041	1.6094

Solution We have from the data, $n = 0, 1, \ldots, 8$, and $h = 0.5$. Now using Simpson's 1/3 rule,

$$\int_1^5 \log x \, dx = \frac{h}{3}[y_0 + y_8 + 4(y_1 + y_3 + y_5 + y_7) + 2(y_2 + y_4 + y_6)]$$

$$= \frac{0.5}{3}[(0 + 1.6094) + 4(4.0787) + 2(3.178)]$$

$$= \frac{0.5}{3}(1.6094 + 16.3148 + 6.356)$$

$$= 4.0467$$

The error in Simpson's rule is given by

$$E = \frac{x_{2N} - x_0}{180} h^4 y^{(iv)}(\xi) \qquad \text{(ignoring the sign)}$$

Since

$$y = \log x, \qquad y' = \frac{1}{x}, \qquad y'' = -\frac{1}{x^2}, \qquad y''' = \frac{2}{x^3}, \qquad y^{(iv)} = -\frac{6}{x^4}$$

$$\max_{1 \le x \le 5} y^{(iv)}(x) = 6, \qquad \min_{1 \le x \le 5} y^{(iv)}(x) = 0.0096$$

Therefore, the error bounds are given by

$$\frac{(0.0096)(4)h^4}{180} < E < \frac{(6)(4)h^4}{180}$$

If the result is to be accurate up to five decimal places, then

$$\frac{24h^4}{180} < 10^{-5}$$

That is, $h^4 < 0.000075$ or $h < 0.09$. It may be noted that the actual value of integral is

$$\int_1^5 \log x\, dx = [x \log x - x]_1^5 = 5 \log 5 - 4$$

Example 7.8 Evaluate the integral

$$I = \int_0^1 \frac{dx}{1+x^2}$$

using (i) trapezoidal rule, (ii) Simpson's 1/3 rule by taking $h = 1/4$. Hence, compute the approximate value of π.

Solution At first, we shall tabulate the function as

x	0	1/4	1/2	3/4	1
$y = \dfrac{1}{1+x^2}$	1	0.9412	0.8000	0.6400	0.5000

using trapezoidal rule, and taking $h = 1/4$

$$I = \frac{h}{2}[y_0 + y_4 + 2(y_1 + y_2 + y_3)] = \frac{1}{8}[1.5 + 2(2.312)] = 0.7828 \quad (1)$$

using Simpson's 1/3 rule, and taking $h = 1/4$, we have

$$I = \frac{h}{3}[y_0 + y_4 + 4(y_1 + y_3) + 2y_2] = \frac{1}{12}[1.5 + 4(1.512) + 1.6] = 0.7854 \quad (2)$$

But the closed form solution to the given integral is

$$\int_0^1 \frac{dx}{1+x^2} + [\tan^{-1} x]_0^1 = \frac{\pi}{4} \quad (3)$$

Equating (2) and (3), we get $\pi = 3.1416$.

Example 7.9 Compute the integral

$$I = \sqrt{\frac{2}{\pi}} \int_0^1 e^{-x^2/2}\, dx$$

using Simpson's 1/3 rule, taking $h = 0.125$.

Solution At the outset, we shall construct the table of the function as required.

x	0	0.125	0.250	0.375	0.5	0.625	0.750	0.875	1.0
$y = \sqrt{\dfrac{2}{\pi}} e^{-x^2/2}$	0.7979	0.7917	0.7733	0.7437	0.7041	0.6563	0.6023	0.5441	0.4839

Using Simpson's 1/3 rule, we have

$$I = \frac{h}{3}[y_0 + y_8 + 4(y_1 + y_3 + y_5 + y_7) + 2(y_2 + y_4 + y_6)]$$

$$= \frac{0.125}{3}[0.7979 + 0.4839 + 4(0.7917 + 0.7437 + 0.6563 + 0.5441)$$

$$+ 2(0.7733 + 0.7041 + 0.6023)]$$

$$= \frac{0.125}{3}(1.2818 + 10.9432 + 4.1594) = 0.6827$$

Hence, $I = 0.6827$.

Example 7.10 A missile is launched from a ground station. The acceleration during its first 80 seconds of flight, as recorded, is given in the following table:

t (s)	0	10	20	30	40	50	60	70	80
a (m/s^2)	30	31.63	33.34	35.47	37.75	40.33	43.25	46.69	50.67

compute the velocity of the missile when t = 80 s, using Simpson's 1/3 rule.

Solution Since acceleration is defined as the rate of change of velocity, we have

$$\frac{dv}{dt} = a \quad \text{or} \quad v = \int_0^{80} a\ dt$$

Using Simpson's 1/3-rule, we have

$$v = \frac{h}{3}[(y_0 + y_8) + 4(y_1 + y_3 + y_5 + y_7) + 2(y_2 + y_4 + y_6)]$$

$$= \frac{10}{3}[(30 + 50.67) + 4(31.63 + 35.47 + 40.33 + 46.69)$$

$$+ 2(33.34 + 37.75 + 43.25)]$$

$$= 3086.1 \text{ m/s}$$

Therefore, the required velocity is given by v = 3.0861 km/s.

7.7 ROMBERG'S INTEGRATION

In Section 7.6, we have observed that the trapezoidal rule of integration of a definite integral is of O(h^2), while that of Simpson's 1/3 and 3/8 rules are of fourth-order accurate. We can improve the accuracy of trapezoidal and Simpson's rules using Richardson's extrapolation procedure as described in Section 7.4, which is also called *Romberg's integration method.* For example, the error in trapezoidal rule of a definite integral

$$I = \int_a^b f(x)\,dx$$

can be written in the form

$$I = I_T + c_1 h^2 + c_2 h^4 + c_3 h^6 + \cdots$$

By applying Richardson's extrapolation procedure to trapezoidal rule, we obtain the following general formula

$$I_{Tm}\left(\frac{h}{2^m}\right) = \frac{4^m I_{T(m-1)}\left(\frac{h}{2^m}\right) - I_{T(m-1)}\left(\frac{h}{2^{m-1}}\right)}{4^m - 1}$$

where $m = 1, 2, \ldots,$ with $I_{T0}(h) = I_T(h)$. (7.55)

For illustration, we consider the following example.

Example 7.11 Using Romberg's integration method, find the value of

$$\int_1^{1.8} y(x)\, dx$$

starting with trapezoidal rule, for the tabular values

x	1.0	1.1	1.2	1.3	1.4	1.5	1.6	1.7	1.8
$y = f(x)$	1.543	1.669	1.811	1.971	2.151	2.352	2.577	2.828	3.107

Solution Taking

$$x_0 = 1, \qquad x_n = 1.8, \qquad h = \frac{1.8 - 1.0}{N}, \qquad x_i = x_0 + ih$$

Let I_T denote the integration by Trapezoidal rule, then for

$$N = 1,\ h = 0.8,\ I_T = \frac{h}{2}(y_0 + y_1) = 0.4(1.543 + 3.107) = 1.8600$$

$$N = 2,\ h = 0.4,\ I_T = \frac{h}{2}(y_0 + 2y_1 + y_2) = 0.2[1.543 + 2(2.151) + 3.107]$$
$$= 1.7904$$

$$N = 4,\ h = 0.2,\ I_T = \frac{h}{2}[y_0 + 2(y_1 + y_2 + y_3) + y_4]$$
$$= 0.1[1.543 + 2(1.811 + 2.151 + 2.577) + 3.107]$$
$$= 1.7728$$

Similarly for

$$N = 8, \qquad h = 0.1, \qquad I_T = 1.7684$$

Now, using Romberg's formula (7.55), we have

$$I_{T1}\left(\frac{h}{2}\right) = \frac{4(1.7904) - 1.8600}{3} = 1.7672$$

$$I_{T2}\left(\frac{h}{2^2}\right) = \frac{4^2(1.7728) - 1.7672}{4^2 - 1} = 1.77317$$

$$I_{T3}\left(\frac{h}{2^3}\right) = \frac{4^3(1.7672) - 1.77317}{4^3 - 1} = 1.7671$$

Thus, after three steps, it is found that the value of the tabulated integral is 1.7671.

7.8 DOUBLE INTEGRATION

To evaluate numerically a double integral of the form

$$I = \int \left[\int f(x, y)\, dx\right] dy \tag{7.56}$$

over a rectangular region bounded by the lines $x = a$, $x = b$, $y = c$, $y = d$ we shall employ either trapezoidal rule or Simpson's rule, developed in Section 7.6, repeatedly with respect to one variable at a time. Noting that, both the integrations are just a linear combination of values of the given function at different values of the independent variable, we divide the interval $[a, b]$ into N equal sub-intervals of size h, such that $h = (b - a)/N$; and the interval (c, d) into M equal sub-intervals of size k, so that $k = (d - c)/M$. Thus, we have

$$x_i = x_0 + ih, \quad x_0 = a, \quad x_N = b, \qquad \text{for } i = 1, 2, \ldots, N - 1$$

$$y_i = y_0 + ik, \quad y_0 = c, \quad y_M = d, \qquad \text{for } i = 1, 2, \ldots, M - 1$$

Thus, we can generate a table of values of the integrand, and the above procedure of integration is illustrated by considering a couple of examples.

Example 7.12 Evaluate the double integral

$$I = \int_1^2 \int_1^2 \frac{dx\, dy}{x + y}$$

by using trapezoidal rule, with $h = k = 0.25$.

Solution Taking $x = 1, 1.25, 1.50, 1.75, 2.0$ and $y = 1, 1.25, 1.50, 1.75, 2.0$, the following table is generated using the integrand

$$f(x, y) = \frac{1}{x + y}$$

x	y				
	1.00	1.25	1.50	1.75	2.00
1.00	0.5	0.4444	0.4	0.3636	0.3333
1.25	0.4444	0.4	0.3636	0.3333	0.3077
1.50	0.4	0.3636	0.3333	0.3077	0.2857
1.75	0.3636	0.3333	0.3077	0.2857	0.2667
2.00	0.3333	0.3077	0.2857	0.2667	0.25

Keeping one variable say x fixed and varying the variable y, the application of trapezoidal rule to each row in the above table gives

$$\int_1^2 f(1, y)\, dy = \frac{0.25}{2}[0.5 + 2(0.4444 + 0.4 + 0.3636) + 0.3333]$$

$$= 0.4062 \tag{1}$$

$$\int_1^2 f(1.25, y)\, dy = \frac{0.25}{2}[0.4444 + 2(0.4 + 0.3636 + 0.3333) + 0.3077]$$

$$= 0.3682 \tag{2}$$

$$\int_1^2 f(1.5, y)\, dy = \frac{0.25}{2}[0.4 + 2(0.3636 + 0.3333 + 0.3077) + 0.2857]$$

$$= 0.3369 \tag{3}$$

$$\int_1^2 f(1.75, y)\, dy = \frac{0.25}{2}[0.3636 + 2(0.3333 + 0.3077 + 0.2857) + 0.2667]$$

$$= 0.3105 \tag{4}$$

and

$$\int_1^2 f(2, y)\, dy = \frac{0.25}{2}[0.3333 + 2(0.3077 + 0.2857 + 0.2667) + 0.25]$$

$$= 0.2879 \tag{5}$$

Therefore,

$$I = \int_1^2 \int_1^2 \frac{dx\, dy}{x + y} = \frac{h}{2}\{f(1, y) + 2[f(1.25, y) + f(1.5, y) + f(1.75, y)] + f(2, y)\} \tag{6}$$

Substituting Eqs. (1)–(5) into Eq. (6), we get the required result as

$$I = \frac{0.25}{2}[0.4062 + 2(0.3682 + 0.3369 + 0.3105) + 0.2879] = 0.3407$$

Example 7.13 Evaluate

$$\int_0^{\pi/2} \int_0^{\pi/2} \sqrt{\sin\,(x + y)}\; dx\, dy$$

by numerical double integration.

Solution Taking $x = y = 0,\ \pi/8,\ \pi/4,\ 3\pi/8,\ \pi/2$, we can generate the following table of the integrand

$$f(x, y) = \sqrt{\sin(x + y)}$$

	y				
x	0	$\pi/8$	$\pi/4$	$3\pi/8$	$\pi/2$
0	0.0	0.6186	0.8409	0.9612	1.0
$\pi/8$	0.6186	0.8409	0.9612	1.0	0.9612
$\pi/4$	0.8409	0.9612	1.0	0.9612	0.8409
$3\pi/8$	0.9612	1.0	0.9612	0.8409	0.6186
$\pi/2$	1.0	0.9612	0.8409	0.6186	0.0

Keeping one variable say x as fixed and y as variable, and applying trapezoidal rule to each row of the above table, we get

$$\int_0^{\pi/2} f(0, y)\, dx = \frac{\pi}{16}[0.0 + 2(0.6186 + 0.8409 + 0.9612) + 1.0] = 1.1469$$

$$\int_0^{\pi/2} f\left(\frac{\pi}{8}, y\right) dx = \frac{\pi}{16}[0.6186 + 2(0.8409 + 0.9612 + 1.0) + 0.9612] = 1.4106$$

Similarly, we get

$$\int_0^{\pi/2} f\left(\frac{\pi}{4}, y\right) dx = 1.4778, \int_0^{\pi/2}\left(\frac{3\pi}{8}, y\right) dx = 1.4106,$$

and

$$\int_0^{\pi/2}\left(\frac{\pi}{2}, y\right) dx = 1.1469$$

Using these results, we finally obtain

$$\int_0^{\pi/2}\int_0^{\pi/2} \sqrt{\sin\,(x+y)}\, dx\, dy = \frac{\pi}{16}\left\{f(0, y) + 2\left[f\left(\frac{\pi}{8}, y\right) + f\left(\frac{\pi}{4}, y\right)\right.\right.$$

$$\left.\left. + f\left(\frac{3\pi}{8}, y\right)\right] + f\left(\frac{\pi}{2}, y\right)\right\}$$

$$= \frac{\pi}{16}[1.1469 + 2(1.4106 + 1.4778 + 1.4106)$$

$$+ 1.1469] = 2.1386$$

7.9 GAUSSIAN QUADRATURE FORMULAE

The numerical integration technique associated with the natural coordinates (ξ, η) used in finite element method for evaluating the element matrices is the Gauss–Legendre quadrature. In Newton–Cotes formulae, the integral of a function is approximated by the sum of its functional values at a set of equally spaced points, multiplied by certain weighting coefficients. However, in Gaussian quadrature, we have the freedom to choose not only the weighting coefficients but also the location of abscissas also called 'sampling points' at which the function is to be evaluated. In fact, they are no longer equally-spaced and the number of functional evaluations are same in both the cases, while we can achieve better accuracy. To illustrate, the Gaussian quadrature, let us consider an integral for which the Gauss formulae is given as

$$\int_a^b W(x)f(x)dx = \sum_{i=1}^{n} W_i f(x_i) \tag{7.57}$$

where W_i are a set of weights and x_i are sampling points. We classify various Gauss formulae based on $W(x)$.*

Gauss–Chebyshev quadrature : $W(x) = (1 - x^2)^{-1/2}$, $-1 < x < 1$
Gauss–Legendre quadrature : $W(x) = 1$, $-1 < x < 1$
Gauss–Leguerre quadrature : $W(x) = x^{\alpha}e^{-x}$, $0 < x < \infty$
Gauss–Hermite quadrature : $W(x) = \exp(-x^2)$, $-\infty < x < \infty$

Thus, Gauss–Legendre quadrature formula can be expressed in the form

$$\int_{-1}^{1} f(x)dx = \sum_{i=1}^{n} W_i f(x_i) \tag{7.58}$$

It may be noted that in case, the limits of integration is from a to b, they can be changed to from -1 to 1, using the transformation

$$x = \frac{1}{2}\xi(b-a) + \frac{1}{2}(b+a) \tag{7.59}$$

Now, consider Gauss–Legendre n-point formula as

$$\int_{-1}^{1} f(\xi)d\xi = \sum_{i=1}^{n} W_i f(\xi_i) = W_1 f(\xi_1) + W_2 f(\xi_2) + \cdots + W_n f(\xi_n) \tag{7.60}$$

where W_i are called weights and ξ_i are the sampling points or Gauss points. From Eq. (7.60), we can observe that there are n Gauss points and n weights, thus in all $2n$ arbitrary parameters. The formula (7.60) will be exact, if $f(\xi)$ is a polynomial of degree $(2n - 1)$ or less. Therefore, $f(\xi)$ is of the form

$$f(\xi) = a_0 + a_1\xi + a_2\xi^2 + a_3\xi^3 + \ldots + a_{2n-1}\xi^{2n-1} \tag{7.61}$$

*For more details, the reader may refer to Stroud and Secrest, 1966.

Substituting Eq. (7.61) into Eq. (7.60), its left-hand side becomes

$$\int_{-1}^{1} f(\xi)d\xi = \int_{-1}^{1} (a_0 + a_1\xi + a_2\xi^2 + \cdots + a_{2n-1}\xi^{2n-1})d\xi$$

$$= 2a_0 + \frac{2}{3}a_2 + \frac{2}{5}a_4 + \cdots \tag{7.62a}$$

By choosing, $\xi = \xi_i$, Eq. (7.61) changes to

$$f(\xi_i) = a_0 + a_1\xi_i + a_2\xi_i^2 + a_3\xi_i^3 + \cdots + a_{2n-1}\xi_i^{2n-1}$$

Using this expression, the right-hand side of Eq. (7.60) becomes.

$$\int_{-1}^{1} f(\xi)d\xi = W_1(a_0 + a_1\xi_1 + a_2\xi_1^2 + a_3\xi_1^3 + \cdots + a_{2n-1}\xi_1^{2n-1})$$

$$+ W_2(a_0 + a_1\xi_2 + a_2\xi_2^2 + a_3\xi_2^3 + \cdots + a_{2n-1}\xi_2^{2n-1}) + \cdots$$

$$+ W_n(a_0 + a_1\xi_n + a_2\xi_n^2 + a_3\xi_n^3 + \cdots + a_{2n-1}\xi_n^{2n-1})$$

which on rewriting yields

$$\int_{-1}^{1} f(\xi)d\xi = a_0(W_1 + W_2 + W_3 + \cdots + W_n)$$

$$+ a_1(W_1\xi_1 + W_2\xi_2 + W_3\xi_3 + \cdots + W_n\xi_n)$$

$$+ a_2(W_1\xi_1^2 + W_2\xi_2^2 + W_3\xi_3^2 + \cdots + W_n\xi_n^2)$$

$$+ \cdots + a_{2n-1}(W_1\xi_1^{2n-1} + W_2\xi_2^{2n-1} + W_3\xi_3^{2n-1} + \cdots + W_n\xi_n^{2n-1}) \tag{7.62b}$$

Now, Eqs. (7.62a) and (7.62b) are one and the same and hence by equating the coefficients of a_i in them, we immediately get the following $2n$ equations:

$$W_1 + W_2 + W_3 + \cdots + W_n = 2$$

$$W_1\xi_1 + W_2\xi_2 + W_3\xi_3 + \cdots + W_n\xi_n = 0$$

$$W_1\xi_1^2 + W_2\xi_2^2 + W_3\xi_3^2 + \cdots + W_n\xi_n^2 = \frac{2}{3}$$

$$\vdots \qquad \vdots \qquad \vdots \qquad \vdots$$

$$W_1\xi_1^{2n-1} + W_2\xi_2^{2n-1} + W_3\xi_3^{2n-1} + \cdots + W_n\xi_n^{2n-1} = 0 \tag{7.62c}$$

The solution of these equations gives us $2n$ unknows such as W_i, the weighting coefficients and ξ_i, the sampling points or Gauss points for $i = 1, 2, \ldots, n$ and they are presented in Table 7.1 for values of n equal to 1 through 6.

Table 7.1 Abscissas and Weights for Gauss–Legendre Quadrature

$$\int_{-1}^{1} f(\xi)d\xi = \sum_{i=1}^{n} W_i f(\xi_i)$$

Number of points n	Location, ξ_i	Weights, W_i
1	0.0	2.0
2	$\pm 1/\sqrt{3} = \pm 0.577350$	1.0
3	0.0	8/9 = 0.888889
	± 0.774597	5/9 = 0.555556
4	± 0.339981	0.652145
	± 0.861136	0.347855
5	0.0	0.568889
	± 0.538469	0.478629
	± 0.906180	0.236927
6	± 0.238619	0.467914
	± 0.661209	0.360762
	± 0.932470	0.171325

It may be observed from the table that Gauss points are located symmetrically about the origin or the mid-point of the interval and that symmetrically-placed points have the same weights.

To have a feel for the method, let us consider one-point and two-point approximations in the following examples:

Example 7.14 Evaluate

$$\int_{-1}^{1} f(\xi)d\xi = W_1 f(\xi_1) \tag{1}$$

Solution This is one-point Gausss–Legendre quadrature formula given by (7.60). Since there are only two parameters W_1 and ξ_1, the formula (1) will be exact if $f(\xi)$ is a polynomial of degree one. Thus, we may write $f(\xi) = a_0 + a_1\xi$. Then it is required that

$$\text{Error} = \int_{-1}^{1} (a_0 + a_1\xi)d\xi - W_1 f(\xi_1) = 0$$

That is,

$$2a_0 - W_1(a_0 + a_1\xi_1) = 0$$

Equivalently,

$$a_0(2 - W_1) - W_1 a_1 \xi_1 = 0 \tag{2}$$

Thus, error is zero if and only if

$$W_1 = 2 \text{ and } \xi_1 = 0 \tag{3}$$

Therefore, for any general $f(\xi)$, we have

$$\int_{-1}^{1} f(\xi)\,d\xi = 2f(0) \tag{4}$$

Example 7.15 Evaluate

$$\int_{-1}^{1} f(\xi)\,d\xi = W_1 f(\xi_1) + W_2 f(\xi_2) \tag{1}$$

Solution This is a two-point Gauss–Legendre formula given by Eq. (7.60). since we have four parameters W_1, W_2, ξ_1 and ξ_2 to be computed, the formula (1) will be exact if $f(\xi)$ is a cubic polynomial. Thus, we may write

$$f(\xi) = a_0 + a_1\xi + a_2\xi^2 + a_3\xi^3$$

Then, it is required that

$$\text{Error} = \int_{-1}^{1} (a_0 + a_1\xi + a_2\xi^2 + a_3\xi^3)\,d\xi - [W_1 f(\xi_1) + W_2 f(\xi_2)] = 0 \tag{2}$$

That is,

$$2a_0 + \frac{2}{3}a_2 - W_1(a_0 + a_1\xi_1 + a_2\xi_1^2 + a_3\xi_1^3)$$
$$-W_2\,(a_0 + a_1\xi_2 + a_2\xi_2^2 + a_3\xi_2^3) = 0$$

Equivalently,

$$a_0(2 - W_1 - W_2) - a_1(W_1\xi_1 + W_2\xi_2)$$
$$+ a_2\left(\frac{2}{3} - W_1\xi_1^2 - W_2\xi_2^2\right) - a_3\left(W_1\xi_1^3 + W_2\xi_2^3\right) = 0 \tag{3}$$

The error will be zero if and only if the following equations are satisfied:

$$\left.\begin{aligned} W_1 + W_2 &= 2 \\ W_1\xi_1 + W_2\xi_2 &= 0 \\ W_1\xi_1^2 + W_2\xi_2^2 &= \frac{2}{3} \\ W_1\xi_1^3 + W_2\xi_2^3 &= 0 \end{aligned}\right\} \tag{4}$$

These are non-linear equations, whose solution is found to be

$$W_1 = W_2 = 1 \quad \text{and} \quad \xi_1 = -\xi_2 = \frac{1}{\sqrt{3}} = 0.577350$$

Example 7.16 Evaluate the integral

$$\int_{-1}^{1} (3\xi^2 + \xi^3)\,d\xi$$

using Gauss–Legendre four-point quadrature formula.

Solution Here, the given data is $n = 4$. Therefore,

$$I = \int_{-1}^{1} (3\xi^2 + \xi^3) d\xi = \sum_{n=1}^{4} W_i f(\xi_i) \tag{1}$$

where

$$f(\xi_i) = 3\xi_i^2 + \xi_i^3 = \xi_i^2 (\xi_i + 3)$$

Taking abscissas and weights from Table 7.1, we have

$$\begin{aligned} I = \int_{-1}^{1} (3\xi^2 + \xi^3) d\xi &= 0.347855\,(-0.861136)^2\,(3 - 0.861136) + 0.652145 \\ &\quad (-0.339981)^2\,(3 - 0.339981) + 0.347855 \\ &\quad (0.861136)^2\,(3 + 0.861136) + 0.652145 \\ &\quad (0.339981)^2\,(3 + 0.339981) \\ &= 0.551728 + 0.200511 + 0.995994 + 0.251766 \\ &= 1.999995 \end{aligned}$$

Integrals in two dimensions

One can easily extend the Gauss–Legendre quadrature formula to two-dimensional integrals of the form

$$I = \int_{-1}^{1} \int_{-1}^{1} f(\xi, \eta)\, d\eta\, d\xi \tag{7.63}$$

The above area integrals in the (ξ, η) coordinate system can be numerically evaluated by first evaluating the inner integral, assuming ξ constant and then evaluating the outer integral. Thus, the inner integral gives

$$\int_{-1}^{1} f(\xi, \eta)\, d\eta = \sum_{j=1}^{n} W_j f(\xi, \eta_j) = g(\xi) \tag{7.64}$$

where η_j and W_j are the Gauss–Legendre sampling points and weighting coefficients given in Table 7.1. Now the outer integral becomes

$$\int_{-1}^{1} g(\xi)\, d\xi = \sum_{i=1}^{m} W_i g(\xi_i) \tag{7.65}$$

Substituing the value of $g(\xi)$ from Eq. (7.64), we get

$$\int_{-1}^{1} \int_{-1}^{1} f(\xi, \eta)\, d\eta\, d\xi = \sum_{i=1}^{m} \sum_{j=1}^{n} W_i W_j f(\xi_i\, \eta_j) \tag{7.66}$$

The implementation of this equation is usually carried out as a single sum over $n \times m$ sampling points with $W_i W_j$-type products.

7.10 MULTIPLE INTEGRALS

For evaluating multiple integrals, we extend below the Gaussian quadrature formula as developed in Section 7.9. Thus, consider a formula of the type

$$I = \int_{-1}^{1}\int_{-1}^{1}\int_{-1}^{1} f(x, y, z)\, dx\, dy\, dz$$

$$= \sum_{i=1}^{n}\sum_{j=1}^{n}\sum_{k=1}^{n} W_i W_j W_k\, f(x_i, y_j, z_k) \tag{7.67}$$

where $f(x, y, z)$ is a polynomial, containing a linear combination of terms of the type $x^r\, y^s\, z^t$. We further assume that r, s and t are non-negative integers. Suppose, we assume that

$$f(x, y, z) = x^r\, y^s\, z^t$$

In view of the fact that the limits of integration are constants and the integrand is factorable, we write

$$I = \int_{-1}^{1}\int_{-1}^{1}\int_{-1}^{1} x^r y^s z^t\, dx\, dy\, dz$$

$$= \left[\int_{-1}^{1} x^r\, dx\right]\left[\int_{-1}^{1} y^s\, dy\right]\left[\int_{-1}^{1} z^t\, dz\right]$$

Now, using Gaussian quadrature formula given by Eq. (7.58) we shall be able to write the above equation as

$$I = \left[\sum_{i=1}^{n} W_i x_i^r\right]\left[\sum_{j=1}^{n} W_j y_j^s\right]\left[\sum_{k=1}^{n} W_k z_k^t\right]$$

or

$$I = \sum_{i=1}^{n}\sum_{j=1}^{n}\sum_{k=1}^{n} W_i W_j W_k\, x_i^r\, y_j^s z_k^t \tag{7.68}$$

We shall illustrate this technique through the following example.

Example 7.17 Evaluate

$$I = \int_{0}^{1}\int_{0}^{2}\int_{-1}^{0} x^3 yz^2\, dx\, dy\, dz$$

Using Gaussian quadrature formula and a two-term formulas for x, y and z.

Solution: We know that any finite range $a \le y \le b$ can be mapped onto the range $-1 \le x \le 1$ using the linear transformation $y = (b - a)x/2 + (b + a)/2$. Thus, the given integral becomes

$$I = \int_{-1}^{1}\int_{-1}^{1}\int_{-1}^{1} \frac{1}{4} \times \frac{1}{8}(u-1)^3(v+1)(w+1)^2 \frac{du}{2} dv \frac{dw}{2}$$

where, we have used

$$x = \frac{1}{2}(u-1), \; y = v+1, \; z = \frac{1}{2}(w+1)$$

or

$$I = \frac{1}{128}\int_{-1}^{1}\int_{-1}^{1}\int_{-1}^{1}(u-1)^3(v+1)(w+1)^2 \, du \, dv \, dw \tag{1}$$

We also know from Section 7.9 that the two and three point Gaussian quadrature formulae are

$$\int_{-1}^{1} f(x)\,dx = [(1)f(-0.5774) + (1)f(0.5774)] \tag{2}$$

and

$$\int_{-1}^{1} f(x)\,dx = \left[\frac{5}{9}f(-0.7746) + \frac{8}{9}f(0) + \frac{5}{9}f(0.7746)\right] \tag{3}$$

Thus, using two-term formula (2) for x, y and z or for u, v and w, the integral (1) becomes

$$I = \frac{1}{128}\sum_{i=1}^{2}\sum_{j=1}^{2}\sum_{k=1}^{2} W_i W_j W_k \,(u_i-1)^3(v_j+1)(W_k+1)^2 \tag{4}$$

observe that $W_1 = 1$, $W_2 = 1$. Now writing down all the terms explicitly, we have

$$\begin{aligned} I = \frac{1}{128}[&(1)(1)(1)(-0.5774-1)^3(-0.5774+1)(-0.5774+1)^2 \\ &+ (1)(1)(1)(-0.5774-1)^3(-0.5774+1)(0.5774+1)^2 \\ &+ (1)(1)(1)(-0.5774-1)^3(0.5774+1)(-0.5774+1)^2 \\ &+ (1)(1)(1)(-0.5774-1)^3(0.5774+1)(0.5774+1)^2 \\ &+ (1)(1)(1)(0.5774-1)^3(-0.5774+1)(-0.5774+1)^2 \\ &+ (1)(1)(1)(0.5774-1)^3(-0.5774+1)(0.5774+1)^2 \\ &+ (1)(1)(1)(0.5774-1)^3(0.5774+1)(-0.5774+1)^2 \\ &+ (1)(1)(1)(0.5774-1)^3(0.5774+1)(0.5774+1)^2] \end{aligned}$$

$$= \frac{1}{128}[-0.2963 - 4.1271 - 1.1057 - 15.4048$$

$$-0.005698 - 0.07939 - 0.02127 - 0.2963]$$

or

$$I = \frac{1}{128}[-21.3366] = -0.16669 \tag{5}$$

However, the exact solution is

$$I = \int_0^1 \int_0^2 \int_{-1}^0 x^3 y z^2 \, dx \, dy \, dz$$

$$= -\frac{1}{4} \int_0^1 \int_0^2 y z^2 \, dy \, dz$$

$$= -\frac{1}{2} \int_0^1 z^2 \, dz = -\frac{1}{6} = -0.1667 \qquad (6)$$

Comparing the numerical solution with the exact solution given by Eqs. (5) and (6) we observe that the numerical solution is four decimal accurate.

EXERCISES

7.1 Define shift operator E, average operator μ and differential operator D. Hence, show that

$$D^2 = \frac{1}{h^2}\left(\Delta^2 - \Delta^3 + \frac{11}{12}\Delta^4 - \frac{5}{6}\Delta^5 + \cdots\right)$$

7.2 Derive the formula

$$D^3 = \mu\left[\delta^3 - \left(\frac{1}{12} + \frac{1}{6}\right)\delta^5 + \cdots\right]$$

7.3 Find the first derivative of $f(x)$ at $x = 0.4$ from the following table:

x	0.1	0.2	0.3	0.4
$f(x)$	1.10517	1.22140	1.34986	1.49182

7.4 From the following table of values, estimate $y'(1.10)$ and $y''(1.10)$:

x	1.00	1.05	1.10	1.15	1.20	1.25	1.30
y	1.0000	1.0247	1.0488	1.0724	1.0954	1.1180	1.1402

7.5 A slider in a machine moves along a fixed straight rod. Its distance x cm along the rod is given below for various values of time t (seconds). Find the velocity of the slider and its acceleration when $t = 0.3$ s.

t	0.0	0.1	0.2	0.3	0.4	0.5	0.6
x	3.013	3.162	3.287	3.364	3.395	3.381	3.324

Use both the forward difference formula and the central difference formula to find the velocity and compare the results.

7.6 Given the table of values, estimate $y''(1.3)$:

x	1.3	1.5	1.7	1.9	2.1	2.3
y	2.9648	2.6599	2.3333	1.9922	1.6442	1.2969

7.7 The following divided difference table is for $y = 1/x$. Use it to find $y'(0.75)$ (i) from a quadratic polynomial fit (ii) from a cubic polynomial fit. What degree polynomial fit gives the most accurate value of $y'(0.75)$.

x	$y = 1/x$	1st divided difference	2nd divided difference	3rd divided difference	4th divided difference
0.25	4.0000				
		−8.0000			
0.50	2.0000		10.6664		
		−2.6668		−14.2219	
0.75	1.3333		2.6672		12.0875
		−1.3332		−2.1344	
1.00	1.0000		1.0664		1.4240
		−0.8000		−0.7104	
1.25	0.8000		0.5336		
		−0.5332			
1.50	0.6667				

7.8 Evaluate the integral

$$\int_{1.0}^{1.8} \frac{e^x + e^{-x}}{2} dx$$

using Simpson's 1/3 rule, by taking $h = 0.2$.

7.9 Evaluate the integral

$$\int_0^6 [f(x)]^2 \, dx$$

using Simpson's 1/3 rule, given that

x	0	1	2	3	4	5	6
$f(x)$	1	0	1	4	9	16	25

7.10 Using Simpson's 1/3 rule, Evaluate the integral

$$\int_0^{\pi/2} \frac{dx}{\sin^2 x + \frac{1}{4}\cos^2 x}$$

7.11 Compute the integral,

$$\int_1^2 \frac{dx}{x}$$

using Simpson's 1/3 rule, and also obtain the error bounds by taking $h = 0.25$.

7.12 Evaluate the integral

$$\int_0^1 e^x \, dx$$

using Simpson's 1/3 rule, by dividing the interval of integration into eight equal parts.

7.13 Using the Richardson extrapolation limit, find $y''(0.6)$ of the following tabulated function by applying the formula

$$F(x) = \frac{1}{h^2}[y(x+h) - 2y(x) + y(x-h)]$$

with $h = 0.4, 0.2, 0.1$.

x	0.2	0.4	0.5	0.6	0.7	0.8	1.0
$y(x)$	1.42007	1.88124	2.12815	2.38676	2.65797	2.94289	3.55975

7.14 Apply Romberg's integration method to evaluate

$$\int_{1.0}^{1.8} \cosh\, dx$$

by applying trapezoidal rule with $h = 0.8, 0.4, 0.2, 0.1$.

7.15 Evaluate the integral

$$\int_1^2 \frac{dx}{x}$$

using Romberg's method of integration starting with trapezoidal rule, taking $h = 1, 0.5, 0.25, 0.0125$.

7.16 Evaluate the double integral

$$I = \int_0^1 \int_0^2 \frac{2xy}{(1+x^2)(1+y^2)}\, dy\, dx$$

using Simpson's 1/3-rule, with step length $h = k = 0.25$.

7.17 Compute numerically

$$I = \iint_D \frac{dx\, dy}{x^2 + y^2}$$

where D is the square with corners at (1, 1), (2, 1), (2, 2), (1, 2).

7.18 Evaluate

$$I = \int_0^1 \int_1^2 (x^2 + y^2)\, dx\, dy$$

using Simpson's 1/3 rule.

7.19 The velocity v m/s of a particle at a time t seconds is given in the following table:

t	0	2	4	6	8	10	12
v	4	6	16	34	60	94	136

Find the distance travelled by the particle in 12 s and also the acceleration at $t = 2$ s.

7.20 Find y' and y'' of the function which is tabulated below at the point $x = 2.03$.

x	1.96	1.98	2.00	2.02	2.04
y	0.7825	0.7739	0.7651	0.7563	0.7473

7.21 A body is in the form of a solid of revolution, whose diameter d in cm of its sections at various distances x cm from one end is given in the table below. Compute the volume of the solid using Simpson's 1/3 rule.

x	0.0	2.5	5.0	7.5	10.0	12.5	15.0
d	5.00	5.50	6.00	6.75	6.25	5.50	4.00

7.22 Evaluate the following triple integral

$$I = \int_0^1 \int_{-1}^0 \int_{-1}^1 yz\, e^x \, dxdydz$$

Using Gaussian quadrature formula and taking a three-term formula for x and two-term formula for y and z.

Chapter 8

Ordinary Differential Equations

8.1 INTRODUCTION

Many problems in science and engineering when formulated mathematically are readily expressed in terms of ordinary differential equations with appropriate initial and or boundary conditions. For example, the trajectory of a ballistic missile, the motion of an artificial satellite in its orbit, are governed by ordinary differential equations. Theories concerning electrical networks, bending of beams, stability of aircraft, etc., are modelled by differential equations. To be more precise, the rate of change of any quantity with respect to another can be modelled by an ordinary differential equation.

Closed form solutions may not be possible to obtain, for every modelled problem, while numerical methods exist, to solve them using computers.

In general, a linear or non-linear ordinary differential equation can be written as

$$\frac{d^n y}{dt^n} = f\left(t, y, \frac{dy}{dt}, \ldots, \frac{d^{n-1} y}{dt^{n-1}}\right) \tag{8.1}$$

However, we shall concentrate our discussion on a system of first order differential equations of the form

$$\frac{dy}{dt} = f(t, y) \tag{8.2}$$

with the initial condition $y(t_0) = y_0$, which is called an *initial value problem* (IVP). It is justified, in view of the fact that any higher order ordinary differential equation can be reduced to a system of first order differential equations by substitution. For example, consider a second order differential equation of the form

$$y'' = f(t, y, y')$$

Introducing the substitution $p = y'$, the above equation reduces to a system of two first order differential equations, such as

$$y' = p, \qquad p' = f(t, y, p)$$

The existence and uniqueness of the solution of an initial value problem described in Eq. (8.2) is stated in the following theorem.

Theorem 8.1 Let $f(t, y)$ be real and continuous in the strip R, defined by

$$t \in [t_0, T], \qquad -\infty \le y \le \infty$$

then for any $t \in [t_0, T]$, and for any y_1, y_2, there exists a constant L, satisfying the inequality.

$$|f(t, y_1) - f(t, y_2)| \le L\,|y_1 - y_2|$$

so that $|f_y(t, y)| \le L$, for every $t, y \in R$. Here, L is called *Lipschitz constant.*

Proof: It follows from the mean value theorem for derivatives that there exists a number η such that η lies between y_1 and y_2 in R and

$$f(t, y_2) - f(t, y_1) = (y_2 - y_1)\left.\frac{\partial f}{\partial y}\right|_{y=\eta}$$

Since (t, η) is in R, we have

$$|f(t, y_2) - f(t, y_1)| \le l.u.b\left|\frac{\partial f}{\partial y}\right|(y_2 - y_1)$$

where $l.u.b$ denotes the least upper bound for all points (t, y) in R. That is,

$$|f(t, y_2) - f(t, y_1)| \le L\,|y_2 - y_1|$$

To illustrate the Lipschitz condition, let us consider $f(t, y) = 2\sqrt{y}$, $y(1) = 0$. If the Lipschitz condition were to satisfy, then L would have to satisfy the inequality

$$|2\sqrt{y_1} - 2\sqrt{y_2}| \le L\,|y_1 - y_2|$$

But

$$\underset{y_1 \to y_2}{Lt}\ \frac{2\sqrt{y_1} - 2\sqrt{y_2}}{y_1 - y_2} = \left.\frac{d}{dy}(2\sqrt{y})\right|_{y=y_2} = \frac{1}{\sqrt{y_2}}$$

Hence, $f(t, y)$ does not satisfy the Lipschitz condition and the function $f_y = \dfrac{\partial f}{\partial y} = \dfrac{1}{\sqrt{y}}$ is unbounded in any rectangular strip containing the point $(1, 0)$. Let us consider another example say $f(t, y) = y + 1$, $y(1) = 0$. Here, we have $f_y = \dfrac{\partial f}{\partial y} = 1$ and application of theorem 8.1, yields, $L = 1$. The Lipschitz condition requires

$$|(y_2 + 1) - (y_1 + 1)| \le L\,|y_2 - y_1|$$

It is trivial to see that $L = 1$ is the smallest value of L for which the condition holds.

If the above conditions are satisfied, then for any y_0, the IVP (8.2) has a unique solution $y(t)$, for $t \in [t_0, T]$. In fact, we assume the existence and uniqueness of a solution to the IVP described by Eq. (8.2). The function f may be linear or non-linear. We also assume that the function $f(t, y)$ is sufficiently differentiable with respect to either t or y. In the following sections, we shall discuss various numerical methods for finding the solution of an initial value problem.

8.2 TAYLOR'S SERIES METHOD

Consider an initial value problem described by Eq. (8.2), that is

$$\frac{dy}{dt} = f(t, y), \qquad y(t_0) = y_0$$

Here, we assume that $f(t, y)$ is sufficiently differentiable with respect to t and y. If $y(t)$ is the exact solution of (8.2), we can expand $y(t)$ by Taylor's series about the point $t = t_0$ and obtain

$$y(t) = y(t_0) + (t - t_0)\, y'(t_0) + \frac{(t - t_0)^2}{2!} y''(t_0) + \frac{(t - t_0)^3}{3!} y'''(t_0) + \frac{(t - t_0)^4}{4!} y^{IV}(t_0) + \cdots \tag{8.3}$$

Since, the solution is not known, the derivatives in the above expansion are not known explicitly. However, f is assumed to be sufficiently differentiable, and therefore, the derivatives can be obtained directly from the given differential equation itself. Noting that f is an implicit function of y, we have

$$\left.\begin{aligned}
y' &= f(t, y) \\
y'' &= \frac{\partial f}{\partial t} + \frac{\partial f}{\partial y}\frac{dy}{dt} = f_t + ff_y \\
y''' &= f_{tt} + ff_{ty} + f(f_{ty} + ff_{yy}) + f_y(f_t + ff_y) \\
&= f_{tt} + 2ff_{ty} + f^2 f_{yy} + f_y(f_t + ff_y) \\
y^{IV} &= f_{ttt} + 3ff_{tty} + 3f^2 f_{tyy} + f_y\,(f_{tt} + 2ff_{ty} + f^2 f_{yy}) \\
&\quad + 3(f_t + ff_y)\,(f_{ty} + ff_{yy}) + f_y^2\,(f_t + ff_y)
\end{aligned}\right\} \tag{8.4}$$

and so on. Continuing in this manner, we can express any derivative of y in terms of $f(t, y)$ and its partial derivatives. We shall demonstrate this method through an example.

Example 8.1 Using Taylor's series method, find the solution of the initial value problem

$$\frac{dy}{dt} = t + y, \qquad y(1) = 0$$

at $t = 1.2$, with $h = 0.1$ and compare the result with the closed form solution.

Solution Let us compute the first few derivatives from the given differential equation as follows:

$$y' = t + y, \qquad y'' = 1 + y', \qquad y''' = y'', \qquad y^{IV} = y''', \qquad y^{V} = y^{IV} \tag{1}$$

Prescribing the initial condition, that is, at $t_0 = 1$, $y_0 = y(t_0) = 0$, we have

$$y_0' = 1, \qquad y_0'' = 2, \qquad y_0''' = y_0^{IV} = y_0^{V} = 2$$

Now, using Taylor's series method, we have

$$y(t) = y_0 + (t - t_0)y_0' + \frac{(t-t_0)^2}{2}y_0'' + \frac{(t-t_0)^3}{6}y_0'''$$

$$+ \frac{(t-t_0)^4}{24}y_0^{\text{IV}} + \frac{(t-t_0)^5}{120}y_0^{\text{V}} + \cdots \qquad (2)$$

Substituting the above values of the derivatives, and the initial condition, into (2), we obtain

$$y(1.1) = 0 + (0.1)(1) + \frac{0.01}{2}(2) + \frac{0.001}{6}(2) + \frac{0.0001}{24}(2) + \frac{0.00001}{120}(2) + \cdots$$

$$= 0.1 + 0.01 + \frac{0.001}{3} + \frac{0.0001}{12} + \frac{0.00001}{60} + \cdots$$

$$= 0.1 + 0.01 + 0.000333 + 0.0000083 + 0.0000001 + \cdots$$

$$\cong 0.1103414.$$

Therefore,

$$y(1.1) = y_1 = 0.1103414 \cong 0.1103$$

Taking $y_1 = 0.1103$ at $t = 1.1$, the values of the derivatives as computed from Eq. (1) are

$$y_1' = 1.1 + 0.1103 = 1.2103$$

$$y_1'' = 1 + 1.2103 = 2.2103$$

$$y_1''' = y_1^{\text{IV}} = y_1^{\text{V}} = 2.2103$$

Substituting the value of y_1 and its derivatives into Taylor's series expansion (2) we get, after retaining terms up to fifth derivative only

$$y(1.2) = y_1 + (t - t_1)y_1' + \frac{(t-t_1)^2}{2}y_1'' + \frac{(t-t_1)^3}{6}y_1''' + \frac{(t-t_1)^4}{24}y_1^{\text{IV}} + \frac{(t-t_1)^5}{120}y_1^{\text{V}}$$

$$= 0.1103 + 0.12103 + 0.0110515 + 0.0003683 + 0.000184 + 0.0000003$$

$$= 0.2429341$$

Hence,

$$y(1.2) = 0.2429341 \cong 0.2429 \qquad (3)$$

To obtain the closed form solution, we rewrite the given IVP as

$$\frac{dy}{dt} - y = t \quad \text{or} \quad d(ye^{-t}) = te^{-t}$$

On integration, we get

$$y = -e^t(te^{-t} + e^{-t}) + ce^t = ce^t - t - 1$$

Using the initial condition, we get

$$0 = ce - 2 \quad \text{or} \quad c = \frac{2}{e}$$

Therefore, the closed form solution is

$$y = -t - 1 + 2e^{t-1}$$

when $t = 1.2$, the closed form solution becomes

$$y(1.2) = -1.2 - 1 + 2(1.2214028) = -2.2 + 2.4428056 = 0.2428 \tag{4}$$

Comparing the results (3) and (4), obtained numerically and in closed form, we observe that they agree up to three decimals.

8.3 EULER METHOD

Euler method is one of the oldest numerical methods used for integrating the ordinary differential equations. Though this method is not used in practice, its understanding will help us to gain insight into the nature of predictor–corrector methods, which we shall develop in later sections.

Consider the differential equation of first order

$$\frac{dy}{dt} = f(t, y) \tag{8.5}$$

with the initial condition $y(t_0) = y_0$. The integral of Eq. (8.5) is a curve in, ty-plane. Here, we find successively $y_1, y_2, \ldots, y_m$; where y_m is the value of y at $t = t_m = t_0 + mh$, $m = 1, 2, \ldots$ and h being small. In this method, we use the property that in a small interval, a curve is nearly a straight line. Thus at (t_0, y_0), we approximate the curve by a tangent at that point. Therefore,

$$\left(\frac{dy}{dt}\right)_{(t_0, y_0)} = \frac{y - y_0}{t - t_0} = f(t_0, y_0)$$

That is,

$$y = y_0 + (t - t_0)\, f(t_0, y_0)$$

Hence, the value of y corresponding to $t = t_1$ is given by

$$y_1 = y_0 + (t_1 - t_0)\, f(t_0, y_0) \tag{8.6}$$

Similarly approximating the solution curve in the next interval (t_1, t_2) by a line through (t_1, y_1) having its slope $f(t_1, y_1)$, we obtain (see Fig. 8.1)

$$y_2 = y_1 + hf(t_1, y_1) \tag{8.7}$$

Thus, we obtain in general, the solution of the given differential equation in the form of a recurrence relation

$$y_{m+1} = y_m + hf(t_m, y_m) \tag{8.8}$$

Geometrically, this method has a very simple meaning. The desired function curve is approximated by a polygon train, where the direction of each part is determined by the value of the function $f(t, y)$ at its starting point. This idea is illustrated in Fig. 8.1.

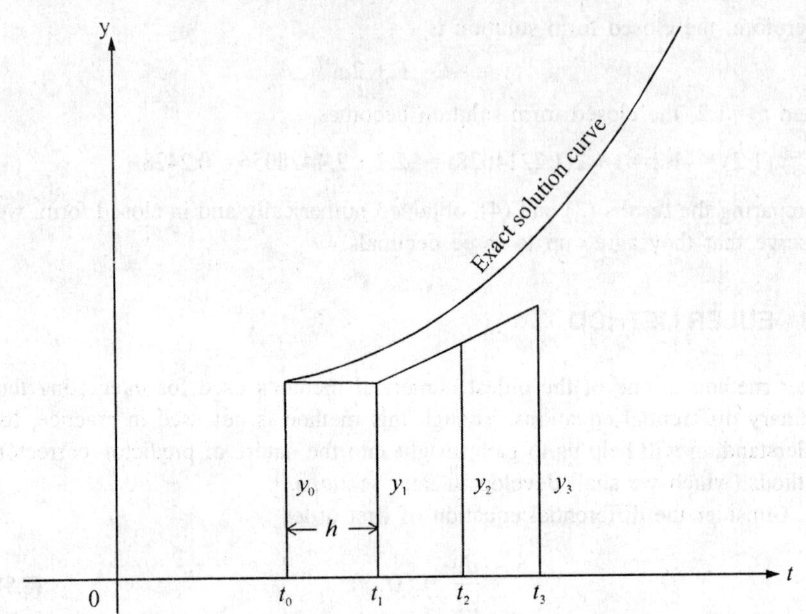

Fig. 8.1 Geometrical meaning of Euler's method.

For illustration, an example follows.

Example 8.2 Given

$$\frac{dy}{dt} = \frac{y - t}{y + t}$$

with the initial condition $y = 1$ at $t = 0$. Find y approximately at $t = 0.1$, in five steps, using Euler's method.

Solution Since the number of steps involved are five, we shall march in steps of $0.1/5 = 0.02$. Therefore, taking step size $h = 0.02$, we shall compute the value of y at $t = 0.02$, 0.04, 0.06, 0.08 and 0.1. Thus,

$$y_1 = y_0 + hf(t_0, y_0), \qquad \text{where } y_0 = 1,\ t_0 = 0$$

Therefore,

$$y_1 = 1 + 0.02\,\frac{1-0}{1+0} = 1.02$$

Similarly

$$y_2 = y_1 + hf(t_1, y_1) = 1.02 + 0.02\,\frac{1.02 - 0.02}{1.02 + 0.02} = 1.0392$$

$$y_3 = y_2 + hf(t_2, y_2) = 1.0392 + 0.02\,\frac{1.0392 - 0.04}{1.0392 + 0.04} = 1.0577$$

$$y_4 = y_3 + hf(t_3, y_3) = 1.0577 + 0.02\,\frac{1.0577 - 0.06}{1.0577 + 0.06} = 1.0756$$

$$y_5 = y_4 + hf(t_4, y_4) = 1.0756 + 0.02\,\frac{1.0756 - 0.08}{1.0756 + 0.08} = 1.0988$$

Hence, the value of y corresponding to $t = 0.1$ is 1.0988.

8.3.1 Modified Euler's Method

The modified Euler's method gives greater improvement in accuracy over the original Euler's method. Here, the core idea is that we use a line through (t_0, y_0) whose slope is the average of the slopes at (t_0, y_0) and $(t_1, y_1^{(1)})$, where $y_1^{(1)} = y_0 + hf(t_0, y_0)$ is the value of y at $t = t_1$ as obtained in Euler's method, which approximates the curve in the interval (t_0, t_1).

Geometrically, from Fig. 8.2, if L_1 is the tangent at (t_0, y_0), L_2 is the line through $(t_1, y_1^{(1)})$ of slope $f(t_1, y_1^{(1)})$ and $\bar{L}$ is the line through $(t_1, y_1^{(1)})$ but with

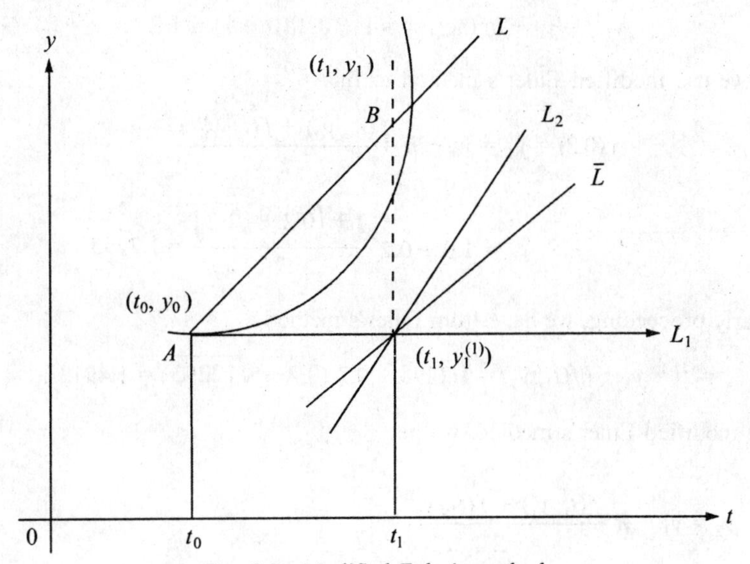

Fig. 8.2 Modified Euler's method.

a slope equal to the average of $f(t_0, y_0)$ and $f(t_1, y_1^{(1)})$, then the line L through (t_0, y_0) and parallel to $\bar{L}$ is used to approximate the curve in the interval (t_0, t_1).

Thus, the ordinate of the point B will give the value of y_1. Now, the equation of the line AL is given by

$$y_1 = y_0 + \left[\frac{f(t_0, y_0) + f(t_1, y_1^{(1)})}{2}\right](t_1 - t_0)$$

$$= y_0 + h\left[\frac{f(t_0, y_0) + f(t_1, y_1^{(1)})}{2}\right] \tag{8.9}$$

Similarly proceeding, we arrive at the recurrence relation

$$y_{m+1} = y_m + h\left[\frac{f(t_m, y_m) + f(t_{m+1}, y_{m+1}^{(1)})}{2}\right] \tag{8.10}$$

This is the *modified Euler's method.*

Example 8.3 Using modified Euler's method, obtain the solution of the differential equation

$$\frac{dy}{dt} = t + \sqrt{y} = f(t, y)$$

with the initial condition $y_0 = 1$ at $t_0 = 0$ for the range $0 \le t \le 0.6$ in steps of 0.2.

Solution At first, we use Euler's method to get

$$y_1^{(1)} = y_0 + hf(t_0, y_0) = 1 + 0.2\,(0 + 1) = 1.2$$

then, we use modified Euler's method to find

$$y(0.2) = y_1 = y_0 + h\,\frac{f(t_0, y_0) + f(t_1, y_1^{(1)})}{2}$$

$$= 1.0 + 0.2\,\frac{1 + \left(0.2 + \sqrt{1.2}\right)}{2} = 1.2295$$

Similarly proceeding, we have from Euler's method

$$y_2^{(1)} = y_1 + hf(t_1, y_1) = 1.2295 + 0.2\,(0.2 + \sqrt{1.2295}) = 1.4913$$

Using modified Euler's method, we get

$$y_2 = y_1 + h\,\frac{f(t_1, y_1) + f(t_2, y_2^{(1)})}{2}$$

$$= 1.2295 + 0.2\,\frac{\left(0.2 + \sqrt{1.2295}\right) + \left(0.4 + \sqrt{1.4913}\right)}{2} = 1.5225$$

Finally

$$y_3^{(1)} = y_2 + hf\,(t_2, y_2) = 1.5225 + 0.2\,\left(0.4 + \sqrt{1.5225}\right) = 1.8493$$

Now, modified Euler's method gives

$$y\,(0.6) = y_3 = y_2 + h\,\frac{f(t_2, y_2) + f(t_3, y_3^{(1)})}{2}$$

$$= 1.5225 + 0.1[(0.4 + \sqrt{1.5225}) + (0.6 + \sqrt{1.8493})] = 1.8819$$

Hence, the solution to the given problem is given by

t	0.2	0.4	0.6
y	1.2295	1.5225	1.8819

8.4 RUNGE–KUTTA METHODS

Computationally, most efficient methods in terms of accuracy were developed by two German mathematicians, Runge and Kutta. These methods are well- known as Runge–Kutta methods. They are distinguished by their orders in the sense that they agree with Taylor's series solution up to terms of h^r, where r is the order of the method. These methods do not demand prior computation of higher derivatives of $y(t)$ as in Taylor's series method. Fourth-order Runge–Kutta methods are widely used for finding the numerical solutions of linear or non-linear ordinary differential equations, the development of which is complicated algebraically. Hence, we convey the basic idea of these methods by developing the second-order Runge–Kutta method which we shall refer hereafter as R–K method.

We may recall that the modified Euler's method described in Eq. (8.10), which can be viewed as

$$y_{n+1} = y_n + h \text{ (average of slopes)} \tag{8.11}$$

This, in fact, is the basic idea of R–K method. Here, we find the slope not only at t_n, but also at several other interior points, and take the weighted average of these slopes and add to y_n to get y_{n+1}. Now, we shall derive the second order R–K method as follows:

Consider the initial value problem described as

$$\frac{dy}{dt} = f(t, y), \qquad y(t_n) = y_n \tag{8.12}$$

we also define

$$k_1 = hf(t_n, y_n), \qquad k_2 = hf(t_n + \alpha h, y_n + \beta k_1)$$

and take the weighted average of k_1 and k_2 and add to y_n to get y_{n+1}. That is, we seek a formula of the form

$$y_{n+1} = y_n + W_1 k_1 + W_2 k_2 \tag{8.13}$$

where α, β, W_1 and W_2 are constants to be determined so that Eq. (8.13) will agree with the Taylor's series expansion as high an order as possible. Thus, using Taylor's series expansion, we have

$$y(t_{n+1}) = y(t_n) + hy'(t_n) + \frac{h^2}{2}y''(t_n) + \frac{h^3}{6}y'''(t_n) + \cdots$$

Rewriting the derivatives of y in terms of f of Eq. (8.12), we get

$$y(t_{n+1}) = y_n + hf(t_n, y_n) + \frac{h^2}{2}(f_t + ff_y)$$

$$+ \frac{h^3}{6}[f_{tt} + 2ff_{ty} + f^2 f_{yy} + f_y(f_t + ff_y)] + O(h^4) \tag{8.14}$$

Here, all derivatives are evaluated at (t_n, y_n).

Next, we shall rewrite Eq. (8.13) after inserting the expressions for k_1 and k_2 from Eq. (8.12) as

$$y_{n+1} = y_n + W_1 hf(t_n, y_n) + W_2 hf(t_n + \alpha h, y_n + \beta k_1)$$

Now, using Taylor's series expansion of two variables, we obtain

$$y_{n+1} = y_n + W_1 hf(t_n, y_n) + W_2 h\Bigg[f(t_n, y_n) + (\alpha h f_t + \beta k_1 f_y)$$

$$+\left(\frac{\alpha^2 h^2}{2} f_{tt} + \alpha h \beta k_1 f_{ty} + \frac{\beta^2 k_1^2}{2} f_{yy}\right) + O(h^3)\Bigg] \tag{8.15}$$

Here again, all derivatives are computed at (t_n, y_n). On inserting the expression for k_1, the above equation becomes

$$y_{n+1} = y_n + (W_1 + W_2)hf + W_2 h\Bigg[(\alpha h f_t + \beta h f f_y)$$

$$+\left(\frac{\alpha^2 h^2}{2} f_{tt} + \alpha\beta h^2 f f_{ty} + \frac{\beta^2 h^2}{2} f^2 f_{yy}\right) + O(h^3)\Bigg]$$

On rearranging in the increasing powers of h, we get

$$y_{n+1} = y_n + (W_1 + W_2)hf + W_2 h^2(\alpha f_t + \beta f f_y)$$

$$+ W_2 h^3 \left(\frac{\alpha^2}{2} f_{tt} + \alpha\beta f f_{ty} + \frac{\beta^2 f^2 f_{yy}}{2}\right) + O(h^4) \tag{8.16}$$

Now, equating coefficients of h and h^2 in Eqs. (8.14) and (8.16), we obtain

$$W_1 + W_2 = 1, \qquad W_2(\alpha f_t + \beta f f_y) = \frac{f_t + f f_y}{2}$$

implying

$$W_1 + W_2 = 1, \qquad W_2 \alpha = W_2 \beta = \frac{1}{2} \tag{8.17}$$

Thus, we have three equations in four unknowns, and therefore, we can choose one value arbitrarily. Solving Eqs. (8.17), we have

$$W_1 = 1 - W_2, \qquad \alpha = \frac{1}{2W_2}, \qquad \beta = \frac{1}{2W_2} \tag{8.18}$$

where W_2 is arbitrary and various values can be assigned to it. We shall consider only two cases which are popular.

Case I: If we choose $W_2 = 1/3$, then $W_1 = 2/3$ and $\alpha = \beta = 3/2$. Equation (8.13) gives

$$y_{n+1} = y_n + \frac{1}{3}(2k_1 + k_2) \tag{8.19}$$

where

$$k_1 = hf(t_n, y_n), \qquad k_2 = hf\left(t_n + \frac{3}{2}h, y_n + \frac{3}{2}k_1\right)$$

Case II: If we consider $W_2 = 1/2$, then $W_1 = 1/2$ and $\alpha = \beta = 1$. Now, Eq. (8.13) gives

$$y_{n+1} = y_n + \frac{k_1 + k_2}{2} \tag{8.20}$$

where,

$$k_1 = hf(t_n, y_n), \qquad k_2 = hf(t_n + h, y_n + k_1)$$

In fact, we can recognize that Eq. (8.20) is the modified Euler's method, and is therefore, a special case of a second order Runge–Kutta method. Equations (8.19) and (8.20) are known as *second-order R–K methods*, since they agree with Taylor's series solution up to terms of h^2.

Defining the local truncation error, TE, as the difference between the exact solution $y(t_{n+1})$ at $t = t_{n+1}$ and the numerical solution y_{n+1}, obtained using the second order R–K method, we have

$$\text{TE} = y(t_{n+1}) - y_{n+1} \tag{8.21}$$

Now, substituting

$$W_2 = \frac{1}{2\alpha}, \qquad W_1 = 1 - \frac{1}{2\alpha}, \qquad \beta = \alpha,$$

into Eq. (8.16), we get

$$y_{n+1} = y_n + hf_n + \frac{h^2}{2}(f_t + ff_y)_{t=t_n} + \frac{h^3\alpha}{4}(f_{tt} + 2ff_{ty} + f^2 f_{yy})_{t=t_n} + \cdots \tag{8.22}$$

Then, using Eqs. (8.14) and (8.22) into Eq. (8.21), we obtain

$$\text{TE} = h^3\left[\left(\frac{1}{6} - \frac{\alpha}{4}\right)(f_{tt} + 2ff_{ty} + f^2 f_{yy}) + \frac{1}{6}f_y(f_t + ff_y)\right] \tag{8.23}$$

Using Eq. (8.4), the above expression further simplifies to

$$\text{TE} = h^3\left[\left(\frac{1}{6} - \frac{\alpha}{4}\right)(y''' - f_y y') + \frac{1}{6}f_y y'\right]$$

Therefore, the expression for local truncation error is given by

$$\text{TE} = h^3\left[\left(\frac{1}{6} - \frac{\alpha}{4}\right)y''' + \frac{\alpha}{4}f_y y'\right] \tag{8.24}$$

It can be easily verified from Eq. (8.24) that the magnitude of the TE in case I is less than that of case II. Therefore, the second order R–K method described by Eq. (8.19) is adopted in practice.

Following similar procedure, Runge–Kutta formulae of any order can be obtained. However, their derivations become exceedingly lengthy and complicated. Amongst them, the most popular and commonly used in practice is the R–K method of fourth-order, which agrees with the Taylor series method up to terms of $O(h^4)$. This well-known fourth-order R–K method is described in the following steps.

$$y_{n+1} = y_n + \frac{1}{6}(k_1 + 2k_2 + 2k_3 + k_4) \tag{8.25}$$

where

$$k_1 = hf(t_n, y_n)$$

$$k_2 = hf\left(t_n + \frac{h}{2}, y_n + \frac{k_1}{2}\right)$$

$$k_3 = hf\left(t_n + \frac{h}{2}, y_n + \frac{k_2}{2}\right)$$

$$k_4 = hf(t_n + h, y_n + k_3)$$

We observe that the second-order Runge–Kutta method described in Eq. (8.19) and (8.20) requires the evaluation of the function twice for each complete step of integration. Similarly, fourth-order Runge–Kutta method requires the evaluation of the function four times. The discussion on optimal order R–K method is beyond the scope of this book. Now, we shall illustrate the Runge–Kutta methods of second- and fourth-order through the following examples:

Example 8.4 Use the following second order Runge–Kutta method described by

$$y_{n+1} = y_n + \frac{1}{3}(2k_1 + k_2)$$

where

$$k_1 = hf(x_n, y_n) \quad \text{and} \quad k_2 = hf\left(x_n + \frac{3}{2}h, y_n + \frac{3}{2}k_1\right)$$

and find the numerical solution of the initial value problem described as

$$\frac{dy}{dx} = \frac{y + x}{y - x}, \quad y(0) = 1$$

at $x = 0.4$ and taking $h = 0.2$.

Solution In the present problem

$$f(x, y) = \frac{y + x}{y - x}, \quad h = 0.2, \quad x_0 = 0, \quad y_0 = 1$$

We calculate

$$k_1 = hf(x_0, y_0) = 0.2\,\frac{1 + 0}{1 - 0} = 0.2$$

$$k_2 = hf[x_0 + 0.3, y_0 + (1.5)(0.2)] = hf(0.3, 1.3) = 0.2\,\frac{1.3 + 0.3}{1.3 - 0.3} = 0.32$$

Now, using the given R–K method, we get

$$y(0.2) = y_1 = 1 + \frac{1}{3}(0.4 + 0.32) = 1.24$$

Now, taking $x_1 = 0.2$, $y_1 = 1.24$, we calculate

$$k_1 = hf(x_1, y_1) = 0.2\,\frac{1.24 + 0.2}{1.24 - 0.2} = 0.2769$$

$$k_2 = hf\left(x_1 + \frac{3}{2}h,\, y_1 + \frac{3}{2}k_1\right) = hf(0.5, 1.6554) = 0.2\,\frac{1.6554 + 0.5}{1.6554 - 0.5} = 0.3731$$

Again using the given R–K method, we obtain

$$y(0.4) = y_2 = 1.24 + \frac{1}{3}[2(0.2769) + 0.3731] = 1.54897$$

Example 8.5 Solve the following differential equation

$$\frac{dy}{dt} = t + y$$

with the initial condition $y(0) = 1$, using fourth-order Runge–Kutta method from

$$t = 0 \text{ to } t = 0.4 \qquad \text{taking } h = 0.1$$

Solution The fourth-order Runge–Kutta method is described as

$$y_{n+1} = y_n + \frac{1}{6}(k_1 + 2k_2 + 2k_3 + k_4) \tag{1}$$

where

$$k_1 = hf(t_n, y_n)$$

$$k_2 = hf\left(t_n + \frac{h}{2},\, y_n + \frac{k_1}{2}\right)$$

$$k_3 = hf\left(t_n + \frac{h}{2},\, y_n + \frac{k_2}{2}\right)$$

$$k_4 = hf(t_n + h,\, y_n + k_3)$$

In this problem, $f(t, y) = t + y$, $h = 0.1$, $t_0 = 0$, $y_0 = 1$. As a first step, we calculate

$k_1 = hf(t_0, y_0) = 0.1\ (1) = 0.1$

$k_2 = hf(t_0 + 0.05, y_0 + 0.05) = hf(0.05, 1.05) = 0.1\ [0.05 + 1.05] = 0.11$

$k_3 = hf(t_0 + 0.05, y_0 + 0.055) = 0.1\ (0.05 + 1.055) = 0.1105$

$k_4 = 0.1\ (0.1 + 1.1105) = 0.12105$

Now, we compute from, Eq. (1) that

$$y_1 = y_0 + \frac{1}{6}(k_1 + 2k_2 + 2k_3 + k_4)$$

$$= 1 + \frac{1}{6}(0.1 + 0.22 + 0.2210 + 0.12105)$$

$$= 1.11034$$

Therefore, $y(0.1) = y_1 = 1.11034$. In the second step, we have to find $y_2 = y(0.2)$. Knowing that $t_1 = 0.1$, $y_1 = 1.11034$, we compute

$$k_1 = hf(t_1, y_1) = 0.1\,(0.1+1.11034) = 0.121034$$

$$k_2 = hf\left(t_1 + \frac{h}{2}, y_1 + \frac{k_1}{2}\right) = 0.1[0.15 + (1.11034 + 0.060517)] = 0.13208$$

$$k_3 = hf\left(t_1 + \frac{h}{2}, y_1 + \frac{k_2}{2}\right) = 0.1[0.15 + (1.11034 + 0.06604)] = 0.132638$$

$$k_4 = hf(t_1 + h, y_1 + k_3) = 0.1[0.2 + (1.11034 + 0.132638)] = 0.1442978$$

and then from Eq. (1), we see that

$$y_2 = 1.11034 + \frac{1}{6}[0.121034 + 2(0.13208) + 2(0.132638) + 0.1442978] = 1.2428$$

Similarly by calculating,

$$k_1 = hf(t_2, y_2) = 0.1\,[0.2 + 1.2428] = 0.14428$$

$$k_2 = hf\left(t_2 + \frac{h}{2}, y_2 + \frac{k_1}{2}\right) = 0.1[0.25 + (1.2428 + 0.07214)] = 0.156494$$

$$k_3 = hf\left(t_2 + \frac{h}{2}, y_2 + \frac{k_2}{2}\right) = 0.1[0.25 + (1.2428 + 0.078247)] = 0.1571047$$

$$k_4 = hf(t_2 + h, y_2 + k_3) = 0.1[0.3 + (1.2428 + 0.1571047)] = 0.16999047$$

Using Eq. (1), we compute

$$y(0.3) = y_3 = y_2 + \frac{1}{6}(k_1 + 2k_2 + 2k_3 + k_4) = 1.399711$$

Finally, we calculate

$$k_1 = hf(t_3, y_3) = 0.1[0.3 + 1.3997] = 0.16997$$

$$k_2 = hf\left(t_3 + \frac{h}{2}, y_3 + \frac{k_1}{2}\right) = 0.1[0.35 + (1.3997 + 0.084985)] = 0.1834685$$

$$k_3 = hf\left(t_3 + \frac{h}{2}, y_3 + \frac{k_2}{2}\right) = 0.1[0.35 + (1.3997 + 0.091734)] = 0.1841434$$

$$k_4 = hf(t_3 + h, y_3 + k_3) = 0.1[0.4 + (1.3997 + 0.1841434)] = 0.19838434$$

Using them in Eq. (1), we get

$$y(0.4) = y_4 = y_3 + \frac{1}{6}(k_1 + 2k_2 + 2k_3 + k_4) = 1.58363$$

which is the required result.

Runge–Kutta method for a system of equations

The fourth-order Runge–Kutta method can be extended to solve numerically the higher-order ordinary differential equations—linear or non-linear. For illustration, let us consider a second order ordinary differential equation of the form

$$\frac{d^2y}{dt^2} = f\left(t, y, \frac{dy}{dt}\right) \tag{8.26}$$

using the substitution

$$\frac{dy}{dt} = p \tag{8.27}$$

Equation (8.26) can be reduced to two first-order simultaneous differential equations as given below.

$$\frac{dy}{dt} = p = f_1(t,y,p), \qquad \frac{dp}{dt} = f_2(t,y,p) \tag{8.28}$$

Now, we can directly write down the Runge–Kutta fourth-order formulae for solving the system (8.28).

Let the initial conditions of the above system be given by

$$y(t_n) = y_n, \qquad y'(t_n) = p\,(t_n) = p_n \tag{8.29}$$

Then, we define

$$\left.\begin{aligned}
k_1 &= hf_1\,(t_n, y_n, p_n), & l_1 &= hf_2\,(t_n, y_n, p_n)\\
k_2 &= hf_1\left(t_n + \frac{h}{2}, y_n + \frac{k_1}{2}, p_n + \frac{l_1}{2}\right) & l_2 &= hf_2\left(t_n + \frac{h}{2}, y_n + \frac{k_1}{2}, p_n + \frac{l_1}{2}\right)\\
k_3 &= hf_1\left(t_n + \frac{h}{2}, y_n + \frac{k_2}{2}, p_n + \frac{l_2}{2}\right), & l_3 &= hf_2\left(t_n + \frac{h}{2}, y_n + \frac{k_2}{2}, p_n + \frac{l_2}{2}\right)\\
k_4 &= hf_1\,(t_n + h, y_n + k_3, p_n + l_3), & l_4 &= hf_2\,(t_n + h, y_n + k_3, p_n + l_3)
\end{aligned}\right\} \tag{8.30}$$

Now, using the initial conditions y_n, p_n and fourth-order R–K formula, we compute

$$\left.\begin{aligned}
y_{n+1} &= y_n + \frac{1}{6}(k_1 + 2k_2 + 2k_3 + k_4)\\
\text{and}\qquad p_{n+1} &= p_n + \frac{1}{6}(l_1 + 2l_2 + 2l_3 + l_4)
\end{aligned}\right\} \tag{8.31}$$

This method can be extended on similar lines to solve system of n first order differential equations. Here follows an example for illustration.

Example 8.6 Solve the following van der Pol's equation

$$y'' - (0.1)\,(1 - y^2)\,y' + y = 0$$

using fourth order Runge–Kutta method for $x = 0.2$, with the initial values $y\,(0) = 1$, $y'(0) = 0$.

Solution Let

$$\frac{dy}{dx} = p = f_1(x,y,p)$$

Then

$$\frac{dp}{dx} = (0.1)(1 - y^2)p - y = f_2(x, y, p)$$

Thus, the given van der Pol's equation reduced to two first-order equations.

In the present problem, we are given that $x_0 = 0$, $y_0 = 1$, $p_0 = y_0' = 0$. Taking $h = 0.2$, we compute

$$k_1 = hf_1(x_0, y_0, p_0) = 0.2(0.0) = 0.0$$

$$l_1 = hf_2(x_0, y_0, p_0) = 0.2(0.0 - 1) = -0.2$$

$$k_2 = hf_1\left(x_0 + \frac{h}{2}, y_0 + \frac{k_1}{2}, p_0 + \frac{l_1}{2}\right) = hf_1(0.1, 1.0, -0.1) = -0.02$$

$$l_2 = hf_2\left(x_0 + \frac{h}{2}, y_0 + \frac{k_1}{2}, p_0 + \frac{l_1}{2}\right) = hf_2(0.1, 1.0, -0.1) = 0.2(0-1)$$

$$= -0.2$$

$$k_3 = hf_1\left(x_0 + \frac{h}{2}, y_0 + \frac{k_2}{2}, p_0 + \frac{l_2}{2}\right) = hf_1(0.1, 0.99, -0.1) = 0.2(-0.1)$$

$$= -0.02$$

$$l_3 = hf_2\left(x_0 + \frac{h}{2}, y_0 + \frac{k_2}{2}, p_0 + \frac{l_2}{2}\right) = hf_2(0.1, 0.99, -0.1)$$

$$= 0.2[(0.1)(0.0199)(-0.1) - 0.99] = -0.1980$$

$$k_4 = hf_1(x_0 + h, y_0 + k_3, p_0 + l_3), = hf_1(0.2, 0.98, -0.1980) = -0.0396$$

$$l_4 = hf_2(x_0 + h, y_0 + k_3, p_0 + l_3) = hf_2(0.2, 0.98, -0.1980)$$

$$= 0.2[(0.1)(1 - 0.9604)(-0.1980) - 0.98] = -0.19616.$$

Now, $y(0.2) = y_1$ is given by

$$y(0.2) = y_1 = y_0 + \frac{1}{6}[k_1 + 2k_2 + 2k_3 + k_4]$$

$$= 1 + \frac{1}{6}[0.0 + 2(-0.02) + 2(-0.02) + (-0.0396)]$$

$$= 1 - 0.019935 = 0.9801$$

and

$$y'(0.2) = p_1 = p_0 + \frac{1}{6}(l_1 + 2l_2 + 2l_3 + l_4)$$

$$= 0 + \frac{1}{6}[-0.2 + 2(-0.2) + 2(-0.1980) + (-0.19616)]$$

$$= -0.19869 \ (= -0.1987)$$

Therefore, the required solution is

$$y(0.2) = 0.9801, \qquad y'(0.2) = -0.1987$$

8.5 PREDICTOR–CORRECTOR METHODS

The methods presented in Sections 8.2 to 8.4 are in general known as single-step methods, where we have seen that the computation of y at t_{n+1}, that is y_{n+1} requires the knowledge of y_n only. In *predictor–corrector methods* which we discuss below, also known as *multi-step methods*, we require to know the solution y at t_n, t_{n-1}, t_{n-2}, etc., to compute the value of y at t_{n+1}. Thus, a *predictor formula* is used to predict the value of y at t_{n+1} and then a *corrector formula* is used to improve the value of y_{n+1}. For example, consider a differential equation

$$\frac{dy}{dt} = f(t, y)$$

with the initial condition $y(t_n) = y_n$. Using simple Euler's and modified Euler's method, we can write down a simple predictor–corrector pair (P–C) as

$$\left.\begin{aligned} \text{P}: y_{n+1}^{(0)} &= y_n + hf(t_n, y_n) \\ \text{C}: y_{n+1}^{(1)} &= y_n + \frac{h}{2}\left[f(t_n, y_n) + f\left(t_{n+1}, y_{n+1}^{(0)}\right)\right] \end{aligned}\right\} \tag{8.32}$$

Here, $y_{n+1}^{(1)}$ is the first corrected value of y_{n+1}. The corrector formula may be used iteratively as defined below:

$$y_{n+1}^{(r)} = y_n + \frac{h}{2}\left[f(t_n, y_n) + f\left(t_{n+1}, y_{n+1}^{(r-1)}\right)\right], \qquad r = 1, 2, \ldots \tag{8.33}$$

The iteration is terminated when two successive iterates agree to the desired accuracy. In this pair, to extrapolate the value of y_{n+1}, we have approximated the solution curve in the interval (t_n, t_{n+1}) by a straight line passing through (t_n, y_n) and (t_{n+1}, y_{n+1}). The accuracy of the predictor formula can be improved by considering a quadratic curve through the equispaced points (t_{n-1}, y_{n-1}), (t_n, y_n), (t_{n+1}, y_{n+1}). Suppose, we fit a quadratic curve of the form

$$y = a + b\,(t - t_{n-1}) + c\,(t - t_n)\,(t - t_{n-1}) \tag{8.34}$$

where a, b, c are constants to be determined. Since the curve passes through (t_{n-1}, y_{n-1}) and (t_n, y_n) and satisfies

$$\left(\frac{dy}{dt}\right)_{(t_n, y_n)} = f(t_n, y_n)$$

we obtain

$$y_{n-1} = a, \qquad y_n = a + bh = y_{n-1} + bh$$

Therefore,

$$b = \frac{y_n - y_{n-1}}{h}$$

and

$$\left(\frac{dy}{dt}\right)_{(t_n, y_n)} = f(t_n, y_n) = \{b + c[(t - t_{n-1}) + (t - t_n)]\}_{(t_n, y_n)}$$

Which gives

$$f(t_n, y_n) = b + c\,(t_n - t_{n-1}) = b + ch$$

or

$$c = \frac{f(t_n, y_n)}{h} - \frac{(y_n - y_{n-1})}{h^2}$$

Substituting these values of a, b and c into the quadratic equation (8.34), we get

$$y_{n+1} = y_{n-1} + 2(y_n - y_{n-1}) + 2[hf(t_n, y_n) - (y_n - y_{n-1})]$$

That is,

$$y_{n+1} = y_{n-1} + 2hf(t_n, y_n) \tag{8.35}$$

Thus, instead of considering the predictor-corrector pair (8.32), we may consider the predictor-corrector pair given by

$$\left.\begin{aligned} \text{P}: y_{n+1} &= y_{n-1} + 2hf(t_n, y_n) \\ \text{C}: y_{n+1} &= y_n + \frac{h}{2}[f(t_n, y_n) + f(t_{n+1}, y_{n+1})] \end{aligned}\right\} \tag{8.36}$$

The essential difference between them is, the predictor of (8.36) is more accurate. However, the predictor of (8.36) cannot be used to predict y_{n+1} for a given initial value problem. The reason being, its use require the knowledge of past two points. In such a situation, a Runge–Kutta method is generally used to start the predictor method.

8.5.1 Milne's Method

It is also a *multi-step method* where we assume that the solution to the given initial value problem is known at the past four equispaced points t_0, t_1, t_2 and t_3. To derive Milne's predictor–corrector pair, we proceed as follows:

Let us consider the typical differential equation

$$\frac{dy}{dt} = f(t, y), \qquad y(t_0) = y_0$$

On integration between the limits t_0 and t_4, we get

$$\int_{t_0}^{t_4} \frac{dy}{dt}\, dt = \int_{t_0}^{t_4} f(t, y)\, dt \tag{8.37}$$

That is,

$$y_4 - y_0 = \int_{t_0}^{t_4} f(t, y)\, dt \tag{8.38}$$

To carry out integration, we employ a quadrature formula such as *Newton's forward difference formula* (6.33), so that

$$f(t, y) = f_0 + s\Delta f_0 + \frac{s(s-1)}{2}\Delta^2 f_0 + \frac{s(s-1)(s-2)}{6}\Delta^3 f_0 + \cdots \tag{8.39}$$

where

$$s = \frac{t - t_0}{h}, \qquad t = t_0 + sh$$

Substituting Eq. (8.39) into Eq. (8.38), we obtain

$$y_4 = y_0 + \int_{t_0}^{t_4}\left[f_0 + s\Delta f_0 + \frac{s(s-1)}{2}\Delta^2 f_0 + \frac{s(s-1)(s-2)}{6}\Delta^3 f_0 \right.$$
$$\left. + \frac{s(s-1)(s-2)(s-3)}{24}\Delta^4 f_0 + \cdots \right] dt$$

Now, by changing the variable of integration (from t to s), the limits of integration also changes (from 0 to 4), and thus the above expression becomes

$$y_4 = y_0 + h\int_0^4\left[f_0 + s\Delta f_0 + \frac{s(s-1)}{2}\Delta^2 f_0 + \frac{s(s-1)(s-2)}{6}\Delta^3 f_0 \right.$$
$$\left. + \frac{s(s-1)(s-2)(s-3)}{24}\Delta^4 f_0 + \cdots \right] ds \qquad (8.40)$$

which simplifies to

$$y_4 = y_0 + h\left[4f_0 + 8\Delta f_0 + \frac{20}{3}\Delta^2 f_0 + \frac{8}{3}\Delta^3 f_0 + \frac{28}{90}\Delta^4 f_0 \right] \qquad (8.41)$$

Substituting the differences, such as $\Delta f_0 = f_1 - f_0$, $\Delta^2 f_0 = f_2 - 2f_1 + f_0$, etc., Eq. (8.41) can be further simplified to

$$y_4 = y_0 + \frac{4h}{3}(2f_1 - f_2 + 2f_3) + \frac{28}{90}h\Delta^4 f_0 \qquad (8.42)$$

Alternatively, it can also be written as

$$y_4 = y_0 + \frac{4h}{3}(2y_1' - y_2' + 2y_3') + \frac{28}{90}h\Delta^4 y_0' \qquad (8.43)$$

This is known as *Milne's predictor formula.*

Similarly, integrating Eq. (8.37) over the interval t_0 to t_2 or $s = 0$ to 2 and repeating the above steps, we get

$$y_2 = y_0 + \frac{h}{3}(y_0' + 4y_1' + y_2') - \frac{1}{90}h\Delta^4 y_0' \qquad (8.44)$$

which is known as *Milne's corrector formula.*

In general, Milne's predictor–corrector pair can be written as

$$\left.\begin{aligned} \text{P}: y_{n+1} &= y_{n-3} + \frac{4h}{3}(2y_{n-2}' - y_{n-1}' + 2y_n') \\ \text{C}: y_{n+1} &= y_{n-1} + \frac{h}{3}(y_{n-1}' + 4y_n' + y_{n+1}') \end{aligned}\right\} \qquad (8.45)$$

From Eqs. (8.43) and (8.44) we observe that the magnitude of the truncation error in corrector formula is $(1/90)h\Delta^4 y_0'$, while the truncation error in predictor formula is $(28/90)h\Delta^4 y_0'$. Thus, we can notice that the truncation error in corrector formula is less than that in the truncation error in predictor formula.

In order to apply this P–C method to solve numerically any initial value problem, we first predict the value of y_{n+1} by means of predictor formula (8.45), where derivatives are computed using the given differential equation itself. Using the predicted value y_{n+1}, we calculate the derivative y'_{n+1} from the given differential equation and then we use the corrector formula of the pair (8.45) to have the corrected value of y_{n+1}. This in turn may be used to obtain improved value of y_{n+1} by using corrector again. This cycle is repeated until we achieve the required accuracy.

To familiarize with the above numerical scheme, we consider the following examples.

Example 8.7 Find $y(2.0)$ if $y(t)$ is the solution of

$$\frac{dy}{dt} = \frac{1}{2}(t + y)$$

assuming $y(0) = 2$, $y(0.5) = 2.636$, $y(1.0) = 3.595$ and $y(1.5) = 4.968$ using Milne's predictor–corrector method.

Solution Taking $t_0 = 0.0$, $t_1 = 0.5$, $t_2 = 1.0$, $t_3 = 1.5$, where we are given y_0, y_1, y_2 and y_3, we have to compute y_4, the solution of the given differential equation corresponding to $t = 2.0$. The Milne's P–C pair is given as

$$\text{P}: y_{n+1} = y_{n-3} + \frac{4h}{3}\left(2y'_{n-2} - y'_{n-1} + 2y'_n\right)$$

$$\text{C}: y_{n+1} = y_{n-1} + \frac{h}{3}\left(y'_{n-1} + 4y'_n + y'_{n+1}\right)$$

From the given differential equation, we have $y' = (t + y)/2$. Therefore,

$$y'_1 = \frac{t_1 + y_1}{2} = \frac{0.5 + 2.636}{2} = 1.5680$$

$$y'_2 = \frac{t_2 + y_2}{2} = \frac{1.0 + 3.595}{2} = 2.2975$$

$$y'_3 = \frac{t_3 + y_3}{2} = \frac{1.5 + 4.968}{2} = 3.2340$$

Now, using predictor formula, we compute

$$y_4 = y_0 + \frac{4h}{3}\left(2y'_1 - y'_2 + 2y'_3\right)$$

$$= 2 + \frac{4(0.5)}{3}\,[2(1.5680) - 2.2975 + 2(3.2340)]$$

$$= 6.8710$$

Using this predicted value, we shall compute the improved value of y_4 from corrector formula

$$y_4 = y_2 + \frac{h}{3}\left(y'_2 + 4y'_3 + y'_4\right)$$

in the following steps. Now using the available predicted value y_4 and the initial values, we compute

$$y'_4 = \frac{t_4 + y_4}{2} = \frac{2 + 6.8710}{2} = 4.4355$$

$$y'_3 = \frac{t_3 + y_3}{2} = \frac{1.5 + 4.968}{2} = 3.2340$$

and

$$y_2' = 2.2975$$

Thus, the first corrected value of y_4 is given by

$$y_4^{(1)} = 3.595 + \frac{0.5}{3}[2.2975 + 4(3.234) + 4.4355] = 6.8731667$$

Suppose, we apply the corrector formula again, then we have

$$y_4^{(2)} = y_2 + \frac{h}{3}\left[y_2' + 4y_3' + \left(y_4^{(1)}\right)'\right]$$

$$= 3.595 + \frac{0.5}{3}\left[2.2975 + 4(3.234) + \frac{2 + 6.8731667}{2}\right]$$

$$= 6.8733467$$

Finally, the value of y at $t = 2.0$ is given by $y\,(2.0) = y_4 = 6.8734$.

Example 8.8 Tabulate the solution of

$$\frac{dy}{dt} = t + y, \qquad y(0) = 1$$

in the interval $0 \le t \le 0.4$, with $h = 0.1$, using Milne's predictor–corrector method.

Solution Milne's P-C method demands the solution at the first four points t_0, t_1, t_2 and t_3. As it is not a self-starting method, we shall use Runge–Kutta method of fourth order (why?) to get the required solution and then switch over to Milne's P-C method. Thus, taking $t_0 = 0$, $t_1 = 0.1$, $t_2 = 0.2$, $t_3 = 0.3$ we get the corresponding y values using Runge–Kutta method of fourth order; that is, $y_0 = 1$, $y_1 = 1.1103$, $y_2 = 1.2428$ and $y_3 = 1.3997$ (as obtained in Example 8.5). Now, we compute

$$y_1' = t_1 + y_1 = 0.1 + 1.1103 = 1.2103$$

$$y_2' = t_2 + y_2 = 0.2 + 1.2428 = 1.4428$$

$$y_3' = t_3 + y_3 = 0.3 + 1.3997 = 1.6997$$

Using Milne's predictor formula

$$\text{P} : y_4 = y_0 + \frac{4h}{3}\left(2y_1' - y_2' + 2y_3'\right)$$

$$= 1 + \frac{4(0.1)}{3}[2(1.2103) - 1.4428 + 2(1.6997)]$$

$$= 1.58363$$

Before using corrector formula, we compute

$$y_4' = t_4 + y_4 \text{ (predicted value)} = 0.4 + 1.5836 = 1.9836$$

Finally, using *Milne's corrector formula*, we compute

$$C : y_4 = y_2 + \frac{h}{3}(y_4' + 4y_3' + y_2')$$

$$= 1.2428 + \frac{0.1}{3}(1.9836 + 6.7988 + 1.4428)$$

$$= 1.5836$$

The required solution is tabulated below:

t	0	0.1	0.2	0.3	0.4
y	1.0	1.1103	1.2428	1.3997	1.5836

8.5.2 Adam–Moulton Method

It is another *predictor–corrector method*, where we use the fact that the solution to the given initial value problem is known at past four equispaced points t_n, t_{n-1}, t_{n-2}, t_{n-3}. The task is to compute the value of y at t_{n+1}.

Consider the differential equation

$$\frac{dy}{dt} = f(t, y)$$

Integrating between the limits t_n to t_{n+1}, we have

$$\int_{t_n}^{t_{n+1}} \frac{dy}{dt}\, dt = \int_{t_n}^{t_{n+1}} f(t, y)\, dt \tag{8.46}$$

That is,

$$y_{n+1} - y_n = \int_{t_n}^{t_{n+1}} f(t, y)\, dt \tag{8.47}$$

To carry out integration, we employ a quadrature formula such as *Newton's backward interpolation formula* (6.35), so that

$$f(t, y) = f_n + s\nabla f_n + \frac{s(s+1)}{2}\nabla^2 f_n + \frac{s(s+1)(s+2)}{6}\nabla^3 f_n + \cdots \tag{8.48}$$

where,

$$s = \frac{t - t_n}{h}, \qquad t = t_n + sh$$

Substituting (8.48) in Eq. (8.47), we obtain

$$y_{n+1} = y_n + \int_{t_n}^{t_{n+1}} \left[f_n + s\nabla f_n + \frac{s(s+1)}{2}\nabla^2 f_n + \frac{s(s+1)(s+2)}{6}\nabla^3 f_n \right.$$

$$\left. + \frac{s(s+1)(s+2)(s+3)}{24}\nabla^4 f_n + \cdots \right] dt$$

Now by changing the variable of integration (from t to s), the limits of integration also changes (from 0 to 1), and thus the above expression becomes

$$y_{n+1} = y_n + h\int_0^1 \left[f_n + s\nabla f_n + \frac{s(s+1)}{2}\nabla^2 f_n + \frac{s(s+1)(s+2)}{6}\nabla^3 f_n \right.$$

$$\left. + \frac{s(s+1)(s+2)(s+3)}{24}\nabla^4 f_n + \cdots \right] ds \quad (8.49)$$

Actual integration reduces the above expression to

$$y_{n+1} = y_n + h\left(f_n + \frac{1}{2}\nabla f_n + \frac{5}{12}\nabla^2 f_n + \frac{3}{8}\nabla^3 f_n + \frac{251}{720}\nabla^4 f_n \right) \quad (8.50)$$

Now substituting the differences such as

$$\nabla f_n = f_n - f_{n-1}$$
$$\nabla^2 f_n = f_n - 2f_{n-1} + f_{n-2}$$
$$\nabla^3 f_n = f_n - 3f_{n-1} + 3f_{n-2} - f_{n-3}$$

Equation (8.50) simplifies to

$$y_{n+1} = y_n + \frac{h}{24}(55f_n - 59f_{n-1} + 37f_{n-2} - 9f_{n-3}) + \frac{251}{720}h\nabla^4 f_n \quad (8.51)$$

Alternatively, it can be written as

$$y_{n+1} = y_n + \frac{h}{24}[55y'_n - 59y'_{n-1} + 37y'_{n-2} - 9y'_{n-3}] + \frac{251}{720}h\nabla^4 y'_n \quad (8.52)$$

This is known as *Adam's predictor formula*. Here, the truncation error is $(251/720)h\nabla^4 y_n'$. To obtain corrector formula, we use *Newton's backward interpolation formula* (6.35) about f_{n+1} instead of f_n. Thus, starting from Eq. (8.47), we obtain

$$y_{n+1} = y_n + h\int_{-1}^0 \left[f_{n+1} + s\nabla f_{n+1} + \frac{s(s+1)}{2}\nabla^2 f_{n+1} + \frac{s(s+1)(s+2)}{6}\nabla^3 f_{n+1} \right.$$

$$\left. + \frac{s(s+1)(s+2)(s+3)}{24}\nabla^4 f_{n+1} + \cdots \right] ds$$

Carrying out the integration and repeating the steps, we get the corrector formula as

$$y_{n+1} = y_n + \frac{h}{24}(9y'_{n+1} + 19y'_n - 5y'_{n-1} + y'_{n-2}) + \left(\frac{-19}{720}\right) h\nabla^4 y'_{n+1} \quad (8.53)$$

Here, the truncation error is $(19/720)\, h\nabla^4\, y'_{n+1}$. The truncation error in Adam's predicator is approximately thirteen times more than that in the corrector, of course with opposite sign. In general, Adam–Moulton predictor–corrector pair can be written as

$$\begin{aligned} \text{P}: y_{n+1} &= y_n + \frac{h}{24}\left(55y'_n - 59y'_{n-1} + 37y'_{n-2} - 9y'_{n-3}\right) \\ \text{C}: y_{n+1} &= y_n + \frac{h}{24}\left(9y'_{n+1} + 19y'_n - 5y'_{n-1} + y'_{n-2}\right) \end{aligned} \tag{8.54}$$

Here follows an example for illustration.

Example 8.9 Using Adam–Moulton predictor–corrector method, find the solution of the initial value problem

$$\frac{dy}{dt} = y - t^2, \qquad y(0) = 1$$

at $t = 1.0$, taking $h = 0.2$. Compare it with the analytical solution.

Solution In order to use Adam's predictor–corrector method, we require the solution of the given differential equation at the past four equispaced points, for which we use Runge–Kutta method of fourth order which is self-starting. Thus, taking $t_0 = 0$, $y_0 = 1$, $h = 0.2$, we compute $k_1 = 0.2$, $k_2 = 0.218$, $k_3 = 0.2198$, $k_4 = 0.23596$, and get

$$y_1 = y_0 + \frac{1}{6}\left(k_1 + 2k_2 + 2k_3 + k_4\right) = 1.21859$$

Taking $t_1 = 0.2$, $y_1 = 1.21859$, $h = 0.2$, we compute $k_1 = 0.23571$, $k_2 = 0.2492$, $k_3 = 0.25064$, $k_4 = 0.26184$, and get

$$y_2 = y_1 + \frac{1}{6}\left(k_1 + 2k_2 + 2k_3 + k_4\right) = 1.46813$$

Now we take $t_2 = 0.4$, $y_2 = 1.46813$, $h = 0.2$, and compute $k_1 = 0.2616$, $k_2 = 0.2697$, $k_3 = 0.2706$, $k_4 = 0.2757$ to get

$$y_3 = y(0.6) = y_2 + \frac{1}{6}\left(k_1 + 2k_2 + 2k_3 + k_4\right) = 1.73779$$

Thus, we have at our disposal

$$\begin{aligned} y_0 &= y(0) = 1 \\ y_1 &= y(0.2) = 1.21859 \\ y_2 &= y(0.4) = 1.46813 \\ y_3 &= y(0.6) = 1.73779 \end{aligned}$$

Now, we use Adam's predictor–corrector pair to calculate $y(0.8)$ and $y(1.0)$ as follows:

$$\text{P}: y_{n+1} = y_n + \frac{h}{24}\left(55y'_n - 59y'_{n-1} + 37y'_{n-2} - 9y'_{n-3}\right)$$

$$\text{C}: y_{n+1} = y_n + \frac{h}{24}\left(9y'_{n+1} + 19y'_n - 5y'_{n-1} + y'_{n-2}\right)$$

Thus

$$y_4^p = y_3 + \frac{h}{24}\left(55y'_3 - 59y'_2 + 37y'_1 - 9y'_0\right) \tag{1}$$

From the given differential equation, we have $y' = y - t^2$. Therefore,

$$y_0' = y_0 - t_0^2 = 1.0$$
$$y_1' = y_1 - t_1^2 = 1.17859$$
$$y_2' = y_2 - t_2^2 = 1.30813$$
$$y_3' = y_3 - t_3^2 = 1.37779$$

Hence, from Eq. (1), we get

$$y(0.8) = y_4^{\mathrm{p}} = 1.73779 + \frac{0.2}{24}(75.77845 - 77.17967 + 43.60783 - 9)$$
$$= 2.01451$$

Now to obtain the corrector value of y at $t = 0.8$, we use

$$y_4^{\mathrm{c}} = y^{\mathrm{c}}(0.8) = y_3 + \frac{h}{24}(9y_4' + 19y_3' - 5y_2' + y_1') \qquad (2)$$

But,

$$9y_4' = 9(y_4^{\mathrm{p}} - t_4^2) = 9[2.01451 - (0.8)^2] = 12.37059$$

Therefore,

$$y_4 = y^{\mathrm{c}}(0.8) = 1.73779 + \frac{0.2}{24}(12.37059 + 26.17801 - 6.54065 + 1.17859)$$
$$= 2.01434 \qquad (3)$$

Proceeding similarly, we get

$$y_5^{\mathrm{p}} = y^{\mathrm{p}}(1.0) = y_4 + \frac{h}{24}(55y_4' - 59y_3' + 37y_2' - 9y_1')$$

Noting that $y_4' = y_4 - t_4^2 = 1.3743$, we calculate

$$y_5^{\mathrm{p}} = 2.01434 + \frac{0.2}{24}(75.5887 - 81.28961 + 48.40081 - 10.60731)$$
$$= 2.28178$$

Now, the corrector formula for computing y_5 is given by

$$y_5^{\mathrm{c}} = y^{\mathrm{c}}(1.0) = y_4 + \frac{h}{24}(9y_5' + 19y_4' - 5y_3' + y_2') \qquad (4)$$

But.

$$9y_5' = 9\left(y_5^p - t_5^2\right) = 11.53602$$

Thus, finally we get

$$y_5 = y(1.0) = 2.01434 + \frac{0.2}{24}(11.53602 + 26.1117 - 6.88895 + 1.30817)$$
$$= 2.28339 \qquad (5)$$

The analytical solution can be seen in the following steps.

$$\frac{dy}{dt} - y = -t^2$$

After finding integrating factor and solving, we get

$$\frac{d}{dt} ye^{-t} = -e^{-t}t^2$$

Integrating, we get

$$ye^{-t} = -\int e^{-t}t^2\, dt = \int t^2 d(e^{-t}) = t^2e^{-t} + 2te^{-t} + 2e^{-t} + c$$

That is,

$$y = t^2 + 2t + 2 + \frac{c}{e^{-t}}$$

Now using the initial condition, $y(0) = 1$, we get $c = -1$. Therefore, the analytical solution is given by

$$y = t^2 + 2t + 2 - e^t$$

from which, we get

$$y(1.0) = 5 - e = 2.2817 \tag{6}$$

8.6 NUMERICAL STABILITY

In this Section, we shall introduce and discuss an important concept called *numerical stability* which is vital for solving differential equations numerically. It is quite possible that the numerical solution of a given differential equation may grow unbounded even though its exact solution is well-behaved. To elaborate this point, let us consider a simple differential equation such as

$$y' = \frac{dy}{dt} = f(t, y) \tag{8.55}$$

and a numerical scheme. We wish to determine the conditions in terms of parameters of the numerical method such as time step or step size 'h', for which the numerical solution remains bounded, thus getting numerically-stable solution. For, consider the two-dimensional Taylor series expansion of $f(t, y)$, that is,

$$f(t, y) = f(t_0, y_0) + (t - t_0)\frac{\partial f}{\partial t}(t_0, y_0) + (y - y_0)\frac{\partial f}{\partial y}(t_0, y_0)$$

$$+ \frac{1}{2!}\left[(t - t_0)^2 \frac{\partial^2 f}{\partial t^2} + 2(t - t_0)(y - y_0)\frac{\partial^2 f}{\partial t \partial y} + (y - y_0)^2 \frac{\partial^2 f}{\partial y^2}\right] + \cdots$$

Considering only the linear terms and substituting in Eq. (8.55) we have formally

$$y' = \lambda y + \alpha_1 + \alpha_2 t + \cdots \tag{8.56}$$

Here, λ, α_1, α_2 are constants. If we denote

$$\lambda = \frac{\partial f}{\partial y}(t_0, y_0)$$

then, we get the linearized form of Eq. (8.55) about (t_0, y_0). Considering only the first term on the right-hand side of Eq. (8.56), we have a model problem as

$$y' = \lambda y \tag{8.57}$$

Instead of considering the general problem described by Eq. (8.55) we consider Eq. (8.57) as the model problem only for convenience and feasibility of analytical treatment of stability analysis. Here, we take the constant λ in general as complex such that

$$\lambda = \lambda_R + i\lambda_I$$

We assume that $\lambda_R \le 0$, to ensure that the solution does not grow with t. The presence of the imaginary part indicate the oscillatory solution of the form $e^{\pm iwt}$

For illustration, consider Euler method given in Eq. (8.8), that is,

$$y_{n+1} = y_n + hf(t_n, y_n)$$

When applied to the model problem given by Eq. (8.57) we find

$$y_{n+1} = y_n + h\lambda y_n = (1 + h\lambda)\, y_n$$

The solution at nth time step can then be written as

$$y_n = (1 + h\lambda)^n\, y_0 \tag{8.58}$$

or

$$y_n = (1 + h\lambda_R + ih\lambda_1)^n y_0 = \sigma^n y_0 \text{ say}$$

Here, $\sigma = (1 + h\lambda_R + ih\lambda_I)$ is called the amplification factor. It may be stated that the numerical solution remains bounded or is stable if

$$|\sigma| \le 1 \tag{8.59}$$

For $\lambda_R \le 0$, the exact solution of the model problem, that is, $y_0\, e^{\lambda t}$ decays and the region of stability is the entire left-hand plane in the $(h\lambda_R, h\lambda_I)$ region as described in Fig. 8.3.

However, in the case of Euler method, the region of stability is only the portion inside the circle given by

$$|\sigma|^2 = (1 + h\lambda_R)^2 + h^2\lambda_I^2 = 1 \tag{8.60}$$

For any value of $(h\lambda)$ in the left-hand plane and outside the above circle, the numerical solution obtained by Euler method blows up, while the exact solution decays. Thus, the Euler method is stable provided that the step size h and that $h\lambda$ lies inside the circle as displayed in Fig. 8.4. Therefore, for getting the numerically-stable solution, we choose h such that

$$|1 + h\lambda| \le 1.$$

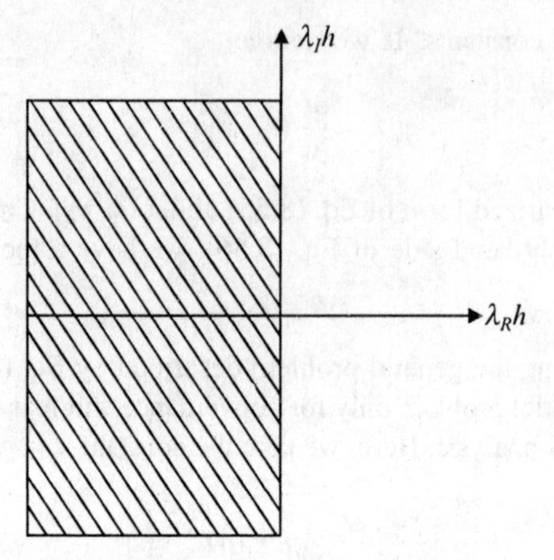

Fig. 8.3 Stability region for the exact solution.

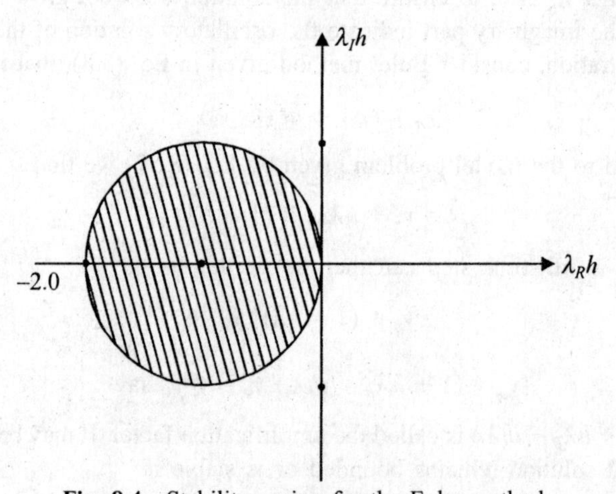

Fig. 8.4 Stability region for the Euler method.

If λ is real and negative, we choose the step size h, so as to satisfy the inequality

$$h \le \frac{2}{|\lambda|} \tag{8.61}$$

From Fig. 8.4, we observe that $(h\lambda)$ is as low as –2. In view of this limitation on h, we require more time steps, and therefore, more computer time to get the solution at the final time. But, if λ is purely imaginary we observe that the circle described by Eq. (8.60) is only tangential to the imaginary axis. Consequently, the Euler method is always unstable.

For example, let us consider the initial value problem described by

$$\frac{dy}{dt} = -\frac{1}{2}y, \; y(0) = 1, \; 0 \le t \le 15$$

In this example, on comparing with the model equation we note that $\lambda = -1/2$, which is real and negative. In order to find the numerical solution of this problem using Euler method, we observe that this method is stable for $h \le 2/|\lambda|$, that is $h \le 4$. Now, let us look at the numerical results obtained for two different time steps say $h = 1.0$ and $h = 4.2$, which are shown graphically in Fig. 8.5 along with the exact solution It can be seen that the numerical solution is stable for

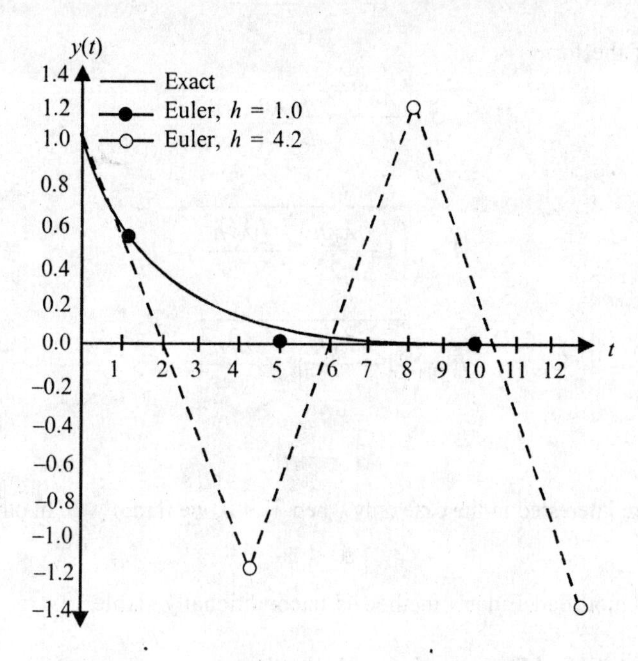

Fig. 8.5 Comparison of numerical solutions—Euler method.

$h = 1.0$, which is also very close to the exact solution. However, the numerical solution for time step $h = 4.2$ is seen to be of oscillatory nature and in fact is unstable, as expected due to violation of the limit prescribed on h by the stability condition.

8.6.1 Stability of Modified Euler's Method

As another example, consider the application of modified Euler's method to the model equation, we obtain

$$y_{n+1} - y_n = \frac{h}{2}[\lambda y_{n+1} + \lambda y_n]$$

or

$$y_{n+1} = \frac{1 + \lambda h/2}{1 - \lambda h/2} y_n$$

Here, the amplification factor is

$$\sigma = \frac{1 + \lambda h/2}{1 - \lambda h/2}$$

For complex $\lambda = \lambda_R + i\lambda_I$, we have

$$\sigma = \frac{1 + \dfrac{\lambda_R h}{2} + \dfrac{i\lambda_I h}{2}}{1 - \dfrac{\lambda_R h}{2} - \dfrac{i\lambda_I h}{2}}$$

which is of the form

$$\sigma = \frac{Ae^{i\theta}}{Be^{i\alpha}} = \frac{A}{B}e^{i(\theta - \alpha)}$$

where

$$A = \sqrt{\left(1 + \frac{\lambda_R h}{2}\right)^2 + \frac{\lambda_I^2 h^2}{4}},$$

$$B = \sqrt{\left(1 - \frac{\lambda_R h}{2}\right)^2 + \frac{\lambda_I^2 h^2}{4}},$$

and

$$|\sigma| = A/B$$

Since we are interested in the case only when $\lambda_R < 0$, we find $A < B$. In other words

$$|\sigma| < 1$$

Hence, the modified Euler's method is unconditionally stable

8.6.2 Stability of Runge–Kutta Methods

In Section 8.4, we have discussed second and fourth order Runge–Kutta methods along with their truncation error estimates. In fact, R–K methods are superior to Euler methods in view of the fact that slopes were introduced not only at t_n but also at other points between t_n and t_{n+1}. Of course, the additional function evaluation, result in additional computation time at each time step. However, R–K methods posseses better stability properties which is seen in the following steps:

Applying the second order R–K method as described in Eq. (8.19) to the model equation we find

$$k_1 = h\lambda y_n$$

$$k_2 = h\left[\lambda y_n + \frac{3}{2}h\lambda^2 y_n\right] = h\lambda\left[1 + \frac{3}{2}h\lambda\right]y_n$$

and

$$y_{n+1} = y_n + \frac{1}{3}\left[3h\lambda + \frac{3}{2}h^2\lambda^2\right]y_n$$

or

$$y_{n+1} = \left[1 + h\lambda + \frac{h^2\lambda^2}{2}\right]y_n = \sigma y_n \tag{8.62}$$

This expression indicates that the method is second order accurate. Now, for stability of second order R–K method, we must have $|\sigma| \le 1$, where

$$\sigma = [1 + h\lambda + h^2\lambda^2/2] \tag{8.63}$$

In order to get the region of stability, it is convenient to set $|\sigma| = |e^{i\theta}| = 1$, for all values of θ. Thus, we have a polynomial in $(h\lambda)$ given by

$$\left[1 + h\lambda + h^2\lambda^2/2\right] - e^{i\theta} = 0 \tag{8.64}$$

The complex roots of the above polynomial can be determined for various values of θ in $0 \le \theta \le \pi$. The resulting stability region is shown in Fig. 8.6.

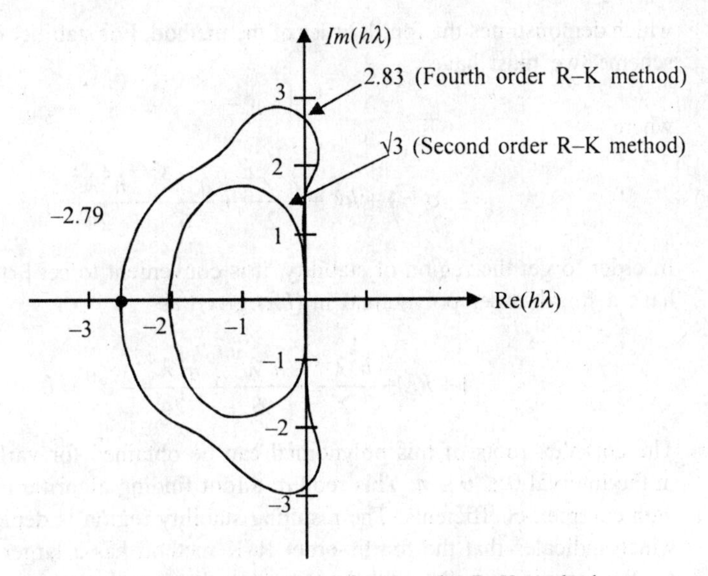

Fig. 8.6 Stability regions for second and fourth order R–K methods.

From this figure, we notice that, on the real axis, the stability boundary is same as that of Euler method. That is, $|h\lambda_R| \le 2$. However, there is a significant improvement in the case of fourth order R–K method. Fourth order Runge–Kutta method is indeed the most popular numerical scheme for solving the initial value problems which is described in Eq. (8.25) and can be presented in the following format:

$$y_{n+1} = y_n + \frac{1}{6}\left[k_1 + 2k_2 + 2k_3 + k_4\right]$$

where

$$k_1 = hf(t_n, y_n)$$

$$k_2 = hf\left(t_n + \frac{h}{2},\ y_n + \frac{k_1}{2}\right)$$

$$k_3 = hf\left(t_n + \frac{h}{2},\ y_n + \frac{k_2}{2}\right)$$

$$k_4 = hf\left(t_n + h, \quad y_n + k_3\right)$$

Observe that four function evaluations are required at each time step. By applying these steps to the model equation $y' = \lambda y$, we arrive at

$$y_{n+1} = \left[1 + h\lambda + \frac{h^2\lambda^2}{2} + \frac{h^3\lambda^3}{6} + \frac{h^4\lambda^4}{24}\right] y_n$$

or

$$y_{n+1} = \sigma y_n \tag{8.65}$$

which demonstrates the fourth order of the method. For stability of this numerical scheme, we must have

$$|\sigma| \le |$$

where

$$\sigma = 1 + h\lambda + \frac{h^2\lambda^2}{2} + \frac{h^3\lambda^3}{6} + \frac{h^4\lambda^4}{24} \tag{8.66}$$

In order to get the region of stability, it is convenient to set $|\sigma| = 1$. Thus, we have a fourth-order polynomial in $(h\lambda)$ given as

$$1 + h\lambda + \frac{h^2\lambda^2}{2} + \frac{h^3\lambda^3}{6} + \frac{h^4\lambda^4}{24} - e^{i\theta} = 0 \tag{8.67}$$

The complex roots of this polynomial can be obtained for various values of θ in the interval $0 \le \theta \le \pi$. This require a root finding algorithm for polynomials with complex coefficients. The resulting stability region is depicted in Fig. 8.6, which indicates that the fourth–order R–K method has a larger stability region on the imaginary axis. We also observe that on the real axis, the stability boundary is given by $|h\lambda_R| \le 2.79$. In practice, to get numerically-stable solutions for a given initial or boundary value problem, we choose the value of h much smaller than the one given by the above condition and also check for the consistancy of the results.

EXERCISES

8.1 Explain Taylor's series method of solving an initial value problem described by

$$\frac{dy}{dx} = f(x, y), \qquad y(x_0) = y_0$$

8.2 Using Taylor's series method, find the solution of the differential equation

$$xy' = x - y \quad \text{given that} \quad y(2) = 2, \text{ at } x = 2.1.$$

8.3 Using Taylor's series method, find the solution of

$$\frac{dy}{dt} = t^2 + y^2$$

with the initial values $t_0 = 1$, $y_0 = 0$, at $t = 1.3$.

8.4 Using Taylor's series method, solve

$$y' = y \sin x + \cos x$$

subject to $x = 0$, $y = 0$ for some x.

8.5 Using modified Euler's method, obtain the solution of

$$\frac{dy}{dt} = 1 - y, \qquad y(0) = 0$$

for the range $0 \le t \le 0.2$, by taking $h = 0.1$.

8.6 Solve the initial value problem

$$yy' = x, \qquad y(0) = 1.5$$

using simple Euler's method, taking $h = 0.1$ and hence find y (0.2).

8.7 Obtain numerically the solution of

$$y' = x^2 + y^2, \qquad y(0) = 0.5$$

using Euler's method to find y at $x = 0.1$ and 0.2

8.8 Use Runge–Kutta method of fourth order to solve numerically the initial value problem

$$10\frac{dy}{dx} = x^2 + y^2, \qquad y(0) = 1$$

and find y in the interval $0 \le x \le 0.4$, taking $h = 0.1$.

8.9 Solve $y'' = x(y')^2 - y^2$, using fourth order Runge–Kutta method for $x = 0.2$ correct to four decimal places with the initial conditions $y(0) = 1$, $y'(0) = 0$.

8.10 Using R–K method of fourth order, solve $y'' = xy' + y^2$, given that $y(0) = 1$, $y'(0) = 2$. Take $h = 0.2$ and find y and y' at $x = 0.2$.

8.11 Use fourth order Runge–Kutta method to solve numerically the following initial value problem

$$\frac{dy}{dt} = y^2 - 100 \exp[-100(t-1)^2], \qquad y(0.8) = 4.9491$$

and find y in the interval $0.8 \le t \le 0.9$ taking $h = 0.01$.

8.12 Find $y(0.8)$ using Milne's P-C method, if $y(x)$ is the solution of the differential equation

$$\frac{dy}{dx} = -xy^2, \qquad y(0) = 2$$

assuming $y(0.2) = 1.92308$, $y(0.4) = 1.72414$, $y(0.6) = 1.47059$.

8.13 Explain the principle of predictor–corrector methods. Derive Milne's predictor–corrector formulae to solve an initial value problem

$$\frac{dy}{dx} = f(x, y), \qquad y(0) = y_0$$

8.14 Using Adam's predictor–corrector method, find y at $t = 4.4$ from the differential equation

$$5t\frac{dy}{dt} + y^2 = 2,$$

given that

t	4.0	4.1	4.2	4.3
y	1.0	1.0049	1.0097	1.0143

8.15 It is well known in the theory of beams that the radius of curvature is given by

$$\frac{\text{EI } y''}{(1 + y'^2)^{3/2}} = M(x)$$

where $M(x)$ is the bending moment. For a cantilever beam, it is known that $y(0) = y'(0) = 0$. Express the above equation into two first order simultaneous differential equations.

8.16 The resonant spring system with a periodic forcing function is given by

$$\frac{d^2y}{dt^2} + 64y = 16 \cos 8t, \qquad y(0) = y'(0) = 0$$

Determine the displacement at $t = 0.1, 0.2, \ldots, 0.5$ using Adam's–Moulton method after getting the required starting values by Runge–Kutta fourth order method.

8.17 Solve the initial value problem

$$\frac{dy}{dx} = 3x^2 + y, \qquad y(0) = 4$$

for the range $0.1 \le x \le 0.5$, using Euler's method by taking $h = 0.1$.

8.18 Using Euler's method, obtain the solution to the initial value problem

$$y' = x + y + xy, \qquad y(0) = 1$$

at $x = 0.1$, by taking $h = 0.025$.

8.19 Solve the initial value problem

$$\frac{dy}{dx} = \log(x + y), \qquad y(0) = 1$$

using modified Euler method and find $y(0.2)$.

8.20 Using fourth order Runge–Kutta method find the solution of the initial value problem

$$y' = 1/(x + y), \qquad y(0) = 1$$

in the range $0.5 \le x \le 2.0$, by taking $h = 0.5$.

8.21 Using fourth order Runge–Kutta method, find the solution of

$$x(dy + dx) = y(dx - dy), \qquad y(0) = 1$$

at $x = 0.1$ and 0.2, by taking $h = 0.1$.

8.22 Find the solution of

$$y' = y(x + y), \qquad y(0) = 1$$

using Milne's P-C method at $x = 0.4$ given that $y(0.1) = 1.11689$, $y(0.2) = 1.27739$ and $y(0.3) = 1.50412$.

8.23 Using Adam's–Moulton P-C method, find the solution of

$$x^2 y' + xy = 1, \qquad y(1) = 1.0$$

at $x = 1.4$, given that $y(1.1) = 0.996$, $y(1.2) = 0.986$, $y(1.3) = 0.972$

8.24 Find the solution of the initial value problem

$$y' = y^2 \sin t \qquad y(0) = 1$$

using Adam's–Moulton P-C method, in the interval (0.2, 0.5), given that $y(0.05) = 1.00125$, $y(0.1) = 1.00502$, $y(0.15) = 1.01136$.

8.25 Solve the following system of differential equations

$$\frac{dx}{dt} = x + 2y, \qquad x(0) = 6$$

$$\frac{dy}{dt} = 3x + 2y, \qquad y(0) = 4$$

over the interval (0.02, 0.06) using Runge–Kutta method, with step size $h = 0.02$.

Chapter 9

Parabolic Partial Differential Equations

9.1 INTRODUCTION

In many applications of science and engineering, there has been a growing desire for numerical answers, for the design of any system and to study its performance. Therefore, in recent years greater emphasis has been shifted from analytical techniques to computer-oriented numerical methods. The principal attraction of numerical methods is that solutions could be obtained for many problems which are not amenable to analytical treatment. The availability of modern digital computers paved the way for the development of efficient and more general finite difference methods for solving particularly partial differential equations.*

In this chapter, we shall present some of the important and powerful finite difference methods for solving parabolic partial differential equations that model heat and mass transfer in fluids and solids, unsteady behaviour of fluid flow past bodies, etc. For example, if the fluid is incompressible and has constant thermal conductivity k, the law of conservation of energy can be expressed mathematically in the form

$$\rho C_p \frac{DT}{Dt} = k\nabla^2 T \tag{9.1}$$

where ρ is the density, C_p is the specific heat of the fluid at constant pressure, T is the temperature and D/Dt is the material derivative. In cartesian coordinates, Eq. (9.1) can be written as

$$\frac{\partial T}{\partial t} + u\frac{\partial T}{\partial x} + v\frac{\partial T}{\partial y} + w\frac{\partial T}{\partial z} = \alpha\left(\frac{\partial^2 T}{\partial x^2} + \frac{\partial^2 T}{\partial y^2} + \frac{\partial^2 T}{\partial z^2}\right) \tag{9.2}$$

Here, $\alpha = k/(\rho C_p)$; u, v and w are the components of fluid velocity at a point (x, y, z) and at time t. Infact, it is a *non-linear parabolic partial differential equation.* For fluids which are at rest, Eq. (9.2) becomes

$$\frac{\partial T}{\partial t} = \alpha\left(\frac{\partial^2 T}{\partial x^2} + \frac{\partial^2 T}{\partial y^2} + \frac{\partial^2 T}{\partial z^2}\right) \tag{9.3}$$

which is also called a *diffusion equation.*

*The broad classification details and analytical solutions of partial differential equations are covered in Sankara Rao (1995).

The finite difference method for solving equations of the type (9.3) though fairly simple is also a powerful technique for solving a variety of heat transfer and fluid flow problems in regular geometries. Most of the concepts associated with the numerical solutions of partial differential equations by finite difference methods can best be illustrated and understood by considering a simple one-dimensional diffusion equation in the form

$$\frac{\partial T}{\partial t} = \alpha \frac{\partial^2 T}{\partial x^2} \tag{9.4}$$

which holds, true in some prescribed region R of the (x, t) space.

9.2 BASIC CONCEPTS IN FINITE DIFFERENCE METHODS

For the purpose of illustration, let us consider a two-dimensional region as shown in Fig. 9.1. It is covered by a rectangular grid formed by two sets of lines drawn parallel to the coordinate axes with grid spacing Δx and Δy in x and y

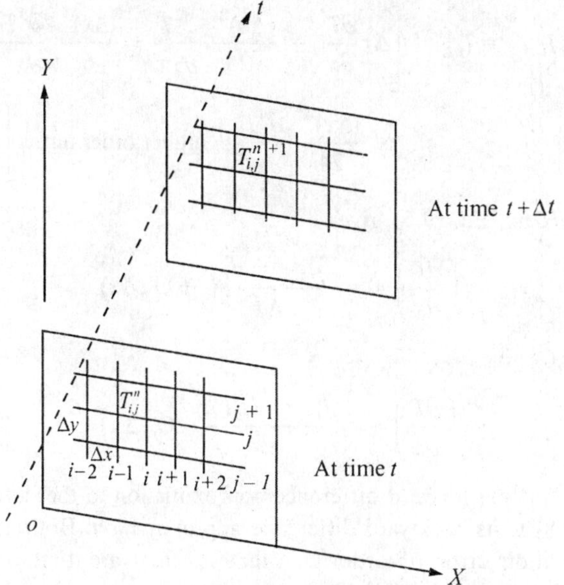

Fig. 9.1 Space–time index notation.

directions respectively. The numerical values of the dependent variables are obtained at the points of intersection of the parallel lines, called *mesh points* or *nodal points*. These values are obtained by discretizing the governing partial differential equations over the region of interest to derive approximately equivalent algebraic equations. The discretization consists of replacing each derivative of the partial differential equation at a mesh point by a finite difference approximation in terms of the values of the dependent variable at the mesh point and at the immediate neighbouring mesh points, and boundary points. In doing so, a set of algebraic equations arise.

Let the temperature T at a representative point be a function of two spacial coordinates x, y and time t. We adopt the following notation. Let the subscripts i and j represent x and y coordinates and superscript, n represents time. Let the mesh spacing in x and y directions are denoted by Δx and Δy and the time step by Δt. Thus, $T(x, y, t)$ can be represented by $T(i\Delta x, j\Delta y, n\Delta t) = T_{i,j}^n$. With this notation, let us assume that the function T and its derivatives are continuous. Then the finite difference approximations to derivatives can be obtained from Taylor's series expansions. For example, the Taylor's series expansion of $T_{i+1,j}$ about the grid point (i, j) gives

$$T_{i+1,j} = T_{i,j} + \left[\Delta x \frac{\partial T}{\partial x} + \frac{(\Delta x)^2}{2} \frac{\partial^2 T}{\partial x^2} + \frac{(\Delta x)^3}{6} \frac{\partial^3 T}{\partial x^3} + \frac{(\Delta x)^4}{24} \frac{\partial^4 T}{\partial x^4} + \text{higher order terms}\right]_{i,j} \tag{9.5}$$

Similarly,

$$T_{i-1,j} = T_{i,j} - \left[\Delta x \frac{\partial T}{\partial x} - \frac{(\Delta x)^2}{2} \frac{\partial^2 T}{\partial x^2} + \frac{(\Delta x)^3}{6} \frac{\partial^3 T}{\partial x^3} - \frac{(\Delta x)^4}{24} \frac{\partial^4 T}{\partial x^4} + \text{higher order terms}\right]_{i,j} \tag{9.6}$$

Solving for $\partial T/\partial x$, Eq. (9.5) gives

$$\left(\frac{\partial T}{\partial x}\right)_{i,j} = \frac{T_{i+1,j} - T_{i,j}}{\Delta x} + \mathrm{O}\,(\Delta x) \tag{9.7}$$

while Eq. (9.6) gives

$$\left(\frac{\partial T}{\partial x}\right)_{i,j} = \frac{T_{i,j} - T_{i-1,j}}{\Delta x} + \mathrm{O}\,(\Delta x) \tag{9.8}$$

Equation (9.7) is the forward difference approximation to the derivative $\partial T/\partial x$, while Eq. (9.8) is its backward difference approximation. Both approximations have a truncation error of order Δx, that is, they are first order accurate. Subtracting Eq. (9.6) from Eq. (9.5), we obtain

$$\left(\frac{\partial T}{\partial x}\right)_{i,j} = \frac{T_{i+1,j} - T_{i-1,j}}{2\Delta x} + \mathrm{O}\,[(\Delta x)^2] \tag{9.9}$$

This is a central difference approximation to the derivative $\partial T/\partial x$, and is second order accurate, in the sense that it has a truncation error of order $\mathrm{O}(\Delta x^2)$.

The central difference approximation to a second order partial derivative $\partial^2 T/\partial x^2$ can be similarly obtained by adding Eqs. (9.5) and (9.6). Thus,

$$\left(\frac{\partial^2 T}{\partial x^2}\right)_{i,j} = \frac{T_{i+1,j} - 2T_{i,j} + T_{i-1,j}}{\Delta x^2} + \mathrm{O}(\Delta x^2) \tag{9.10}$$

which is second order accurate. Similar expressions can be written for y derivatives.

$$\left(\frac{\partial^2 T}{\partial y^2}\right)_{i,j} = \frac{T_{i,j+1} - 2T_{i,j} + T_{i,j-1}}{\Delta y^2} + O(\Delta y^2) \tag{9.11}$$

The expressions for mixed derivatives can be obtained by differentiating with respect to each variable in turn. Thus, for example,

$$\left(\frac{\partial^2 T}{\partial x \partial y}\right)_{i,j} = \frac{\partial}{\partial x}\left(\frac{\partial T}{\partial y}\right)_{i,j} = \frac{\left(\frac{\partial T}{\partial y}\right)_{i+1,j} - \left(\frac{\partial T}{\partial y}\right)_{i-1,j}}{2\Delta x}$$

$$= \frac{(T_{i+1,j+1} - T_{i+1,j-1})/(2\Delta y) - (T_{i-1,j+1} - T_{i-1,j-1})/(2\Delta y)}{2\Delta x}$$

Therefore,

$$\left(\frac{\partial^2 T}{\partial x \partial y}\right)_{i,j} = \frac{T_{i+1,j+1} - T_{i+1,j-1} - T_{i-1,j+1} + T_{i-1,j-1}}{4\Delta x \Delta y} \tag{9.12}$$

Proceeding in a similar manner, the central difference approximation to the third derivative is found to be

$$\left(\frac{\partial^3 T}{\partial x^3}\right)_{i,j} = \frac{T_{i+2,j} - 2T_{i+1,j} + 2T_{i-1,j} - T_{i-2,j}}{2(\Delta x)^3} \tag{9.13}$$

Similar approximations can be obtained even to higher order derivatives. It may be observed that, as the order of the derivative increases, the number of adjacent nodes involved in the approximation also increases. When we are computing adjacent to the boundaries, the usefulness of these approximations become restricted. In fact, while computing near the boundaries, one-sided formulae involving more adjacent points become necessary. For instance, to implement derivative boundary condition, a second order accurate difference formula can be derived involving simultaneously the nodes $(i-2, j)$, $(i-1, j)$ and (i, j) in the form

$$\left(\frac{\partial T}{\partial x}\right)_{i,j} = \frac{aT_{i,j} + bT_{i-1,j} + cT_{i-2,j}}{\Delta x} \tag{9.14}$$

where the coefficients a, b, c can be obtained from Taylor's series expansion of $T_{i-2,j}$ and $T_{i-1,j}$ as shown below:

$$T_{i-2,j} = T_{i,j} - (2\Delta x)\left(\frac{\partial T}{\partial x}\right)_{i,j} + \frac{(2\Delta x)^2}{2}\left(\frac{\partial^2 T}{\partial x^2}\right)_{i,j} - \frac{(2\Delta x)^3}{6}\left(\frac{\partial^3 T}{\partial x^3}\right)_{i,j} + \text{H.O.T.} \tag{9.15}$$

$$T_{i-1,j} = T_{i,j} - \Delta x\left(\frac{\partial T}{\partial x}\right)_{i,j} + \frac{(\Delta x)^2}{2}\left(\frac{\partial^2 T}{\partial x^2}\right)_{i,j} - \frac{(\Delta x)^3}{6}\left(\frac{\partial^3 T}{\partial x^3}\right)_{i,j} + \text{H.O.T.} \tag{9.16}$$

Thus, we have

$$aT_{i,j} + bT_{i-1,j} + cT_{i-2,j} = (a + b + c)T_{i,j} - \Delta x(2c + b)\left(\frac{\partial T}{\partial x}\right)_{i,j}$$

$$+ \frac{(\Delta x)^2}{2}(4c + b)\left(\frac{\partial^2 T}{\partial x^2}\right)_{i,j} + O(\Delta x^3) \quad (9.17)$$

Comparing Eqs. (9.14) and (9.17), we get

$$a + b + c = 0, \qquad b + 2c = -1, \qquad b + 4c = 0.$$

Solving these equations, we obtain

$$a = \frac{3}{2}, \quad b = -\frac{4}{2}, \quad c = \frac{1}{2}$$

Thus,

$$\left(\frac{\partial T}{\partial x}\right)_{i,j} = \frac{3T_{i,j} - 4T_{i-1,j} + T_{i-2,j}}{2\Delta x} + O(\Delta x^2) \quad (9.18)$$

Similarly, one-sided formulae can be derived even for higher order derivatives.

Alternatively, in Chapter 6, we have defined various finite difference operators Δ, ∇, δ, E, μ and D and obtained various relations connecting them. Those relations can also be used to derive finite difference approximations to derivatives of any order to the desired accuracy (see Mitchell and Griffith, 1980).

Thus, the application of a finite difference method to a given physical problem consists of the following three basic steps:

(i) Division of spatial domain into an orthogonal computational grid.
(ii) Discretization of the governing equations and boundary conditions in space and time to derive approximately equivalent algebraic equations for each node.
(iii) Solving the resulting equations by a suitable matrix inversion or iterative technique.

These steps are further explained in the next section with more details.

Example 9.1 Apply Taylor's series expansion to the form

$$\left(\frac{\partial^4 T}{\partial x^4}\right)_{i,j} = aT_{i+2,j} + bT_{i+1,j} + cT_{i,j} + dT_{i-1,j} + eT_{i-2,j}$$

and hence show that

$$\left(\frac{\partial^4 T}{\partial x^4}\right)_{i,j} = \frac{T_{i+2,j} - 4T_{i+1,j} + 6T_{i,j} - 4T_{i-1,j} + T_{i-2,j}}{\Delta x^4} - \frac{(\Delta x)^2}{6}\left(\frac{\partial^6 T}{\partial x^6}\right)_{i,j}$$

Solution Writing down the Taylor's series expansions

$$T_{i\pm 2,j} = T_{i,j} \pm (2\Delta x)\left(\frac{\partial T}{\partial x}\right)_{i,j} + \frac{(2\Delta x)^2}{2}\left(\frac{\partial^2 T}{\partial x^2}\right)_{i,j}$$

$$\pm \frac{(2\Delta x)^3}{6}\left(\frac{\partial^3 T}{\partial x^3}\right)_{i,j} + \frac{(2\Delta x)^4}{24}\left(\frac{\partial^4 T}{\partial x^4}\right)_{i,j}$$

$$\pm \frac{(2\Delta x)^5}{120}\left(\frac{\partial^5 T}{\partial x^5}\right)_{i,j} + \frac{(2\Delta x)^6}{720}\left(\frac{\partial^6 T}{\partial x^6}\right)_{i,j} \pm \cdots$$

and

$$T_{i\pm 1,j} = T_{i,j} \pm (\Delta x)\left(\frac{\partial T}{\partial x}\right)_{i,j} + \frac{(\Delta x)^2}{2}\left(\frac{\partial^2 T}{\partial x^2}\right)_{i,j}$$

$$\pm \frac{(\Delta x)^3}{6}\left(\frac{\partial^3 T}{\partial x^3}\right)_{i,j} + \frac{(\Delta x)^4}{24}\left(\frac{\partial^4 T}{\partial x^4}\right)_{i,j}$$

$$\pm \frac{(\Delta x)^5}{120}\left(\frac{\partial^5 T}{\partial x^5}\right)_{i,j} + \frac{(\Delta x)^6}{720}\left(\frac{\partial^6 T}{\partial x^6}\right)_{i,j} \pm \cdots$$

Substituting these expansions into the form

$$\left(\frac{\partial^4 T}{\partial x^4}\right)_{i,j} = aT_{i+2,j} + bT_{i+1,j} + cT_{i,j} + dT_{i-1,j} + eT_{i-2,j}$$

We obtain the following equations:

$$a + b + c + d + e = 0$$

$$2a + b - d - 2e = 0$$

$$2a + \frac{b}{2} + \frac{d}{2} + 2e = 0$$

$$\frac{4}{3}a + \frac{b}{6} - \frac{d}{6} - \frac{4}{3}e = 0$$

$$\frac{2}{3}a + \frac{b}{24} + \frac{d}{24} + \frac{2}{3}e = 1$$

Whose solution is found to be

$$a = 1, \quad b = -4, \quad c = 6, \quad d = -4, \quad e = 1$$

Thus, we have

$$T_{i+2,j} - 4T_{i+1,j} + 6T_{i,j} - 4T_{i-1,j} + T_{i-2,j} = (\Delta x)^4\left(\frac{\partial^4 T}{\partial x^4}\right)_{i,j} + \frac{\Delta x^6}{6}\left(\frac{\partial^6 T}{\partial x^6}\right)_{i,j} + \cdots$$

which immediately gives

$$\left(\frac{\partial^4 T}{\partial x^4}\right)_{i,j} = \frac{T_{i+2,j} - 4T_{i+1,j} + 6T_{i,j} - 4T_{i-1,j} + T_{i-2,j}}{\Delta x^4} - \frac{(\Delta x)^2}{6}\left(\frac{\partial^6 T}{\partial x^6}\right)_{i,j}$$

9.3 EXPLICIT METHODS

These are basic finite difference methods for solving partial differential equations. Here, our approach in solving such equations is to replace the derivatives of the given differential equation by their finite difference approximations. In this process, we develop various finite difference formulae.

A formula which expresses one unknown value at a given node in terms of the known preceding values is called an *explicit formula*. Some of the well-known explicit methods are presented below for solving parabolic partial differential equations.

9.3.1 Schmidt Method

This is a simple two-level explicit method. For illustration, let us consider a one-dimensional diffusion equation (9.4) at the (i, n) grid point. Thus, we have

$$\left(\frac{\partial T}{\partial t}\right)_i^n = \alpha \left(\frac{\partial^2 T}{\partial x^2}\right)_i^n \tag{9.19}$$

Introducing Forward difference approximation to the Time derivative and Central difference approximation to Space derivative (FTCS), the above equation becomes

$$\frac{T_i^{n+1} - T_i^n}{\Delta t} = \alpha \frac{T_{i+1}^n - 2T_i^n + T_{i-1}^n}{(\Delta x)^2} \tag{9.20}$$

That is,

$$T_i^{n+1} = T_i^n + \frac{\alpha \Delta t}{(\Delta x)^2}\left(T_{i+1}^n - 2T_i^n + T_{i-1}^n\right)$$

We now define

$$\frac{\alpha \Delta t}{(\Delta x)^2} = r \tag{9.21}$$

then the above equation becomes

$$T_i^{n+1} = rT_{i-1}^n + (1 - 2r)\, T_i^n + rT_{i+1}^n \tag{9.22}$$

This is the *Schmidt explicit* formula which gives the approximate value of T_i^{n+1} at the $(n + 1)$th time level in terms of the known values T_{i-1}^n, T_i^n and T_{i+1}^n at the n th time level. This is also called a *two-level explicit formula*.

For $n = 0$, Eq. (9.22) gives approximate values of T_i^1 at grid points along the first time row $t = \Delta t$, in terms of known initial and boundary values T_i^0 $(i = 1, 2, \ldots, I)$. Taking $n = 1$, we can compute approximate values of T_i^2 at the grid points along the second time level, in terms of just computed values T_i^1, and so on. This process is continued until a predetermined time is reached. We shall illustrate this method through the following examples.

Example 9.2 Solve the following initial boundary value problem using an explicit finite difference method:

$$\frac{\partial T}{\partial t} = \frac{\partial^2 T}{\partial x^2}, \quad 0 \le x \le 1$$

Given

$$T = \sin \pi x \quad \text{when } t = 0,\ 0 \le x \le 1$$

$$T = 0 \quad \text{at } x = 0 \text{ and } x = 1 \quad \text{for } t > 0$$

and hence examine the accuracy of the numerical solution at $t = 0.006$ with its analytical solution.

Solution Let us choose $\Delta x = 0.1$. To get a numerically-stable solution, we choose Δt in such a way as to satisfy the stability condition (see also Section 9.5)

$$r = \frac{\Delta t}{\Delta x^2} \le 0.5$$

which gives $\Delta t \le 0.005$. In this example, we take $\Delta t = 0.002$, so that $r = 0.2$. Thus using the FTCS approximation, the given differential equation reduces to

$$T_i^{n+1} = rT_{i-1}^n + (1 - 2r)\, T_i^n + rT_{i+1}^n \tag{1}$$

Note that, $n = 0$ corresponds to the initial condition at $t = 0$ and the end values of T are given by the boundary conditions. Thus,

$$T_1^0 = 0.3090, \qquad T_2^0 = \sin\frac{\pi}{5} = 0.5878, \qquad T_3^0 = 0.8090$$

$$T_4^0 = 0.9511, \qquad T_5^0 = \sin\frac{\pi}{2} = 1.0, \qquad T_6^0 = 0.9511$$

$$T_7^0 = 0.8090, \qquad T_8^0 = 0.5878, \qquad T_9^0 = 0.3090$$

The boundary conditions give $T_0^0 = T_{10}^0 = 0.0$. Now the plan of computation, the initial and boundary conditions are shown in Fig. 9.2.

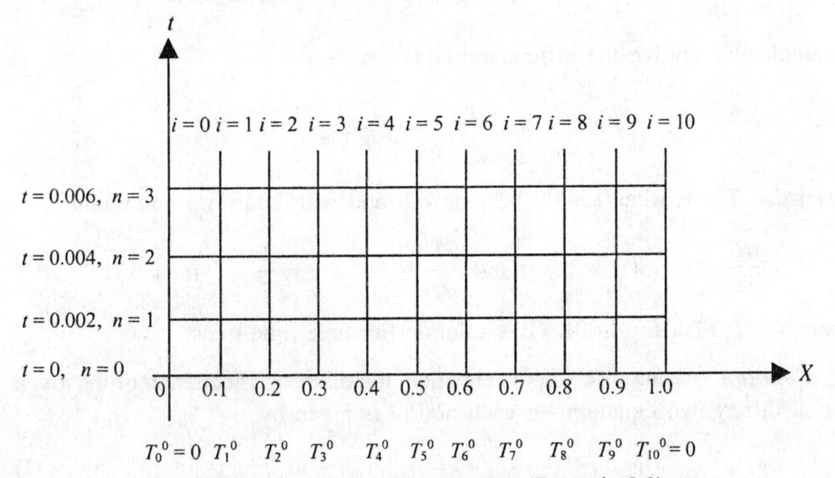

Fig. 9.2 Scheme of computation (Example 9.2).

We observe the symmetry of the solution on either side of the centre line $x = 0.5$, and therefore, the value of T at $x = 0.6$ is the same as at $x = 0.4$ and so on. In other words, $T_6^0 = T_4^0$ etc. These entries are shown as the first row in Table 9.1. For $i = 1$, $n = 0$, using the numerical scheme given by Eq. (1), we get

$$T_1^1 = 0.2T_0^0 + 0.6T_1^0 + 0.2T_2^0 = 0.3030$$

Similarly, for $i = 2, 3, 4$ and 5, $n = 0$, we can compute T_2^1, T_3^1, T_4^1 and T_5^1 which is the solution of the given initial boundary value problem for $t = \Delta t = 0.002$. Thus, we can use the simple algorithm described by Eq. (1) and compute further the values of T for subsequent time steps at all the nodes, which are conveniently tabulated in Table 9.1.

Table 9.1

n	t	x: 0.0	0.1	0.2	0.3	0.4	0.5	0.6	0.7	0.8	0.9	1.0
0	0.0	0.0	0.3090	0.5878	0.8090	0.9511	1.0	0.9511	0.8090	0.5878	0.3090	0.0
1	0.002	0.0	0.3030	0.5763	0.7932	0.9325	0.9804	0.9325	0.7932	0.5763	0.3030	0.0
2	0.004	0.0	0.2970	0.5650	0.7777	0.9142	0.9612	0.9142	0.7777	0.5650	0.2970	0.0
3	0.006	0.0	0.2912	0.5540	0.7625	0.8963	0.9424	0.8963	0.7625	0.5540	0.2912	0.0
	*.	0.0	0.2912	0.5540	0.7625	0.8964	0.9425	0.8964	0.7625	0.5540	0.2912	0.0

The analytical solution to the present problem can be easily found to be (see Sankara Rao, 1995),

$$T(x, t) = e^{-\pi^2 t} \sin \pi x$$

Thus,

$$T(x, 0.006) = 0.9425 \sin \pi x$$

The exact solution (* mark) and the computed solution at $t = 0.006$ is shown in Table 9.1 and both solutions found to agree up to third decimal place. This result demonstrates the power of even simple explicit numerical method.

Example 9.3 Solve the differential equation

$$\frac{\partial T}{\partial t} = \frac{\partial^2 T}{\partial x^2}, \qquad 0 \le x \le \frac{1}{2}$$

given that $T = 0$ when $t = 0$, $0 \le x \le 1/2$, and with boundary conditions

$$\frac{\partial T}{\partial x} = 0 \ \text{ at } x = 0 \qquad \text{and} \ \frac{\partial T}{\partial x} = 1 \ \text{ at } x = \frac{1}{2} \qquad \text{for } t > 0$$

taking $\Delta x = 0.1$, $\Delta t = 0.001$. Give solution for three-time steps.

Solution Using FTCS approximation, the discretized equation of the given partial differential equation for each node i is given by

$$T_i^{n+1} = rT_{i-1}^n + (1 - 2r)T_i^n + rT_{i+1}^n \tag{1}$$

where $r = \Delta t/\Delta x^2$. The data of the present problem gives $r = 0.1$. The required computation and the initial conditions are shown in Fig. 9.3.

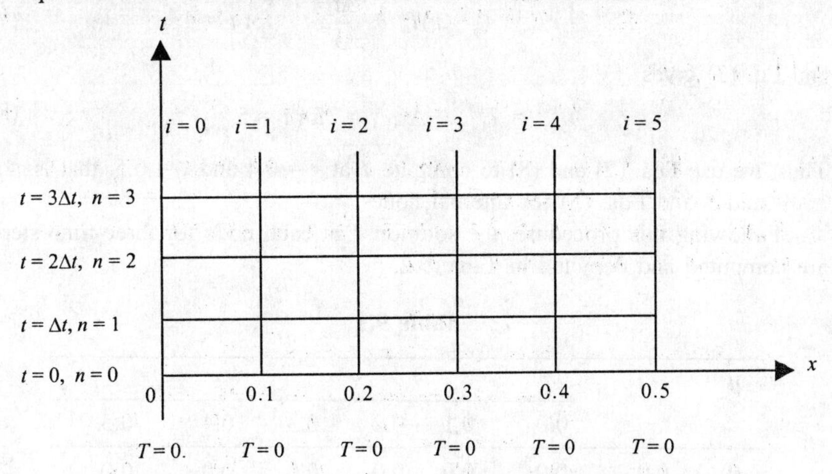

Fig. 9.3 Scheme of computation (Example 9.3).

In this problem, derivative boundary conditions are specified, which require special care, the implementation of which can be explained as follows:

At the left end, that is at $x = 0$, $\partial T/\partial x = 0$ for $t > 0$. Using forward difference approximation, it can be equivalently written as

$$\frac{T_{i+1}^n - T_i^n}{\Delta x} = 0 \text{ implying } T_i^n = T_{i+1}^n, \text{ therefore, } T_0^n = T_1^n \tag{2}$$

At the right end, that is, at $x = 1/2$, the boundary condition is $\partial T/\partial x = 1$, $t > 0$. Using backward difference approximation, its equivalent is

$$\frac{T_i^n - T_{i-1}^n}{\Delta x} = 1 \text{ implying } T_i^n = T_{i-1}^n + \Delta x, \text{ therefore, } T_5^n = T_4^n + \Delta x \tag{3}$$

When $i = 1$, Eq. (1) gives

$$T_1^{n+1} = rT_0^n + (1 - 2r)T_1^n + rT_2^n \tag{4}$$

Eliminating T_0^n from Eqs. (2) and (4), we get

$$T_1^{n+1} = (1 - r)T_1^n + rT_2^n$$

Thus, we use the following equations from $x = 0$ to 0.3, that is, for $i = 0, 1, 2$ and 3.

$$\left.\begin{aligned} T_1^{n+1} &= (1 - r)T_1^n + rT_2^n \\ T_0^{n+1} &= T_1^{n+1} \\ T_i^{n+1} &= rT_{i-1}^n + (1 - 2r)T_i^n + rT_{i+1}^n, \qquad \text{for } i = 2, 3 \end{aligned}\right\} \tag{5}$$

Also, for $i = 4$, Eq. (1) becomes

$$T_4^{n+1} = rT_3^n + (1 - 2r)T_4^n + rT_5^n \tag{6}$$

From Eqs. (3) and (6), we eliminate T_5^n to get

$$T_4^{n+1} = rT_3^n + (1 - r)T_4^n + \frac{\Delta t}{\Delta x}, \qquad \text{for } i = 4 \tag{7}$$

and Eq. (3) gives

$$T_5^{n+1} = T_4^{n+1} + \Delta x, \qquad \text{for } i = 5 \tag{8}$$

Thus, we use Eqs. (7) and (8) to compute T at $x = 0.4$ and $x = 0.5$, that is, for $i = 4$ and 5 and Eqs. (5) for internal nodes.

Following this procedure, the solution T at each node for three-time steps are computed and depicted in Table 9.2.

Table 9.2

n	t	x					
		0.0	0.1	0.2	0.3	0.4	0.5
0	0.0	0.0	0.0	0.0	0.0	0.0	0.0
1	0.001	0.0	0.0	0.0	0.0	0.01	0.11
2	0.002	0.0	0.0	0.0	0.001	0.019	0.119
3	0.003	0.0	0.0	0.0001	0.0027	0.0272	0.1272

9.3.2 Durfort–Frankel Method (1953)

This is another explicit method developed for solving diffusion equation

$$T_t = \alpha\, T_{xx} \tag{9.23}$$

In this method, both time and space derivatives are replaced by their central difference approximations. Thus, Eq. (9.23) is replaced as

$$\frac{T_i^{n+1} - T_i^{n-1}}{2\Delta t} = \alpha \frac{T_{i-1}^n - 2T_i^n + T_{i+1}^n}{\Delta x^2} \tag{9.24}$$

On rearrangement, it simplifies to

$$T_i^{n+1} = T_i^{n-1} + 2r\left(T_{i-1}^n - 2T_i^n + T_{i+1}^n\right) \tag{9.25}$$

where

$$r = \alpha \frac{\Delta t}{\Delta x^2}$$

Further, in Eq. (9.25), T_i^n is replaced by its average value at $(n + 1)$th and $(n - 1)$th time levels. Thus, if we substitute

$$T_i^n = \frac{T_i^{n+1} + T_i^{n-1}}{2}$$

into Eq. (9.25), we obtain

$$T_i^{n+1} = \frac{2r}{1 + 2r}\left(T_{i-1}^n + T_{i+1}^n\right) + \frac{1 - 2r}{1 + 2r} T_i^{n-1} \tag{9.26}$$

Since, it involves three time levels, it is called an *explicit* three-level finite difference scheme. To initiate computation using Durfort–Frankel scheme, we should know the solution at two-time levels, that is when $n = 0$ and $n = 1$ or at $t = 0$ and $t = \Delta t$.

To implement this scheme, from the given initial and boundary conditions, we shall compute the solution at $t = \Delta t$, using two-level explicit method and then use Eq. (9.26) to obtain the numerical solution at subsequent time steps, at all nodes covering the region of computation.

9.4 IMPLICIT METHODS

These are also basic finite difference methods for solving parabolic partial differential equations. A method in which the computation of many present unknown values necessitates the solution of a set of simultaneous equations is called an *implicit method.* For a given problem, the choice of a particular method depends on the computational accuracy required, the affordable computational time and storage memory. Some of the well-known implicit methods are discussed in the following sections.

9.4.1 Classical Implicit Method

This method was developed by O'Brien et al. to solve the diffusion equation (9.23). Here, we replace the time derivative by its backward difference approximation and the space derivative is replaced by its central difference approximation. Thus, using BTCS approximation, the diffusion equation evaluated at $(i, n + 1)$ grid point namely

$$\left(\frac{\partial T}{\partial t}\right)_i^{n+1} = \alpha \left(\frac{\partial^2 T}{\partial x^2}\right)_i^{n+1}$$

gives

$$\frac{T_i^{n+1} - T_i^n}{\Delta t} = \alpha \left[\frac{T_{i-1}^{n+1} - 2T_i^{n+1} + T_{i+1}^{n+1}}{(\Delta x)^2}\right]$$

on simplification, we get the following finite difference scheme

$$-rT_{i-1}^{n+1} + (1 + 2r)T_i^{n+1} - rT_{i+1}^{n+1} = T_i^n \tag{9.27}$$

for $i = 1, 2, \ldots, N - 1$, the number of unknowns in x-direction. This equation is implicit in the sense that there are three unknown values of T at $(n + 1)$th time level. For $n = 0$ and for $i = 1, 2, \ldots, N - 1$, Eq. (9.27) gives $(N - 1)$ simultaneous linear algebraic equations for $(N - 1)$ unknown values of T at internal grid points along the first time level in terms of known initial and boundary values. These equations can be solved to get the values of T_i^1, for $i = 1, 2, \ldots, N - 1$. Similarly by taking $n = 1$, we get $(N - 1)$ equations for $(N - 1)$ unknown values of T at the second time level, in terms of boundary values and the recently computed values of T at the first-time level and so on. Thus, for example, if the values of

T are known at nth time level, by substituting $i = 1, 2, \ldots, N-1$ in Eq. (9.27), we get the following set of simultaneous linear algebraic equations in tridiagonal form as given below:

$$\left.\begin{aligned}(1+2r)T_1^{n+1} - rT_2^{n+1} &= d_1^n \\ -rT_1^{n+1} + (1+2r)T_2^{n+1} - rT_3^{n+1} &= d_2^n \\ -rT_2^{n+1} + (1+2r)T_3^{n+1} - rT_4^{n+1} &= d_3^n \\ &\vdots \\ -rT_{N-2}^{n+1} + (1+2r)T_{N-1}^{n+1} &= d_{N-1}^n\end{aligned}\right\} \quad (9.28)$$

where,

$$d_1^n = T_1^n + rT_0^{n+1}, \qquad d_i^n = T_i^n, \qquad i = 2, 3, \ldots, N-2$$

and

$$d_{N-1}^n = T_{N-1}^n + rT_N^{n+1}$$

the subscript 0 and N correspond to the boundary value. This tridiagonal system can be solved exactly using Crout's LU reduction technique, as explained in Section 3.4 or Thomas Algorithm as explained in Appendix.

9.4.2 Crank–Nicolson Method (1947)

In this method, the diffusion equation is approximated by replacing both time and space derivatives by their central difference approximations (CTCS) at a point

$$\left[i\Delta x, \left(n + \frac{1}{2}\right)\Delta t\right]$$

which is half-way between the grid points (i, n) and $(i, n+1)$. Thus, the diffusion equation can be written as

$$\left(\frac{\partial T}{\partial t}\right)_i^{n+(1/2)} = \frac{\alpha}{2}\left[\left(\frac{\partial^2 T}{\partial x^2}\right)_i^n + \left(\frac{\partial^2 T}{\partial x^2}\right)_i^{n+1}\right] \quad (9.29)$$

Now, using central difference approximation to both the space and time derivatives, we get

$$\frac{T_i^{n+1} - T_i^n}{\Delta t} = \frac{\alpha}{2}\left(\frac{T_{i+1}^n - 2T_i^n + T_{i-1}^n}{\Delta x^2} + \frac{T_{i+1}^{n+1} - 2T_i^{n+1} + T_{i-1}^{n+1}}{\Delta x^2}\right)$$

or

$$-\frac{r}{2}T_{i-1}^{n+1} + (1+r)T_i^{n+1} - \frac{r}{2}T_{i+1}^{n+1} = \frac{r}{2}T_{i-1}^n + (1-r)T_i^n + \frac{r}{2}T_{i+1}^n \quad (9.30)$$

for $i = 0, 1, 2, \ldots, N-1, N$, where N is the number of divisions, into which the interval of integration is subdivided, thus, $i = 0$ and $i = N$ corresponds to the left and right boundaries respectively.

It can be observed from Eq. (9.30) that the new value of temperature T_i^{n+1} is not given directly in terms of known T_i^n at one-time step earlier but is also a function of unknown temperatures at adjacent positions, and is therefore, called an *implicit method*.

Thus, assuming, that the values of T are known at the nth time level and taking $i = 1, 2, \ldots, N - 1$, Eq. (9.30) yields a set of linear simultaneous equations in terms of unknowns T_i^{n+1} as given below in a tridiagonal form:

$$\begin{bmatrix} 1+r & -\frac{r}{2} & & & & \\ -\frac{r}{2} & 1+r & -\frac{r}{2} & & & \\ & -\frac{r}{2} & 1+r & -\frac{r}{2} & & \\ & \vdots & \vdots & \vdots & & \\ & & -\frac{r}{2} & 1+r & -\frac{r}{2} & \\ & & & -\frac{r}{2} & 1+r \end{bmatrix} \begin{pmatrix} T_1^{n+1} \\ T_2^{n+1} \\ T_3^{n+1} \\ \vdots \\ T_{N-2}^{n+1} \\ T_{N-1}^{n+1} \end{pmatrix} = \begin{pmatrix} d_1^n \\ d_2^n \\ d_3^n \\ \vdots \\ d_{N-2}^n \\ d_{N-1}^n \end{pmatrix} \tag{9.31}$$

where

$$\left.\begin{aligned} d_1^n &= \frac{r}{2}T_0^n + (1+r)T_1^n + \frac{r}{2}T_2^n + \frac{r}{2}T_0^{n+1} \\ d_i^n &= \frac{r}{2}T_{i-1}^n + (1-r)T_i^n + \frac{r}{2}T_{i+1}^n, \qquad i = 2, 3, \ldots, N-2 \\ d_{N-1}^n &= \frac{r}{2}T_{N-2}^n + (1-r)T_{N-1}^n + \frac{r}{2}T_N^n + \frac{r}{2}T_N^{n+1} \end{aligned}\right\} \tag{9.32}$$

all of which are known *quantities*. This resulting tridiagonal system can be solved exactly using Crout's reduction technique as explained in Section 3.4. In this method, not only the computational effort, but also the required computer memory space can be minimized. Also, since we use the same coefficient tridiagonal matrix to compute the solution at subsequent time steps, there will be a considerable saving in computer time as well, particularly while handling higher order systems. An example follows for illustration of this method.

Example 9.4 Use Crank–Nicolson method to find the numerical solution of the following parabolic partial differential equation after one-time step:

$$T_t = T_{xx}, \qquad 0 < x < 1$$

subject to the initial condition

$$T(x, 0) = 1, \qquad 0 < x < 1$$

and the boundary conditions

$$T(0, t) = T(1, t) = 0, \qquad t \geq 0$$

Compute the solution by taking $\Delta x = 1/4$ and $\Delta t = 1/32$.

Solution The Crank–Nicolson implicit finite difference scheme for the given differential equation is given by Eq. (9.30). That is

$$-\frac{r}{2}T_{i-1}^{n+1} + (1+r)T_i^{n+1} - \frac{r}{2}T_{i+1}^{n+1} = \frac{r}{2}T_{i-1}^n + (1-r)T_i^n + \frac{r}{2}T_{i+1}^n \tag{1}$$

Noting that $N = 1/\Delta x = 4$, $r = \Delta t/\Delta x^2 = 1/2$ and $N = 0, 4$ corresponds to the left and right boundaries respectively, for $i = 1, 2, 3$ and $n = 0$, Eq. (1) yields

$$
\begin{aligned}
-\frac{1}{4}T_0^1 + \frac{3}{4}T_1^1 - \frac{1}{4}T_2^1 &= \frac{1}{4}T_0^0 + \frac{1}{2}T_1^0 + \frac{1}{4}T_2^0 \\
-\frac{1}{4}T_1^1 + \frac{3}{4}T_2^1 - \frac{1}{4}T_3^1 &= \frac{1}{4}T_1^0 + \frac{1}{2}T_2^0 + \frac{1}{4}T_3^0 \\
-\frac{1}{4}T_2^1 + \frac{3}{4}T_3^1 - \frac{1}{4}T_4^1 &= \frac{1}{4}T_2^0 + \frac{1}{2}T_3^0 + \frac{1}{4}T_4^0
\end{aligned}
\tag{2}
$$

From the boundary conditions, it is clear that

$$T_0^n = T_4^n = 0 \qquad \text{for all } n = 0, 1, 2, \ldots$$

Thus, using the boundary conditions, we get

$$
\begin{aligned}
0 + 6T_1^1 - T_2^1 &= 0 + 2 + 1 \\
-T_1^1 + 6T_2^1 - T_3^1 &= 1 + 2 + 1 \\
-T_2^1 + 6T_3^1 - 0 &= 1 + 2 + 0
\end{aligned}
$$

which in matrix notation reduces to the tridiagonal system

$$
\begin{bmatrix} 6 & -1 & \\ -1 & 6 & -1 \\ & -1 & 6 \end{bmatrix}
\begin{pmatrix} T_1^1 \\ T_2^1 \\ T_3^1 \end{pmatrix} = \begin{pmatrix} 3 \\ 4 \\ 3 \end{pmatrix}
\tag{3}
$$

To solve this system, we preferably use Crout's reduction technique, where we decompose the coefficient matrix of Eq. (3) into the form

$$
A = LU = \begin{bmatrix} l_{11} & 0 & 0 \\ l_{21} & l_{22} & 0 \\ l_{31} & l_{32} & l_{33} \end{bmatrix}
\begin{bmatrix} 1 & u_{12} & u_{13} \\ 0 & 1 & u_{23} \\ 0 & 0 & 1 \end{bmatrix}
$$

Thus, we get

$$l_{11} = 6, \qquad l_{21} = -1, \qquad l_{31} = 0$$

$$l_{11}u_{12} = -1, \quad l_{11}u_{13} = 0 \quad \text{implying } u_{12} = -\frac{1}{6}, \quad u_{13} = 0$$

$$l_{21}u_{12} + l_{22} = 6 \text{ gives } l_{22} = 6 - \frac{1}{6} = \frac{35}{6}$$

$$l_{31}u_{12} + l_{32} = -1 \text{ gives } l_{32} = -1$$

and

$$l_{31}u_{13} + l_{32}\,u_{23} + l_{33} = 6 \qquad \text{or} \qquad l_{33} = \frac{204}{35}$$

Now, the system can be rewritten as

$$
\begin{bmatrix} 6 & 0 & 0 \\ -1 & \frac{35}{6} & 0 \\ 0 & -1 & \frac{204}{35} \end{bmatrix}
\begin{bmatrix} 1 & -\frac{1}{6} & 0 \\ 0 & 1 & -\frac{6}{35} \\ 0 & 0 & 1 \end{bmatrix}
\begin{pmatrix} T_1^1 \\ T_2^1 \\ T_3^1 \end{pmatrix} = \begin{pmatrix} 3 \\ 4 \\ 3 \end{pmatrix}
\tag{4}
$$

the solution of which is given by

$$(T_1^1,\ T_2^1,\ T_3^1) = \left(\frac{11}{17}, \frac{15}{17}, \frac{11}{17}\right) \tag{5}$$

Similarly, we can advance the solution one-time step at a time, by solving the same tridiagonal system (1) repeatedly with appropriate values of T_i^n for $n = 1, 2, \ldots$

To solve this problem by an explicit method and to get a result of same accuracy, we should choose Δt such that

$$\Delta t < \frac{1}{32}$$

Example 9.5 Solve the unsteady heat equation $T_t = T_{xx}$, $0 \le x \le 4$, subject to the initial conditions $T(x, 0) = 1000$,
and the boundary conditions

$$\frac{\partial T}{\partial x}(0,t) = 0.15(T-70), \qquad \frac{\partial T}{\partial x}(4,t) = 0.25(T-70)$$

using Crank–Nicolson finite difference method by taking, $\alpha = 0.132$, $\Delta x = 1.0$, $\Delta t = 1.893$. Give solution after one-time step.

Solution From the given data, we find that

$$r = \frac{\alpha \Delta t}{\Delta x^2} = 0.25$$

and the initial conditions are shown below:

$i = 1$	$i = 2$	$i = 3$	$i = 4$	$i = 5$
$x = 0$	$x = 1$	$x = 2$	$x = 3$	$x = 4$
$T_1^0 = 1000$	$T_2^0 = 1000$	$T_3^0 = 1000$	$T_4^0 = 1000$	$T_5^0 = 1000$

using Crank–Nicolson implicit scheme, we have from Eq. (9.30)

$$-0.25T_{i-1}^{n+1} + 2.5T_i^{n+1} - 0.25T_{i+1}^{n+1} = 0.25T_{i-1}^n + 1.5T_i^n + 0.25T_{i+1}^n \tag{1}$$

Let $n = 0$, $i = 1, 2, \ldots, 5$, then Eq. (1) gives

$$\left.\begin{aligned}
-0.25T_0^1 + 2.5T_1^1 - 0.25T_2^1 &= 0.25T_0^0 + 1.5T_1^0 + 0.25T_2^0 \\
-0.25T_1^1 + 2.5T_2^1 - 0.25T_3^1 &= 0.25T_1^0 + 1.5T_2^0 + 0.25T_3^0 \\
-0.25T_2^1 + 2.5T_3^1 - 0.25T_4^1 &= 0.25T_2^0 + 1.5T_3^0 + 0.25T_4^0 \\
-0.25T_3^1 + 2.5T_4^1 - 0.25T_5^1 &= 0.25T_3^0 + 1.5T_4^0 + 0.25T_5^0 \\
-0.25T_4^1 + 2.5T_5^1 - 0.25T_6^1 &= 0.25T_4^0 + 1.5T_5^0 + 0.25T_6^0
\end{aligned}\right\} \tag{2}$$

In this system of equations, T_0 and T_6 are fictitious points which we shall eliminate using the boundary conditions. Using central difference approximation to the left end boundary condition, we have

$$\frac{T_2^n - T_0^n}{\Delta x} = 0.15(T_1^n - 70)$$

Thus,

$$T_0^1 = T_2^1 - 0.15(T_1^1 - 70) \quad \text{and} \quad T_0^0 = T_2^0 - 0.15(T_1^0 - 70)$$

Therefore,

$$0.25(T_0^0 + T_0^1) = 0.25(T_2^0 + T_2^1) - 0.0375(T_1^0 + T_1^1) + 5.25$$

Hence, the first of Eq. (2) reduces to

$$2.5375T_1^1 - 0.5T_2^1 = 1.4625T_1^0 + 0.5T_2^0 + 5.25 \tag{3}$$

Similarly, using central difference approximation to the right-end boundary condition, we have

$$\frac{T_6^n - T_4^n}{\Delta x} = 0.25(T_5^n - 70)$$

Thus,

$$T_6^1 = T_4^1 + 0.25(T_5^1 - 70) \quad \text{and} \quad T_6^0 = T_4^0 + 0.25(T_5^0 - 70)$$

Therefore,

$$0.25(T_6^0 + T_6^1) = 0.25(T_4^0 + T_4^1) + 0.0625(T_5^0 + T_5^1) - 8.75$$

Hence, the last of Eq. (2) becomes

$$-0.5T_4^1 + 2.4375T_5^1 = 0.5T_4^0 + 1.5625T_5^0 - 8.75 \tag{4}$$

Using Eqs. (3) and (4) into Eq. (2), it can be written in matrix notation, after using the initial conditions in the form

$$\begin{bmatrix} 2.5375 & -0.5 & 0 & 0 & 0 \\ -0.25 & 2.5 & -0.25 & 0 & 0 \\ 0 & -0.25 & 2.5 & -0.25 & 0 \\ 0 & 0 & -0.25 & 2.5 & -0.25 \\ 0 & 0 & 0 & -0.5 & 2.4375 \end{bmatrix} \begin{pmatrix} T_1^1 \\ T_2^1 \\ T_3^1 \\ T_4^1 \\ T_5^1 \end{pmatrix} = \begin{pmatrix} 1967.75 \\ 2000 \\ 2000 \\ 2000 \\ 2053.75 \end{pmatrix}$$

Applying Crout's reduction technique, this matrix equation can be rewritten as

$$\begin{bmatrix} 2.5375 & 0 & 0 & 0 & 0 \\ -0.25 & 2.4507 & 0 & 0 & 0 \\ 0 & -0.25 & 2.475 & 0 & 0 \\ 0 & 0 & -0.25 & 2.4748 & 0 \\ 0 & 0 & 0 & -0.5 & 2.3971 \end{bmatrix} \begin{bmatrix} 1 & -0.1970 & 0 & 0 & 0 \\ 0 & 1 & -0.1020 & 0 & 0 \\ 0 & 0 & 1 & -0.1010 & 0 \\ 0 & 0 & 0 & 1 & -0.0808 \\ 0 & 0 & 0 & 0 & 1 \end{bmatrix}$$

$$\times \begin{pmatrix} T_1^1 \\ T_2^1 \\ T_3^1 \\ T_4^1 \\ T_5^1 \end{pmatrix} = \begin{pmatrix} 1967.75 \\ 2000 \\ 2000 \\ 2000 \\ 2053.75 \end{pmatrix} \tag{5}$$

Now, the solution of Eq. (5) can be easily found as

$$(T_1^1\ T_2^1\ T_3^1\ T_4^1\ T_5^1) = (971.8725\ 996.9774\ 997.8171\ 983.2880\ 1044.2641)$$

This is the heat distribution after one-time step. At subsequent time steps, the right-hand side values of Eq. (5) changes, while the coefficient matrix remains unchanged and hence the solution of the given problem can be obtained at subsequent time steps with little effort and so on.

9.4.3 Weighted Average Implicit Method

To explain this method, we shall recall the heat conduction equation

$$T_t = \alpha T_{xx} \tag{9.33}$$

In this method, we shall replace the time derivative by its backward difference approximation and space derivative by its central difference approximation. Thus, using BTCS approximation, Eq. (9.33) is first evaluated at the grid point $(i, n + 1)$ to get

$$T_i^{n+1} - T_i^n = r(T_{i-1}^{n+1} - 2T_i^{n+1} + T_{i+1}^{n+1}) \tag{9.34}$$

where

$$r = \frac{\alpha \Delta t}{\Delta x^2}$$

Now, we introduce the weighing factor λ such that $0 \le \lambda \le 1$, and rewrite the above expression in the form

$$T_i^{n+1} - T_i^n = r[\lambda (T_{i-1}^{n+1} - 2T_i^{n+1} + T_{i+1}^{n+1}) + (1 - \lambda)(T_{i-1}^n - 2T_i^n + T_{i+1}^n)]$$

as a weighted average at $(n + 1)$th and nth time levels. On rewriting the above equation, we obtain

$$\begin{aligned} -r\lambda T_{i-1}^{n+1} + (1 + 2r\lambda)T_i^{n+1} - r\lambda T_{i+1}^{n+1} &= r(1 - \lambda)T_{i-1}^n \\ &+ [1 - 2r(1 - \lambda)]\, T_i^n + r(1 - \lambda)T_{i+1}^n \end{aligned} \tag{9.35}$$

From Eq. (9.35), it can be observed that when $\lambda = 1$, we can recover O' Brien et. al., formula (9.27); when $\lambda = 1/2$, we obtain Crank–Nicolson formula (9.30); and when $\lambda = 0$, we get the explicit formula represented by Eq. (9.22). Hence, this weighted average method can be considered as the most general implicit method, which combines many methods so far discussed.

9.5 THE CONCEPT OF STABILITY

In Sections 9.3 and 9.4, we have introduced various finite difference methods for solving parabolic partial differential equation which models heat conduction equation. All those methods essentially reduce the solution of the partial differential equation to the solution of a set of algebraic equations. We are also aware that numerical computations carried out on any computer can be accurate only to a finite number of decimal places. At each stage of calculation, some round-off error will be introduced, no matter how small. Let e be an error introduced into the calculations at each mesh point (i, n) and $\bar{u}_i^n$, the resulting numerical solution of the differential equation

$$u_t = \alpha u_{xx} \tag{9.36}$$

with finite differences approximation. If u_i^n is the corresponding solution of the differential equation with a finite difference method, assume no round-off error is involved in each calculation. Then, we define

$$e_i^n = u_i^n - \bar{u}_i^n \tag{9.37}$$

Stability considerations are very important in getting the numerical solution of a differential equation, using finite difference methods. The solution of the finite difference equation is said to be stable, if the error do not grow exponentially as we progress from one-time step to another. That is, e_i^n should remain uniformly bounded as $n \to \infty$, while Δt, $\Delta x \to 0$. There are two standard methods for investigating the growth of errors, in solving the finite difference equations: matrix method and Von-Neumann fourier series expansion method. However, we shall discuss only matrix method in the following section to show the important aspects of theoretical study of numerical stability. Consistency and convergence of the finite difference method are other important theoretical aspects, for which the reader is advised to consult advanced books on the subject (see Richtmyer, 1957).

The matrix method

Let us consider one-dimensional heat conduction problem described by the differential equation

$$u_t = \alpha u_{xx}, \qquad 0 \le x \le L \tag{9.38}$$

subject to the boundary conditions

$$u = 0 \quad \text{at} \quad x = 0 \quad \text{and} \quad x = L, \quad t \ge 0$$

and the initial condition

$$u = f(x), \qquad 0 < x < L, \qquad t = 0$$

Using simple explicit finite difference scheme FTCS, the given differential equation reduces to

$$u_i^{n+1} = ru_{i-1}^n + (1 - 2r)u_i^n + ru_{i+1}^n \tag{9.39}$$

for $n = 0, 1, 2, \ldots$ and for $i = 0, 1, 2, \ldots, N$ such that $\Delta x = L/N$. The subscripts $0, N$ corresponds to $x = 0, L$ the left- and right-end boundaries, where u is known for all t. Expanding Eq. (9.39) for $i = 1, 2, \ldots, (N - 1)$, we have

$$u_1^{n+1} = ru_0^n + (1 - 2r)u_1^n + ru_2^n$$

$$u_2^{n+1} = ru_1^n + (1 - 2r)u_2^n + ru_3^n$$

$$\vdots$$

$$u_{N-1}^{n+1} = ru_{N-2}^n + (1 - 2r)u_{N-1}^n + ru_N^n$$

using the boundary conditions $u_0^n = u_N^n = 0$ for all t, the above system written in matrix notation takes the following form

$$\begin{pmatrix} u_1^{n+1} \\ u_2^{n+1} \\ u_3^{n+1} \\ \vdots \\ u_{N-1}^{n+1} \end{pmatrix} = \begin{bmatrix} 1-2r & r & & & \\ r & 1-2r & r & & \\ & r & 1-2r & r & \\ & & & \ddots & \\ & & & r & 1-2r \end{bmatrix} \begin{pmatrix} u_1^n \\ u_2^n \\ u_3^n \\ \vdots \\ u_{N-1}^n \end{pmatrix} \tag{9.40}$$

equivalently

$$U^{n+1} = [A]U^n \tag{9.41}$$

where U is a column vector, $[A]$ is an $(N-1) \times (N-1)$ square matrix, which happened to be a tridiagonal matrix. Here, $[A]$ is called an *amplification* matrix, U^0 is a vector of initial values, with which we have initiated our computation. The successive rows of computations with each-time step is given as

$$U^1 = [A]\, U^0$$

$$U^2 = [A]\, U^1 = [A]\, [A]\, U^0 = [A^2]\, U^0$$

$$\vdots$$

$$U^{n+1} = [A]\, U^n = [A^2]\, U^{n-1} = \cdots = [A^{n+1}]\, U^0$$

Now, suppose we have introduced errors at every node along $n = 0$ row or at $t = 0$ and start the computation with the vector of values $\overline{U}^0$, we then compute successive rows as follows:

$$\overline{U}^1 = [A]\, \overline{U}^0$$

$$\overline{U}^2 = [A]\, \overline{U}^1 = [A^2]\, \overline{U}^0$$

$$\vdots$$

$$\overline{U}^{n+1} = [A^{n+1}]\, \overline{U}^0$$

Then the error vector e can be written as

$$e^{n+1} = U^{n+1} - \overline{U}^{n+1} = [A^{n+1}]\,(U^0 - \overline{U}^0) = [A^{n+1}]\, e^0 \tag{9.42}$$

indicating that the formula for error propagation is similar to the one used in the computation of U. Since the difference equation is linear, superposition principle is applicable and hence it is enough to consider the propagation of only one line of errors. A finite difference scheme is known to be stable if e^n remains bounded as $n \to \infty$, which is ensured, provided that the eigenvalue with largest modulus of the amplification matrix $[A]$ is less than or equal to unity.

Now, from Eq. (9.40), the matrix $[A]$ can be rewritten as

$$[A] = \begin{bmatrix} 1 & 0 & 0 & & \\ 0 & 1 & 0 & & \\ & & \ddots & & \\ & & 0 & 1 & 0 \\ & & & 0 & 1 \end{bmatrix} + r \begin{bmatrix} -2 & 1 & & & \\ 1 & -2 & 1 & & \\ & & \ddots & & \\ & & 1 & -2 & 1 \\ & & & 1 & -2 \end{bmatrix}$$

Alternatively

$$[A] = [I_{N-1}] + r\,[D_{N-1}]$$

where $[I_{N-1}]$ is a unit matrix of order $(N-1)$ and $[D_{N-1}]$ is an $(N-1) \times (N-1)$ tridiagonal matrix. We may recall from linear algebra that the eigenvalues of an $N \times N$ tridiagonal matrix

$$\begin{bmatrix} b & c & & & & \\ a & b & c & & & \\ & a & b & c & & \\ & & & \ddots & & \\ & & & a & b & c \\ & & & & a & b \end{bmatrix} \tag{9.43}$$

are given by the formula

$$\lambda_s = b + 2\sqrt{ca}\,\cos\frac{s\pi}{N+1}, \qquad s = 1, 2, \ldots, N \tag{9.44}$$

Thus, the eigenvalues of $[D_{N-1}]$ are

$$\lambda_s = -2 + 2\cos\frac{s\pi}{N} = -4\sin^2\frac{s\pi}{2N}, \qquad s = 1, 2, \ldots, (N-1)$$

For the matrix $[A]$, the eigenvalues are

$$1 + r\left(-4\sin^2\frac{s\pi}{2N}\right)$$

Hence, the condition for the stability of simple explicit scheme (9.39) to solve heat conduction equation (9.38) is

$$\left|\, 1 - 4r\sin^2\frac{s\pi}{2N} \,\right| \le 1$$

Therefore,

$$-1 \le -4r\sin^2\frac{s\pi}{2N} \le 1$$

Which gives the limiting value of r as

$$-1 \le 1 - 4r\sin^2\frac{s\pi}{2N}$$

or

$$r \le \frac{1}{2\sin^2\dfrac{s\pi}{2N}}$$

Thus, the explicit scheme is stable if

$$\left(r = \alpha\frac{\Delta t}{\Delta x^2} \le \frac{1}{2}\right) \tag{9.45}$$

Example 9.6 Show that the Crank–Nicolson scheme for the diffusion equation

$$T_t = T_{xx}, \qquad 0 \le x \le L$$

satisfying the conditions

$$T = 0 \quad \text{at} \quad x = 0, \qquad T = L \quad \text{for} \quad t > 0$$

and

$$T = f(x), \quad 0 \le x \le L \qquad \text{for } t = 0$$

has unrestricted stability.

Solution We may recall the Crank–Nicolson finite difference scheme, for the given diffusion equation, given by Eq. (9.30)

$$-rT_{i-1}^{n+1} + 2(1+r)T_i^{n+1} - rT_{i+1}^{n+1} = rT_{i-1}^n + 2(1-r)T_i^n + rT_{i+1}^n, \qquad i = 0, 1, 2, \ldots, N \tag{1}$$

$i = 0, N$ correspond to the left and right boundaries respectively. For $i = 1, 2, \ldots, (N-1)$, Eq. (1) can be expanded and written in the following matrix form:

$$\begin{bmatrix} 2(1+r) & -r & & & \\ -r & 2(1+r) & -r & & \\ & & \ddots & & \\ & & -r & 2(1+r) & -r \\ & & & -r & 2(1+r) \end{bmatrix} \begin{pmatrix} T_1^{n+1} \\ T_2^{n+1} \\ \vdots \\ T_{N-2}^{n+1} \\ T_{N-1}^{n+1} \end{pmatrix}$$

$$= \begin{bmatrix} 2(1-r) & r & & & \\ r & 2(1-r) & r & & \\ & & \ddots & & \\ & & r & 2(1-r) & r \\ & & & r & 2(1-r) \end{bmatrix} \begin{pmatrix} T_1^n \\ T_2^n \\ \vdots \\ T_{N-2}^n \\ T_{N-1}^n \end{pmatrix} \tag{2}$$

That is,

$$(2I - rD_{N-1})\overline{T}^{n+1} = (2I + rD_{N-1})\overline{T}^n \tag{3}$$

where,

$$D_{N-1} = \begin{bmatrix} -2 & 1 & & & \\ 1 & -2 & 1 & & \\ & & \ddots & & \\ & & 1 & -2 & 1 \\ & & & 1 & -2 \end{bmatrix} \tag{4}$$

Thus,

$$\overline{T}^{n+1} = (2I - rD_{N-1})^{-1} (2I + rD_{N-1})\overline{T}^n$$

Alternatively, it can be put in the form $\overline{T}^{n+1} = [A]\overline{T}^n$, where

$$[A] = (2I - rD_{N-1})^{-1} (2I + rD_{N-1})$$

For stability, the modulus of the eigenvalues of $[A]$ should each be less than unity. However, the eigenvalues of D_{N-1} are known to be given by

$$\lambda_s = -2 + 2\cos\frac{s\pi}{N} = -4\sin^2\frac{s\pi}{2N}$$

Hence, the eigenvalues of $[A]$ are

$$\frac{2 - 4r\sin^2\dfrac{s\pi}{2N}}{2 + 4r\sin^2\dfrac{s\pi}{2N}}, \quad s = 1, 2, \ldots, N-1 \tag{5}$$

These are clearly less than one for all positive values of r. This proves that Crank–Nicolson scheme for the given problem is stable for all positive values of r, so has unrestricted stability.

9.6 METHODS FOR TWO-DIMENSIONAL EQUATIONS

Consider the finite difference methods of solution for the equation:

$$\frac{\partial u}{\partial t} = \alpha\left(\frac{\partial^2 u}{\partial x^2} + \frac{\partial^2 u}{\partial y^2}\right) \tag{9.46}$$

in a rectangular region bounded by $\overline{R} = R \times [0 \le t \le T]$, where $R = [0 \le x, y \le 1]$, with appropriate initial and boundary conditions. Here again, the region $\overline{R}$ in (x, y, t) space is covered by a rectangular grid with sides parallel to the axes, as shown in Fig. 9.1, with $\Delta x = \Delta y = h$ and $\Delta t = k$ with grid spacings in the space and time directions respectively. The grid points (x, y, t) are given by $x = i\Delta x$, $y = j\Delta y$, $t = n\Delta t$; where i, j, n are all integers and $i = j = 0$ correspond to the origin. The function u satisfying the difference equation at the grid points is denoted by $u_{i,j}^n$. With this notation, we shall extend the discussion of Sections 9.3 and 9.4 to two-dimensional parabolic Eq. (9.46).

9.6.1 Explicit Methods

Straightforward use of FTCS approximation to Eq. (9.46) yields

$$\frac{u_{i,j}^{n+1} - u_{i,j}^n}{\Delta t} = \alpha\left(\frac{u_{i-1,j}^n - 2u_{i,j}^n + u_{i+1,j}^n}{\Delta x^2} + \frac{u_{i,j-1}^n - 2u_{i,j}^n + u_{i,j+1}^n}{\Delta y^2}\right)$$

Now, we define $r = \alpha\Delta t/h^2$, where $\Delta x = \Delta y = h$. Then, the above equation reduces to

$$u_{i,j}^{n+1} = u_{i,j}^{n} + r\left(u_{i-1,j}^{n} - 2u_{i,j}^{n} + u_{i+1,j}^{n}\right) + r\left(u_{i,j-1}^{n} - 2u_{i,j}^{n} + u_{i,j+1}^{n}\right) \quad (9.47)$$

This is the standard explicit formula involving five points at the time level $t = n\Delta t = nk$. Similarly, the simplest three level scheme for the two-dimensional diffusion Eq. (9.46) can be written as

$$\frac{u_{i,j}^{n+1} - u_{i,j}^{n-1}}{2\Delta t} = \alpha \frac{u_{i-1,j}^{n} + u_{i+1,j}^{n} + u_{i,j-1}^{n} + u_{i,j+1}^{n} - 4u_{i,j}^{n}}{h^2}$$

where $\Delta x = \Delta y = h$. If we define $r = \alpha\Delta t/h^2$, the above equation can be written as

$$u_{i,j}^{n+1} = u_{i,j}^{n-1} + 2r\left(u_{i-1,j}^{n} + u_{i+1,j}^{n} + u_{i,j-1}^{n} + u_{i,j+1}^{n} - 4u_{i,j}^{n}\right) \quad (9.48)$$

This is known as *Richardson's formula*. If $u_{i,j}^{n}$ is replaced by

$$\frac{u_{i,j}^{n+1} + u_{i,j}^{n-1}}{2}$$

Then Eq. (9.48) reduces to

$$u_{i,j}^{n+1} = u_{i,j}^{n-1} + 2r[u_{i-1,j}^{n} + u_{i+1,j}^{n} + u_{i,j-1}^{n} + u_{i,j+1}^{n} + 2(u_{i,j}^{n+1} + u_{i,j}^{n-1})]$$

or

$$u_{i,j}^{n+1} = \frac{1-4r}{1+4r}u_{i,j}^{n-1} + \frac{2r}{1+4r}\left(u_{i-1,j}^{n} + u_{i+1,j}^{n} + u_{i,j-1}^{n} + u_{i,j+1}^{n}\right) \quad (9.49)$$

This formula is called *Durfort–Frankel* scheme.

These explicit methods can be easily extended to solve non-linear partial differential equations and to study numerical simulations in many branches of fluid mechanics (The interested reader may refer to Jain and Sankara Rao, 1969).

9.6.2 Implicit Methods

Taking $\Delta x = \Delta y = h$, if we apply Crank–Nicolson scheme (Section 9.4) to Eq. (9.46), we obtain

$$u_{i,j}^{n+1} = u_{i,j}^{n} + \frac{r}{2}\left(u_{i-1,j}^{n+1} + u_{i+1,j}^{n+1} + u_{i,j-1}^{n+1} + u_{i,j+1}^{n+1} - 4u_{i,j}^{n+1}\right)$$

$$+ \frac{r}{2}\left(u_{i-1,j}^{n} + u_{i+1,j}^{n} + u_{i,j-1}^{n} + u_{i,j+1}^{n} - 4u_{i,j}^{n}\right) \quad (9.50)$$

where, $r = \alpha\Delta t/h^2$. Further, if we define $a = -r/2$, $c = (1 + 2r)$, and

$$b_{i,j}^{n} = u_{i,j}^{n} + \frac{r}{2}\left(u_{i-1,j}^{n} + u_{i+1,j}^{n} + u_{i,j-1}^{n} + u_{i,j+1}^{n} - 4u_{i,j}^{n}\right)$$

We can rewrite Eq. (9.50) as

$$au_{i-1,j}^{n+1} + cu_{i,j}^{n+1} + au_{i+1,j}^{n+1} + au_{i,j-1}^{n+1} + au_{i,j+1}^{n+1} = b_{i,j}^{n} \tag{9.51}$$

This resulting linear algebraic system is no longer tridiagonal, in view of five unknowns $u_{i,j}^{n+1}$, $u_{i-1,j}^{n+1}$, $u_{i+1,j}^{n+1}$, $u_{i,j-1}^{n+1}$, $u_{i,j+1}^{n+1}$ in Eq. (9.51). Such a system is in general solved by an iterative method.

9.6.3 Alternate Direction Implicit Method

While solving time-dependent partial differential equations, in more than one space variable, it is advantageous to break the numerical process into one of sequentially solving simpler problems. Such a process is called *splitting*. For example, we will discover that much is to be gained by *splitting* a two-dimensional problem into a sequence of two one-dimensional problems. This procedure leads to the development of so-called alternate direction implicit (ADI) method. This method was first developed by Peaceman and Rachford in 1955.

It is a two-step process involving the solution of tridiagonal system of equations using a very efficient and direct solver known as Thomas algorithm (For details see Appendix) along lines parallel to x and y axes at the first and second steps respectively. When ADI method is applied to Eq. (9.46), in the first step, we use row-wise traverse, that is, we advance from $n\Delta t$ to $(n + 1)\Delta t$ by approximating the space derivatives using central difference approximation at $(n + 1)$th time level in the x-direction and at n th time level in the y-direction. After row-wise traverse, we proceed column-wise traverse as a second step. In this case, we advance from $(n + 1)$th time level to $(n + 2)$th time level, using central difference approximation to both the space derivatives, this time at $(n + 1)$th level in the x-direction and at $(n + 2)$th time level in the y-direction. Thus, the resulting expressions by taking $\Delta x = \Delta y = h$ and $r = \alpha\Delta t/h^2$ can be written down as

$$u_{i,j}^{n+1} - u_{i,j}^{n} = r\left(u_{i-1,j}^{n+1} - 2u_{i,j}^{n+1} + u_{i+1,j}^{n+1}\right) + r\left(u_{i,j-1}^{n} + u_{i,j+1}^{n} - 2u_{i,j}^{n}\right) \tag{9.52}$$

and

$$u_{i,j}^{n+2} - u_{i,j}^{n+1} = r\left(u_{i-1,j}^{n+1} - 2u_{i,j}^{n+1} + u_{i+1,j}^{n+1}\right) + r\left(u_{i,j-1}^{n+2} - 2u_{i,j}^{n+2} + u_{i,j+1}^{n+2}\right) \tag{9.53}$$

This split pair (9.52) and (9.53) was first introduced by Peaceman and Rachford and is known as *Peaceman–Rachford formula*. Let us apply this formula to a two-dimensional problem as follows.

Example 9.7 A rectangular plate is of 8 units by 6 units. Initially, all the points on the plate are at 50°C. The edges are suddenly brought to the temperatures shown in Fig. 9.4 and held at these temperatures. Use ADI method to trace the history of temperatures at the internal nodes spaced two units apart, assuming that heat flows only in x and y directions (Gerald et al., 1994).

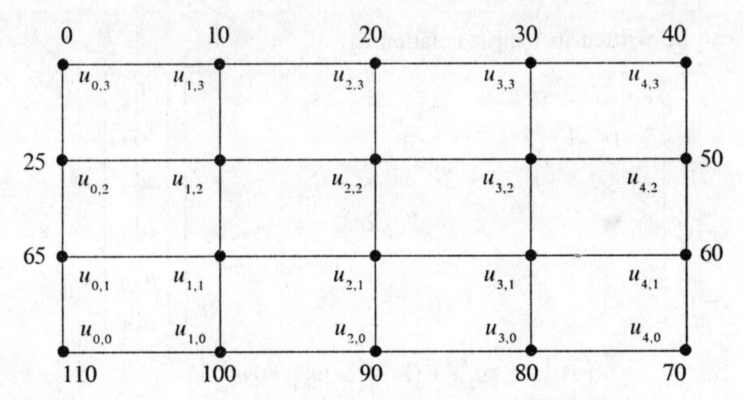

Fig. 9.4 Scheme of computation (Example 9.7).

Solution The governing equation of heat transfer is given by $u_t = u_{xx} + u_{yy}$.

Taking $\Delta x = \Delta y = h$, the discretized form of the above equation is

$$\frac{u_{i,j}^{n+1} - u_{i,j}^{n}}{\Delta t} = \frac{u_{i-1,j} - 2u_{i,j} + u_{i+1,j}}{h^2} + \frac{u_{i,j-1} - 2u_{i,j} + u_{i,j+1}}{h^2} \tag{1}$$

Let $r = \Delta t/\Delta x^2$, then we apply ADI method, where, as a first step we use row-wise traverse and Eq. (1) can be rearranged in the form

$$u_{i,j}^{n+1} - u_{i,j}^{n} = r\left(u_{i-1,j}^{n+1} - 2u_{i,j}^{n+1} + u_{i+1,j}^{n+1}\right) + r\left(u_{i,j-1}^{n} - 2u_{i,j}^{n} + u_{i,j+1}^{n}\right) \tag{2}$$

After the row-wise traverse, we proceed column-wise traverse as a second step and the corresponding finite difference form is

$$u_{i,j}^{n+2} - u_{i,j}^{n+1} = r\left(u_{i-1,j}^{n+1} - 2u_{i,j}^{n+1} + u_{i+1,j}^{n+1}\right) + r\left(u_{i,j-1}^{n+2} - 2u_{i,j}^{n+2} + u_{i,j+1}^{n+2}\right) \tag{3}$$

For row-wise traverse, we consider $j = 1, 2$ and vary $i = 1, 2, 3$. Thus, Eq. (2) gives

$$u_{1,1}^{1} - u_{1,1}^{0} = r\left(u_{0,1}^{1} - 2u_{1,1}^{1} + u_{2,1}^{1}\right) + r\left(u_{1,0}^{0} - 2u_{1,1}^{0} + u_{1,2}^{0}\right)$$

$$u_{2,1}^{1} - u_{2,1}^{0} = r\left(u_{1,1}^{1} - 2u_{2,1}^{1} + u_{3,1}^{1}\right) + r\left(u_{2,0}^{0} - 2u_{2,1}^{0} + u_{2,2}^{0}\right)$$

$$u_{3,1}^{1} - u_{3,1}^{0} = r\left(u_{2,1}^{1} - 2u_{3,1}^{1} + u_{4,1}^{1}\right) + r\left(u_{3,0}^{0} - 2u_{3,1}^{0} + u_{3,2}^{0}\right)$$

$$u_{1,2}^{1} - u_{1,2}^{0} = r\left(u_{0,2}^{1} - 2u_{1,2}^{1} + u_{2,2}^{1}\right) + r\left(u_{1,1}^{0} - 2u_{1,2}^{0} + u_{1,3}^{0}\right)$$

$$u_{2,2}^{1} - u_{2,2}^{0} = r\left(u_{1,2}^{1} - 2u_{2,2}^{1} + u_{3,2}^{1}\right) + r\left(u_{2,1}^{0} - 2u_{2,2}^{0} + u_{2,3}^{0}\right)$$

$$u_{3,2}^{1} - u_{3,2}^{0} = r\left(u_{2,2}^{1} - 2u_{3,2}^{1} + u_{4,2}^{1}\right) + r\left(u_{3,1}^{0} - 2u_{3,2}^{0} + u_{3,3}^{0}\right)$$

which can be written in matrix notation as

$$\begin{bmatrix} 1+2r & -r & & & & \\ -r & 1+2r & -r & & & \\ & -r & 1+2r & & & \\ & & & 1+2r & -r & \\ & & & -r & 1+2r & -r \\ & & & & -r & 1+2r \end{bmatrix} \begin{pmatrix} u_{1,1}^1 \\ u_{2,1}^1 \\ u_{3,1}^1 \\ u_{1,2}^1 \\ u_{2,2}^1 \\ u_{3,2}^1 \end{pmatrix}$$

$$= \begin{pmatrix} ru_{0,1}^1 + ru_{1,0}^0 + (1-2r)u_{1,1}^0 + ru_{1,2}^0 \\ ru_{2,0}^0 + (1-2r)u_{2,1}^0 + ru_{2,2}^0 \\ ru_{4,1}^1 + ru_{3,0}^0 + (1-2r)u_{3,1}^0 + ru_{3,2}^0 \\ ru_{0,2}^1 + ru_{1,1}^0 + (1-2r)u_{1,2}^0 + ru_{1,3}^0 \\ ru_{2,1}^0 + (1-2r)u_{2,2}^0 + ru_{2,3}^0 \\ ru_{4,2}^1 + ru_{3,1}^0 + (1-2r)u_{3,2}^0 + ru_{3,3}^0 \end{pmatrix}$$

Taking $r = 1$, which imply $\Delta t = r\Delta x^2 = \Delta x^2 = 4$ and using the given boundary conditions, we have

$$\begin{bmatrix} 3 & -1 & & & & \\ -1 & 3 & -1 & & & \\ & -1 & 3 & & & \\ & & & 3 & -1 & \\ & & & -1 & 3 & -1 \\ & & & & -1 & 3 \end{bmatrix} \begin{pmatrix} u_{1,1}^1 \\ u_{2,1}^1 \\ u_{3,1}^1 \\ u_{1,2}^1 \\ u_{2,2}^1 \\ u_{3,2}^1 \end{pmatrix} = \begin{pmatrix} 65+50-50+50 \\ 50-50+50 \\ 60+50-50+50 \\ 25+50-50+50 \\ 50-50+50 \\ 50+50-50+50 \end{pmatrix} = \begin{pmatrix} 115 \\ 50 \\ 110 \\ 75 \\ 50 \\ 100 \end{pmatrix}$$

On solving this tridiagonal system, the first row-wise traverse gives result at $t = 4$. Thus, we have

$$\begin{aligned} u_{1,1}^1 &= 56.1904, & u_{1,2}^1 &= 40.4762 \\ u_{2,1}^1 &= 53.5714, & u_{2,2}^1 &= 46.4286 \\ u_{3,1}^1 &= 54.5238, & u_{3,2}^1 &= 48.8095 \end{aligned}$$

These results are used to build up for the next computation. Now as a second step, a column-wise traverse, that is for $i = 1, 2, 3$ varying $j = 1, 2$ we get from Eq. (3)

$$u_{1,1}^2 - u_{1,1}^1 = r\left(u_{0,1}^1 - 2u_{1,1}^1 + u_{2,1}^1\right) + r\left(u_{1,0}^2 - 2u_{1,1}^2 + u_{1,2}^2\right)$$

$$u_{1,2}^2 - u_{1,2}^1 = r\left(u_{0,2}^1 - 2u_{1,2}^1 + u_{2,2}^1\right) + r\left(u_{1,1}^2 - 2u_{1,2}^2 + u_{1,3}^2\right)$$

$$u_{2,1}^2 - u_{2,1}^1 = r\left(u_{1,1}^1 - 2u_{2,1}^1 + u_{3,1}^1\right) + r\left(u_{2,0}^2 - 2u_{2,1}^2 + u_{2,2}^2\right)$$

$$u_{2,2}^2 - u_{2,2}^1 = r\left(u_{1,2}^1 - 2u_{2,2}^1 + u_{3,2}^1\right) + r\left(u_{2,1}^2 - 2u_{2,2}^2 + u_{2,3}^2\right)$$

$$u_{3,1}^2 - u_{3,1}^1 = r\left(u_{2,1}^1 - 2u_{3,1}^1 + u_{4,1}^1\right) + r\left(u_{3,0}^2 - 2u_{3,1}^2 + u_{3,2}^2\right)$$

$$u_{3,2}^2 - u_{3,2}^1 = r\left(u_{2,2}^1 - 2u_{3,2}^1 + u_{4,2}^1\right) + r\left(u_{3,1}^2 - 2u_{3,2}^2 + u_{3,3}^2\right)$$

Taking $r = 1$ and using the computed values in the first step, the above equations can be written in matrix notation as

$$\begin{bmatrix} 3 & -1 & & & & \\ -1 & 3 & & & & \\ & & 3 & -1 & & \\ & & -1 & 3 & & \\ & & & & 3 & -1 \\ & & & & -1 & 3 \end{bmatrix} \begin{pmatrix} u_{1,1}^2 \\ u_{1,2}^2 \\ u_{2,1}^2 \\ u_{2,2}^2 \\ u_{3,1}^2 \\ u_{3,2}^2 \end{pmatrix} = \begin{pmatrix} -56.1904 + 65 + 53.5714 + 100 \\ -40.4762 + 25 + 46.4286 + 10 \\ -53.5714 + 56.1904 + 54.5238 + 90 \\ -46.4286 + 40.4762 + 48.8095 + 20 \\ 54.5238 + 53.5714 + 60 + 80 \\ -48.8095 + 46.4286 + 50 + 30 \end{pmatrix} = \begin{pmatrix} 162.381 \\ 40.9524 \\ 147.1428 \\ 62.8571 \\ 139.0476 \\ 77.6191 \end{pmatrix}$$

whose solution is given by

$$u_{1,1}^2 = 66.0119, \quad u_{2,2}^2 = 41.9643$$
$$u_{1,2}^2 = 35.6548, \quad u_{3,1}^2 = 61.8452$$
$$u_{2,1}^2 = 63.0357, \quad u_{3,2}^2 = 46.4881$$

This is the ADI method of solution at $t = 8$.

This procedure can be continued alternately between row-wise and column-wise traverse until steady state solution is obtained. This ADI technique is more convenient for computer adaptation than for hand computation.

EXERCISES

9.1 Derive central difference formula for third derivative in the following form:

$$\left(\frac{\partial^3 T}{\partial x^3}\right)_{i,j} = \frac{T_{i+2,j} - 2T_{i+1,j} + 2T_{i-1,j} - T_{i-2,j}}{2(\Delta x)^3} - \frac{(\Delta x)^2}{4}\left(\frac{\partial^5 T}{\partial x^5}\right)_{i,j}$$

9.2 Derive one-sided formula for the second derivative in the form

$$\left(\frac{\partial^2 T}{\partial x^2}\right) = \frac{T_{i,j} - 2T_{i-1,j} + T_{i-2,j}}{(\Delta x)^2} + \Delta x\left(\frac{\partial^3 T}{\partial x^3}\right) + \cdots$$

9.3 Obtain a finite difference solution to the following heat conduction problem described by the differential equation $T_t = T_{xx}$, $0 \le x \le 1$, subject to the initial condition at $t = 0$,

$$T = \begin{cases} 2x \text{ for } 0 \le x \le \dfrac{1}{2} \\ 2(1 - x) \text{ for } \dfrac{1}{2} \le x \le 1 \end{cases}$$

and the boundary conditions $T = 0$ when $x = 0$, and $x = 1$ for all $t > 0$, using an explicit method for the first five-time steps, taking $\Delta x = 0.1$, $\Delta t = 0.002$.

9.4 Solve the parabolic partial differential equation $T_t = 4\ T_{xx}$, $0 \le x \le 8$, subject to the boundary conditions $T(0, t) = T(8, t) = 0$, $t > 0$ and the initial condition

$$T(x, 0) = 4x - \frac{x^2}{2}$$

using an explicit finite difference method. Carry out computations for five-time levels, taking $\Delta x = 1$ and $\Delta t = 1/8$.

9.5 Solve the following heat conduction equation $T_t = T_{xx}$, $0 \le x \le 1$ subject to the initial condition $T = 1$ for $0 \le x \le 1$ at $t = 0$, and the boundary conditions $T_x = T$ at $x = 0$ for $t > 0$ and $T_x = -T$ at $x = 1$ for $t > 0$ using an explicit finite difference method, by taking $\Delta x = 0.1$ and $\Delta t = 0.0025$ up to $t = 0.0125$.

9.6 Solve the following initial value problem using Durfort–Frankel method for two time-steps described by the partial differential equation $T_t = T_{xx}$, $0 \le x \le 1$ subject to the initial condition

$$T(x, 0) = \cos \frac{\pi x}{2}$$

and the boundary conditions $T(0, t) = 1$, $T(1, t) = 0$ for $t > 0$, taking

$$\Delta x = \frac{1}{3}, \qquad r = \frac{\Delta t}{\Delta x^2} = \frac{1}{3}$$

9.7 The function T satisfies the non-linear partial differential equation

$$T_t = T_{xx} + T_x^{\,2}, \qquad 0 < x < 1, \qquad t < 0$$

The initial and boundary conditions are given by $T = 0$ when $t = 0$, $0 \le x \le 1$ and $T_x = 1$ at $x = 0$, $T = 0$ at $x = 1$ for $t > 0$. Introduce the transformation $T = \log u$, $u \ne 0$ and hence show that the above problem can be reduced to the problem of finding the solution of $u_t = u_{xx}$, $0 < x < 1$, $t > 0$ satisfying the conditions $u = 1$ when $t = 0$, $0 \le x \le 1$, $u_x = u$ at $x = 0$, $u = 1$ at $x = 1$ for $t > 0$. Solve the resulting problem using an explicit scheme for the first three time steps by taking $\Delta x = 0.2$ and $\Delta t = 0.0025$. Also obtain the solution to the original problem.

9.8 Using Crank–Nicolson finite difference method, find the solution of the partial differential equation afteı one-time step: $T_t = T_{xx}$, $0 < x < \pi$, $t > 0$ subject to the initial condition $T(x, 0) = \sin x$, $0 \le x \le \pi$ and the boundary conditions $T(0, t) = T(\pi, t) = 0$, $t > 0$ by taking

$$\Delta x = \frac{\pi}{4}, \ \Delta t = \frac{\pi^2}{32}$$

9.9 Solve the diffusion equation using Crank–Nicolson method: $T_t = T_{xx}$, $0 \le x \le 1$, subject to the initial condition $T(x, 0) = 0$ and the boundary conditions $T(0, t) = 0$ and $T(1, t) = t$ for $t > 0$ after one-time step, by choosing

$$\Delta x = \frac{1}{4}, \ \Delta t = \frac{1}{8}$$

9.10 Discuss the stability of Durfort–Frankel scheme given by Eq. (9.26) for one-dimensional diffusion equation.

9.11 Solve the unsteady heat equation $T_t = \alpha T_{xx}$, $0 \le x \le 1$, subject to the initial condition $T(x, 0) = x(x - 1)$ and the boundary conditions $T(0, t) = 0$, $T(1, t) = 0$, using Crank–Nicolson method, taking

$$\Delta x = 0.2, \ \alpha = 0.8545, \ r = \alpha \frac{\Delta t}{\Delta x^2} = 1$$

Compute the solution at two-time steps.

9.12 Solve the parabolic partial differential equation $T_t = T_{xx}$, $0 \le x \le 1$ subject to the boundary conditions $T = 0$ at $x = 0$ and $x = 1$ for $t > 0$ and the initial conditions

$$T(x, 0) = \begin{cases} 2x & \text{for } 0 \le x \le \frac{1}{2} \\ 2\,(1 - x) & \text{for } \frac{1}{2} \le x \le 1 \end{cases}$$

using (i) explicit method and (ii) Durfort–Frankel implicit method. The analytical solution to this problem is found to be (see Sankara Rao, 1995, Example 3.4)

$$T(x, t) = \frac{8}{\pi^2} \sum \frac{1}{n^2} \sin \frac{n\pi}{2} \sin (n\pi x)\, e^{-n^2\pi^2 t}$$

Chapter 10

Elliptic Partial Differential Equations

10.1 INTRODUCTION

Steady state problems in two and three dimensions give rise to elliptic partial differential equations. These equations are due to Laplace and Poisson. In vector notation, they can be written as

$$\nabla^2 u = 0 \tag{10.1}$$

$$\nabla^2 u = f \tag{10.2}$$

Typically, the Laplace equation is the very core of the theory of analytic functions of a complex variable. It describes the temperature distribution in an isotropic medium. It summarizes the gravitational or electrostatic potentials at points of empty space. It also characterizes the slow motion of an incompressible viscous fluid. The unsteady Navier–Stokes equations is a mixed parabolic–elliptic type of partial differential equation, which occurs in many important problems of fluid dynamics and heat transfer. Poisson's equation arise in the case of St. Venant theory of torsion of a rectangular bar subject to twisting. Many more physical situations can be cited where these equations occur. For a well-posed problem in an elliptic system, a closed boundary is desirable.

10.2 DERIVATION OF FINITE DIFFERENCE APPROXIMATIONS

Consider the two-dimensional Laplace equation

$$u_{xx} + u_{yy} = 0 \tag{10.3}$$

Following the notation and approximations to derivatives as described in Section 9.2, and if we use central difference approximation to both the space derivatives, the finite difference approximation of Eq. (10.3) is given by

$$\frac{u_{i-1,j} - 2u_{i,j} + u_{i+1,j}}{\Delta x^2} + \frac{u_{i,j-1} - 2u_{i,j} + u_{i,j+1}}{\Delta y^2} = 0$$

If we assume uniform grid spacing in both x and y directions, that is, if we take $\Delta x = \Delta y = h$, it results in considerable simplification. Hence,

$$u_{xx} + u_{yy} = \frac{u_{i-1,j} + u_{i+1,j} + u_{i,j-1} + u_{i,j+1} - 4u_{i,j}}{h^2} = 0$$

which on rewriting becomes

$$u_{i,j} = \frac{u_{i-1,j} + u_{i+1,j} + u_{i,j-1} + u_{i,j+1}}{4} \tag{10.4}$$

Thus, the value of u at any nodal point (i, j) is the mean of its values at its four — north, east, south and west — neighbours, which is shown in the following computational molecule (Fig. 10.1), and is known as *standard five-point formula*.

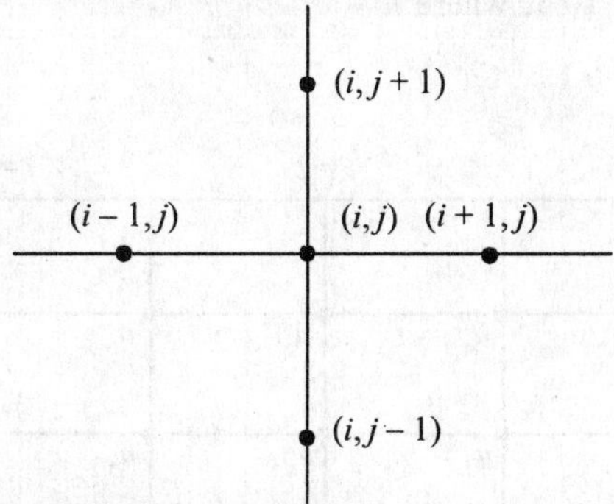

Fig. 10.1 Standard five-point molecule.

It is also well-known that Laplace equation remains invariant under rotation of coordinate axes through 45° (see Fig. 10.2), and hence Eq. (10.4) can also be expressed in the form

$$u_{i,j} = \frac{u_{i-1,j-1} + u_{i+1,j-1} + u_{i+1,j+1} + u_{i-1,j+1}}{4} \tag{10.5}$$

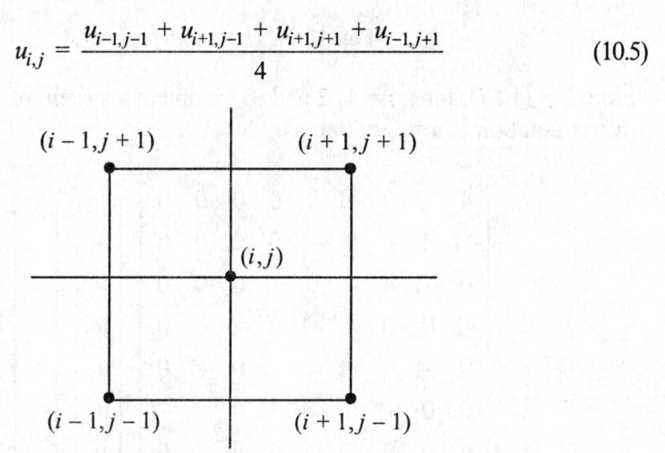

Fig. 10.2 Standard diagonal five-point molecule.

This is called *standard diagonal five-point formula*. Its molecular representation is given in Fig. 10.2

As a second example, let us consider the Poisson's equation in two dimensions in the form

$$u_{xx} + u_{yy} = -2 \tag{10.6}$$

with $u(x, y) = 0$ on the boundary of a unit square $0 \le x \le 1$, $0 \le y \le 1$. Using central difference approximation to both the derivatives and using a uniform grid spacing, that is, $\Delta x = \Delta y = h$, we can approximate Eq. (10.6), and the corresponding finite difference approximation is given by

$$u_{i,j} = \frac{u_{i-1,j} + u_{i+1,j} + u_{i,j-1} + u_{i,j+1} + 2h^2}{4} \tag{10.7}$$

If we choose $h = 0.25$ and denote the internal net with mesh points by $x_i = ih$, $y_j = jh$, where $i, j = 0, 1, 2, 3, 4$; then $i, j = 0, 4$ corresponds to the boundary as depicted in Fig. 10.3, where $u = 0$.

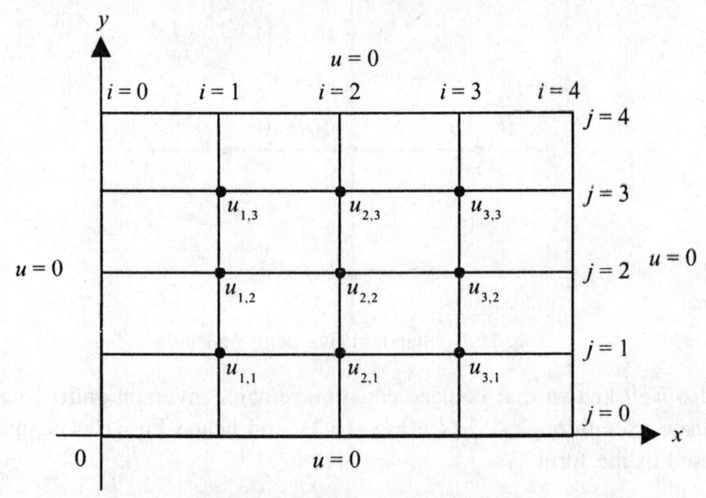

Fig. 10.3 Finite difference grid.

From Eq. (10.7), for $i, j = 1, 2$ and 3, we obtain a system of equations, which in matrix notation can be written as

$$\begin{bmatrix} 4 & -1 & 0 & -1 & 0 & 0 & 0 & 0 & 0 \\ -1 & 4 & -1 & 0 & -1 & 0 & 0 & 0 & 0 \\ 0 & -1 & 4 & 0 & 0 & -1 & 0 & 0 & 0 \\ -1 & 0 & 0 & 4 & -1 & 0 & -1 & 0 & 0 \\ 0 & -1 & 0 & -1 & 4 & -1 & 0 & -1 & 0 \\ 0 & 0 & -1 & 0 & -1 & 4 & 0 & 0 & -1 \\ 0 & 0 & 0 & -1 & 0 & 0 & 4 & -1 & 0 \\ 0 & 0 & 0 & 0 & -1 & 0 & -1 & 4 & -1 \\ 0 & 0 & 0 & 0 & 0 & -1 & 0 & -1 & 4 \end{bmatrix} \begin{pmatrix} u_{1,1} \\ u_{1,2} \\ u_{1,3} \\ u_{2,1} \\ u_{2,2} \\ u_{2,3} \\ u_{3,1} \\ u_{3,2} \\ u_{3,3} \end{pmatrix} = \begin{pmatrix} 2h^2 \\ 2h^2 \\ 2h^2 \\ 2h^2 \\ 2h^2 \\ 2h^2 \\ 2h^2 \\ 2h^2 \\ 2h^2 \end{pmatrix} \tag{10.8}$$

In compact form, Eq. (10.8) can be written as

$$[A](\bar{u}) = (\bar{b}) \tag{10.9}$$

It can be observed that the above coefficient matrix is symmetric, positive definite and sparse. Thus, an elliptic partial differential equation when discretized yields a linear system with large number of zero coefficients. This suggests that an iterative method is most suitable rather than a direct method to find its solution. For illustration of standard five-point formulae, we now consider a number of examples.

Example 10.1 Solve the Laplace equation $u_{xx} + u_{yy} = 0$ by employing five-point formulae, which satisfy the following Dirichlet boundary conditions:

$$u(0, y) = 0, \qquad u(x, 0) = 0$$
$$u(x, 1) = 100x, \qquad u(1, y) = 100y$$

Solution It is clear from the boundary conditions that the region of computation is a square, bounded by $0 \leq x, y \leq 1$. Taking uniform grid spacing $\Delta x = \Delta y = h = 1/4$, the computational grid points are shown in Fig. 10.4. In this problem

$$x_i = x_0 + ih, \quad y_i = y_0 + ih, \qquad i = 0, 1, 2, 3, 4$$

The suffixes 0, 4 corresponds to the boundary.

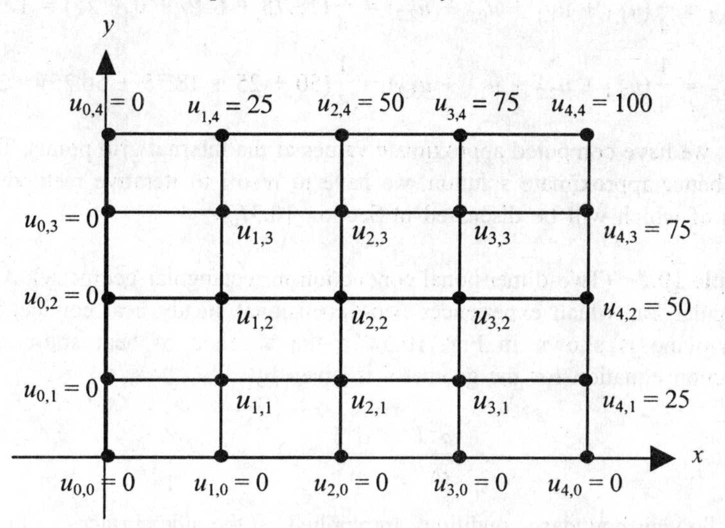

Fig. 10.4 Computational grid (Example 10.1).

The following steps show how the value of u at the interior grid points have been obtained.

Step I: Since, we know the diagonal values, we first use the standard diagonal five-point formula to compute $u_{2,2}$, thus

$$u_{2,2} = \frac{1}{4}(u_{0,0} + u_{4,4} + u_{0,4} + u_{4,0}) = \frac{1}{4}(0 + 100 + 0 + 0) = 25$$

Similarly, we can compute $u_{1,1}$, $u_{3,1}$, $u_{3,3}$, $u_{1,3}$ using the same standard diagonal five-point formula. Therefore,

$$u_{1,1} = \frac{1}{4}(u_{0,0} + u_{2,2} + u_{2,0} + u_{0,2}) = \frac{1}{4}(0 + 25 + 0 + 0) = 6.25$$

$$u_{3,1} = \frac{1}{4}(u_{4,2} + u_{2,0} + u_{4,0} + u_{2,2}) = \frac{1}{4}(50 + 0 + 0 + 25) = 18.75$$

$$u_{3,3} = \frac{1}{4}(u_{4,4} + u_{2,2} + u_{4,2} + u_{2,4}) = \frac{1}{4}(100 + 25 + 50 + 50) = 56.25$$

$$u_{1,3} = \frac{1}{4}(u_{2,4} + u_{0,2} + u_{2,2} + u_{0,4}) = \frac{1}{4}(50 + 0 + 25 + 0) = 18.75$$

Step II: The solution at the remaining nodal points $u_{2,1}$, $u_{3,2}$, $u_{1,2}$ and $u_{2,3}$ can then be obtained using the standard five-point formula. Thus, we get

$$u_{2,1} = \frac{1}{4}(u_{1,1} + u_{3,1} + u_{2,2} + u_{2,0}) = \frac{1}{4}(6.25 + 18.75 + 25 + 0) = 12.50$$

$$u_{3,2} = \frac{1}{4}(u_{3,3} + u_{3,1} + u_{4,2} + u_{2,2}) = \frac{1}{4}(56.25 + 18.75 + 50 + 25) = 37.50$$

$$u_{1,2} = \frac{1}{4}(u_{1,3} + u_{1,1} + u_{0,2} + u_{2,2}) = \frac{1}{4}(18.75 + 6.25 + 0 + 25) = 12.50$$

$$u_{2,3} = \frac{1}{4}(u_{2,4} + u_{2,2} + u_{1,3} + u_{3,3}) = \frac{1}{4}(50 + 25 + 18.75 + 56.25) = 37.50$$

Hence, we have computed approximate values at the internal grid points. To get much better approximate solution, we have to resort to iterative methods, the details of which will be discussed in Section 10.3.

Example 10.2 (Two-dimensional conduction in rectangular geometry): A long rectangular bar which experiences two-dimensional steady heat conduction in the xy-plane is shown in Fig. 10.5. In the absence of heat sources, the conduction equation for the geometry is given by

$$\frac{\partial^2 T}{\partial x^2} + \frac{\partial^2 T}{\partial y^2} = 0 \tag{1}$$

The following boundary conditions are applied on the side surfaces of the bar:

$T = T_0$ at $x = 0$ (prescribed temperature)

$k\dfrac{\partial T}{\partial y} = 0$ at $y = 0$ (insulated boundary)

$-k\dfrac{\partial T}{\partial x} = q$ at $x = L$ (applied heat flux)

$-k\dfrac{\partial T}{\partial y} = h(T - T_f)$ at $y = H$ (convective heat transfer)

Develop the algebraic equations for determining the temperature at the interior and boundary grid points or nodes.

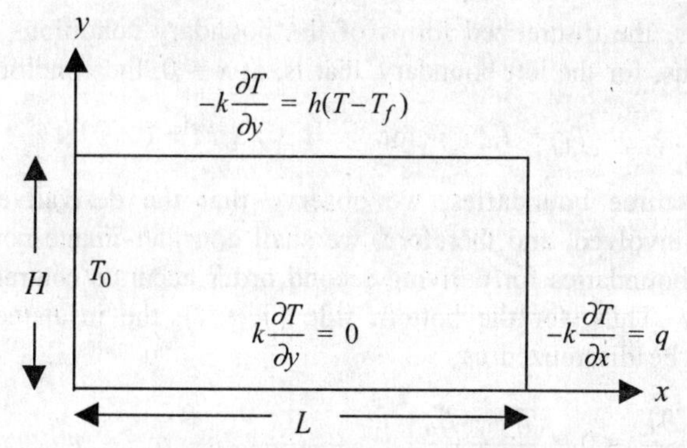

Fig. 10.5 Geometry for Example 10.2.

Solution Dividing the length L of the plate into N equal parts and breadth H into M equal parts, we have

$$\Delta x = \frac{L}{N}, \qquad \Delta y = \frac{H}{M}$$

Thus in the x-direction, i varies from 1 to $N + 1$ (or 0 to N) and $i = 1$ and $N + 1$ correspond to the left and right boundaries; while j varies from 1 to $M + 1$ and $j = 1$ and $M + 1$ correspond to the bottom and top boundaries. Thus, the equi-spaced rectangular grid is shown as in Fig. 10.6:

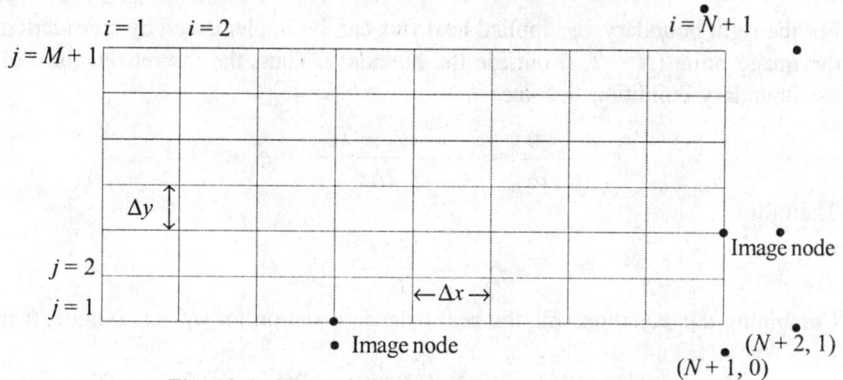

Fig. 10.6 Finite difference grid for Example 10.2.

The discretized form of the heat balance equation (1) using central difference approximation is given by

$$\frac{T_{i-1,j} - 2T_{i,j} + T_{i+1,j}}{\Delta x^2} + \frac{T_{i,j-1} - 2T_{i,j} + T_{i,j+1}}{\Delta y^2} = 0$$

Defining the grid aspect ratio $(\Delta x/\Delta y) = \beta$, the above equation simplifies to

$$T_{i-1,j} + T_{i+1,j} + \beta^2 T_{i,j-1} + \beta^2 T_{i,j+1} - 2(1+\beta^2)T_{i,j} = 0 \tag{2}$$

The nodal equation for all the interior grid points can be obtained from Eq. (2) by setting $i = 2, 3, \ldots, N$ and $j = 2, 3, \ldots, M$. To derive the nodal equation for boundary nodes, the discretized forms of the boundary conditions have to be considered. Thus, for the left boundary, that is, at $x = 0$, the condition becomes

$$T_{1,j} = T_0, \qquad \text{for} \qquad 1 \le j \le M + 1 \tag{3}$$

For the other three boundaries, we observe that the derivative boundary conditions are involved, and therefore, we shall consider image nodes outside the respective boundaries for deriving second order accurate central difference approximations. Thus, for the bottom side ($j = 1$), the insulated boundary conditions can be discretized as

$$\frac{\partial T}{\partial y} = 0 = \frac{T_{i,2} - T_{i,0}}{2\Delta y}, \qquad \text{implying} \qquad T_{i,0} = T_{i,2} \tag{4}$$

Here $(i, 0)$ is an image node located at a distance of Δy below the bottom boundary for each i as shown in Fig. 10.6. Now, the heat balance equation for the bottom boundary can be obtained from Eq. (2) by setting $j = 1$ as

$$T_{i-1,1} + T_{i+1,1} + \beta^2 T_{i,0} + \beta^2 T_{i,2} - 2(1 + \beta^2)T_{i,1} = 0 \tag{5}$$

Eliminating the image node $T_{i,0}$ from Eqs. (4) and (5), the final form of the nodal equation for the bottom boundary is given by

$$T_{i-1,1} + T_{i+1,1} + 2\beta^2 T_{i,2} - 2(1 + \beta^2)T_{i,1} = 0, \qquad 2 \le i \le N \tag{6}$$

For the right boundary, the applied heat flux can be implemented by considering the image point $(N + 2, j)$ outside the boundary. Thus, the discretized form of the boundary condition becomes

$$-k\frac{\partial T}{\partial x} = -k\frac{T_{N+2,j} - T_{N,j}}{2\Delta x} = q$$

Therefore,

$$T_{N+2,j} = -\frac{2q\,\Delta x}{k} + T_{N,j}$$

Combining this equation with the heat balance equation for $(N + 1, j)$ node, that is,

$$T_{N,j} + T_{N+2,j} + \beta^2 T_{N+1,j-1} + \beta^2 T_{N+1,j+1} - 2(1 + \beta^2)T_{N+1,j} = 0$$

Eliminating the image node $T_{N+2,j}$ from the above two equations, we obtain

$$2T_{N,j} + \beta^2 T_{N+1,j-1} + \beta^2 T_{N+1,j+1} - 2(1 + \beta^2)T_{N+1,j} = \frac{2q\,\Delta x}{k} \tag{7}$$

for $2 \le j \le M$. Similarly, for the top boundary, the discretization of the boundary condition gives

$$-k\,\frac{T_{i,M+2} - T_{i,M}}{2\Delta y} = h\,(T_{i,M+1} - T_f).$$

Therefore,

$$T_{i,M+2} = -\frac{2h\,\Delta y}{k}\,(T_{i,M+1} - T_f) + T_{i,M}$$

Combining this equation with the heat balance equation for the boundary node $(i, M + 1)$, that is,

$$T_{i-1,M+1} + T_{i+1,M+1} + \beta^2 T_{i,M} + \beta^2 T_{i,M+2} - 2(1 + \beta^2)T_{i,M+1} = 0$$

we can eliminate the image node $(i, M + 2)$ and get the nodal equation for the top boundary in the form

$$T_{i-1,M+1} + T_{i+1,M+1} + 2\beta^2 T_{i,M} - 2\left(1 + \beta^2 + \frac{\beta^2 h\Delta y}{k}\right)T_{i,M+1} = -\frac{2h\beta^2\Delta y}{k}T_f \quad (8)$$

for $2 \le i \le N$. Finally, we shall consider the corner nodes of the rectangular domain. The nodes (1, 1) and $(1, M + 1)$ are already included in Eq. (3). For the corner node $(N + 1, 1)$, two image points $(N + 1, 0)$ and $(N + 2, 1)$ are considered as shown in Fig. 10.6. The boundary conditions on the bottom and the right surfaces are discretized using these image nodes. After combining the discretized boundary conditions with the heat balance equation (2), the following nodal equation results:

$$T_{N,1} + \beta^2 T_{N+1,2} - (1 + \beta^2)T_{N+1,1} = \frac{q\Delta x}{k} \quad (9)$$

Similarly, by incorporating the boundary conditions of the applied heat flux and convective heat transfer with the help of the two image points $(N + 2, M + 1)$ and $(N + 1, M + 2)$, the equation for the node $(N + 1, M + 1)$ is obtained as

$$T_{N,M+1} + \beta^2 T_{N+1,M} - \left(1 + \beta^2 + \frac{\beta^2 h\Delta y}{k}\right)T_{N+1,M+1} = -\frac{\beta^2 h\Delta y}{k}T_f + \frac{q\Delta x}{k} \quad (10)$$

Thus, Eqs. (2), (3) and (6)–(10), constitute the complete set of simultaneous algebraic equations for computing the temperatures at the interior and boundary nodes. The local truncation error is of $O(\Delta x^2, \Delta y^2)$.

10.3 ITERATIVE METHODS

We have observed in Section 10.2 that space discretization of either Laplace equation or Poisson's equation gives us a system of linear algebraic equations, when written in matrix notation lead to sparse matrices. This situation naturally favours the use of iterative methods rather than the direct methods for their solution; as the total number of arithmetic operations in an iterative method are very few when compared with those involved in direct methods. This subject has been treated extensively in Chapter 3, and is still a current area of research either to improve the existing algorithms or to reduce the total number of arithmetic operations.

The basic idea of any *iterative method* is to perform a few arithmetic operations on the elements of a matrix of an algebraic system iteratively with the aim of achieving the near-exact solution to the prescribed accuracy, hopefully in a small number of iterations. There are a large number of iterative methods with different rates of convergence. Some of the classical methods are:

(a) Jacobi method

(b) Liebmann's method or Gauss–Seidel method

(c) Relaxation method.

There is another class of iterative methods known as *alternate direction implicit methods* referred as ADI method, which will be discussed in the next section.

We shall illustrate below, different classical methods by considering the Poisson's equation (10.2), which when discretized using second order five-point scheme given by

$$u_{i,j} = \frac{1}{4}\left(u_{i-1,j} + u_{i+1,j} + u_{i,j-1} + u_{i,j+1} + h^2 f_{i,j}\right) \tag{10.10}$$

In Jacobi method, if $u_{i,j}^{(r)}$ denotes the rth iterative value of $u_{i,j}$, then the iterative procedure for solving Poisson's equation (10.10), at the interior mesh points is defined by the latest $(r + 1)$th iteration as

$$u_{i,j}^{(r+1)} = \frac{1}{4}\left[u_{i-1,j}^{(r)} + u_{i+1,j}^{(r)} + u_{i,j-1}^{(r)} + u_{i,j+1}^{(r)} + h^2 f_{i,j}\right] \tag{10.11}$$

This is the point Jacobi method.

In Liebmann's method or Gauss–Seidel method, we make use of the recently computed values as soon as they are available and the values of u along each row are computed moving from left to right. Thus, the Liebmann's iterative formula is

$$u_{i,j}^{(r+1)} = \frac{1}{4}\left[u_{i-1,j}^{(r+1)} + u_{i+1,j}^{(r)} + u_{i,j-1}^{(r+1)} + u_{i,j+1}^{(r)} + h^2 f_{i,j}\right] \tag{10.12}$$

Finally, in over-relaxation method, the convergence rate of an iterative process is accelerated by propagating the corrections $\left[u_{i,j}^{(r+1)} - u_{i,j}^{(r)}\right]$, faster through the mesh as follows: $\bar{u}_{i,j}^{(r+1)}$ is the value obtained from a basic iterative method, then the value at the next iteration is defined by

$$u_{i,j}^{(r+1)} = \omega \bar{u}_{i,j}^{(r+1)} + (1 - \omega)u_{i,j}^{(r)} \tag{10.13}$$

Where ω is the over-relaxation factor. Thus, the Jacobi over-relaxation method for the Poisson's equation (10.2) becomes

$$u_{i,j}^{(r+1)} = \frac{1}{4}\omega\left[u_{i-1,j}^{(r)} + u_{i+1,j}^{(r)} + u_{i,j-1}^{(r)} + u_{i,j+1}^{(r)} + h^2 f_{i,j}\right] + (1 - \omega)\, u_{i,j}^{(r)} \tag{10.14}$$

Similarly, the Liebmann's over-relaxation method leads to the following iterative scheme:

$$u_{i,j}^{(r+1)} = \frac{1}{4}\omega\left[u_{i-1,j}^{(r+1)} + u_{i+1,j}^{(r)} + u_{i,j-1}^{(r+1)} + u_{i,j+1}^{(r)} + h^2 f_{i,j}\right] + (1 - \omega)u_{i,j}^{(r)} \tag{10.15}$$

Mitchell and Griffiths, (1980) have shown that the over-relaxation factor ω

lies between 1 and 2. A reasonable optimal value of ω has been estimated by the smaller root of the equation

$$\left(\cos\frac{\pi}{p} + \cos\frac{\pi}{q}\right)^2 \omega^2 - 16\omega + 16 = 0 \tag{10.16}$$

where p and q are the number of mesh divisions in x and y directions respectively. Solving the above equation for ω, we obtain

$$\omega_{\text{opt}} = \frac{4}{2 + \sqrt{4 - c^2}} \tag{10.17}$$

where

$$c = \cos\frac{\pi}{p} + \cos\frac{\pi}{q} \tag{10.18}$$

More details regarding the estimate of the optimal value of the over-relaxation factor ω can be found in the works of Smith (1965), Mitchell, and Griffiths (1980). To have a better understanding of the above iterative schemes, we shall consider below few numerical examples.

Example 10.3 Solve the Laplace equation $\nabla^2 u = 0$ in two-dimensional region for the boundary value problem described in Fig. 10.7 given below:

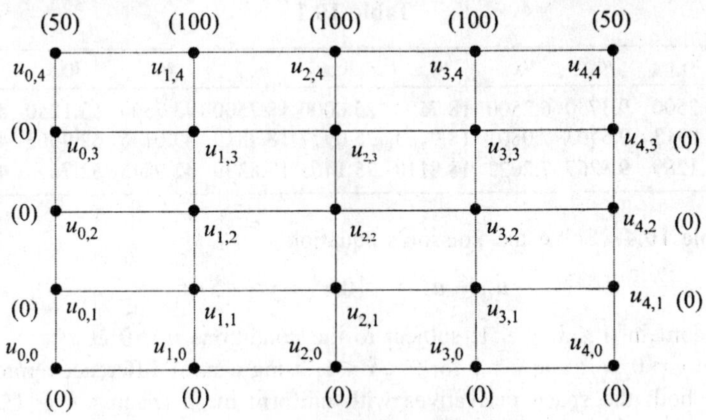

Fig. 10.7 Example 10.3.

Solution Using standard five-point diagonal formula (10.5) for approximating the Laplace equation, we calculate the value of u at the interior nodes as follows:

$$u_{2,2} = \frac{1}{4}(u_{4,4} + u_{0,0} + u_{4,0} + u_{0,4}) = \frac{1}{4}(50 + 0 + 0 + 50) = 25.0$$

$$u_{1,3} = \frac{1}{4}(u_{0,4} + u_{2,2} + u_{2,4} + u_{0,2}) = \frac{1}{4}(50 + 25 + 100 + 0) = 43.75$$

$$u_{3,3} = \frac{1}{4}(u_{4,4} + u_{2,2} + u_{2,4} + u_{4,2}) = \frac{1}{4}(50 + 25 + 100 + 0) = 43.75$$

$$u_{1,1} = \frac{1}{4}(u_{2,2} + u_{0,0} + u_{2,0} + u_{0,2}) = \frac{1}{4}(25 + 0 + 0 + 0) = 6.25$$

$$u_{3,1} = \frac{1}{4}(u_{2,2} + u_{4,0} + u_{4,2} + u_{2,0}) = \frac{1}{4}(25 + 0 + 0 + 0) = 6.25$$

Now, applying standard five-point formula (10.4), we can compute the value of u at the remaining nodal points. Thus,

$$u_{2,3} = \frac{1}{4}(u_{2,2} + u_{2,4} + u_{3,3} + u_{1,3}) = \frac{1}{4}(25 + 100 + 43.75 + 43.75) = 53.125$$

$$u_{1,2} = \frac{1}{4}(u_{1,1} + u_{1,3} + u_{2,2} + u_{0,2}) = \frac{1}{4}(6.25 + 43.75 + 25 + 0) = 18.75$$

$$u_{2,1} = \frac{1}{4}(u_{2,0} + u_{2,2} + u_{3,1} + u_{1,1}) = \frac{1}{4}(0 + 25 + 6.25 + 6.25) = 9.375$$

$$u_{3,2} = \frac{1}{4}(u_{3,1} + u_{3,3} + u_{2,2} + u_{4,2}) = \frac{1}{4}(6.25 + 43.75 + 25 + 0) = 18.75$$

Taking these computed values as first approximations, we can compute a better solution using Gauss-Seidel iterative method as described in Eq. (10.12). The first few iterations are presented in Table 10.1 as given below:

Table 10.1

r	$u_{1,1}$	$u_{2,1}$	$u_{3,1}$	$u_{1,2}$	$u_{2,2}$	$u_{3,2}$	$u_{1,3}$	$u_{2,3}$	$u_{3,3}$
1	6.2500	9.3750	6.2500	18.7500	25.0000	18.7500	43.7500	53.1250	43.7500
2	7.0313	9.5703	7.0801	18.9453	25.0977	18.9819	43.0176	52.9663	42.9871
3	7.1289	9.8267	7.2022	18.8110	25.1465	18.8339	42.9443	52.7695	42.9009

Example 10.4 Solve the Poisson's equation

$$u_{xx} + u_{yy} = -10(x^2 + y^2 + 5)$$

in the domain $0 \le x, y \le 1$; subject to the conditions $u = 0$ at $x = 0$, $x = 1$; $u = 0$ at $y = 0$; $u = 1$ at $y = 1$ for $0 < x < 1$, using central difference approximation to both the space derivatives with uniform mesh spacing $h = 1/3$. Use Liebmann iterative method to find the solution of the resulting system.

Solution After discretization, the given partial differential equation becomes

$$\frac{u_{i-1,j} - 2u_{i,j} + u_{i+1,j}}{(1/3)^2} + \frac{u_{i,j-1} - 2u_{i,j} + u_{i,j+1}}{(1/3)^2} = -10\left[\left(\frac{i}{3}\right)^2 + \left(\frac{j}{3}\right)^2 + 5\right]$$

which gives

$$u_{i,j} = \frac{1}{4}\left[u_{i-1,j} + u_{i+1,j} + u_{i,j-1} + u_{i,j+1} + \frac{10}{9}\left(\frac{i^2}{9} + \frac{j^2}{9} + 5\right)\right]$$

From Fig. 10.8, it can be noted that there are four internal nodes marked by • at which, we seek the solution subject to the given boundary conditions, using standard five-point formula.

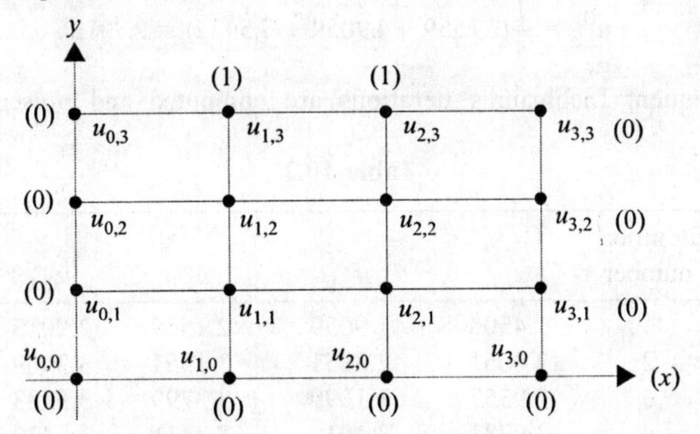

Fig. 10.8 Discretization of Example 10.4.

Thus, we have

$$u_{1,1} = \frac{1}{4}\left[u_{2,1} + u_{1,2} + 0 + 0 + \frac{10}{81}(1 + 1 + 45)\right] = \frac{1}{4}\left(u_{2,1} + u_{1,2} + \frac{470}{81}\right)$$

$$u_{2,1} = \frac{1}{4}\left[u_{1,1} + u_{2,2} + \frac{10}{81}(4 + 1 + 45)\right] = \frac{1}{4}\left(u_{1,1} + u_{2,2} + \frac{500}{81}\right)$$

$$u_{1,2} = \frac{1}{4}\left[u_{2,2} + u_{1,1} + 1 + \frac{10}{81}(1 + 4 + 45)\right] = \frac{1}{4}\left(u_{2,2} + u_{1,1} + \frac{581}{81}\right)$$

$$u_{2,2} = \frac{1}{4}\left[u_{1,2} + u_{2,1} + 1 + \frac{10}{81}(4 + 4 + 45)\right] = \frac{1}{4}\left(u_{1,2} + u_{2,1} + \frac{611}{81}\right)$$

The Liebmann's iteration for the above system can be written as

$$u_{1,1}^{(r+1)} = \frac{1}{4}\left[u_{2,1}^{(r)} + u_{1,2}^{(r)} + 5.8025\right]$$

$$u_{2,1}^{(r+1)} = \frac{1}{4}\left[u_{1,1}^{(r+1)} + u_{2,2}^{(r)} + 6.1728\right]$$

$$u_{1,2}^{(r+1)} = \frac{1}{4}\left[u_{1,1}^{(r+1)} + u_{2,2}^{(r)} + 7.1728\right]$$

$$u_{2,2}^{(r+1)} = \frac{1}{4}\left[u_{1,2}^{(r+1)} + u_{2,1}^{(r+1)} + 7.5432\right]$$

As a first approximation, we take $u_{2,1} = u_{1,2} = 0$ and $u_{2,2} = 0$. Thus for $r = 0$, we obtain

$$u_{1,1}^{(1)} = \frac{1}{4}(0 + 0 + 5.8025) = 1.4506$$

$$u_{2,1}^{(1)} = \frac{1}{4}(1.4506 + 0 + 6.1728) = 1.9059$$

$$u_{1,2}^{(1)} = \frac{1}{4}(1.4506 + 0 + 7.1728) = 2.1559$$

$$u_{2,2}^{(1)} = \frac{1}{4}(2.1559 + 1.9059 + 7.5432) = 2.9013$$

The subsequent Liebmann's iterations are computed and presented as in Table 10.2:

Table 10.2

Iteration number	$u_{1,1}$	$u_{2,1}$	$u_{1,2}$	$u_{2,2}$
1	1.4506	1.9059	2.1559	2.9013
2	2.4661	2.8851	3.1351	3.3909
3	2.9557	3.1299	3.3799	3.5133
4	3.0781	3.1911	3.4411	3.5439
5	3.1087	3.2064	3.4564	3.5515

Example 10.5 The temperature u in a steady heat flow in a square plate bounded by $x = 0$, $y = 0$, $x = 4$, $y = 4$ satisfies Laplace equation $u_{xx} + u_{yy} = 0$. Taking a square mesh of size unity and the boundary values given by $u = 0$ along $x = 0$ and $y = 0$, $u = x^3$ along $y = 4$, $u = 16y$ along $x = 4$. Solve the given Laplace equation, using Liebmann's over-relaxation method.

Solution The initial values at the internal nodes (*see* Fig. 10.9) can be obtained using diagonal and standard formulae. Thus, we get

$$u_{2,2} = \frac{1}{4}(0 + 0 + 0 + 64) = \frac{64}{4} = 16.0$$

$$u_{1,3} = \frac{1}{4}(0 + 16 + 8 + 0) = \frac{24}{4} = 6.0$$

$$u_{3,3} = \frac{1}{4}(8 + 32 + 64 + 16) = \frac{120}{4} = 30.0$$

$$u_{1,1} = \frac{1}{4}(0 + 0 + 16 + 0) = \frac{16}{4} = 4.0$$

$$u_{3,1} = \frac{1}{4}(16 + 0 + 32 + 0) = \frac{48}{4} = 12.0$$

$$u_{2,3} = \frac{1}{4}(6 + 30 + 8 + 16) = \frac{60}{4} = 15.0$$

$$u_{1,2} = \frac{1}{4}(0 + 16 + 6 + 4) = \frac{26}{4} = 6.5$$

$$u_{3,2} = \frac{1}{4}(16 + 32 + 30 + 12) = \frac{90}{4} = 22.5$$

$$u_{2,1} = \frac{1}{4}(4 + 12 + 16 + 0) = \frac{32}{4} = 8.0$$

Using Liebmann's over-relaxation method for Laplace equation, we have

$$u_{i,j}^{(r+1)} = \frac{1}{4}\omega\left[u_{i-1,j}^{(r+1)} + u_{i+1,j}^{(r)} + u_{i,j-1}^{(r+1)} + u_{i,j+1}^{(r)}\right] + (1-\omega)u_{i,j}^{(r)} \tag{1}$$

The optimal value of ω is obtained from Eqs. (10.17) and (10.18). In this problem, we note that $p = q = 4$. Hence,

$$\omega_{\text{opt}} = \frac{4}{2 + \sqrt{4 - c^2}} \quad \text{where} \quad c = \cos\frac{\pi}{p} + \cos\frac{\pi}{q} = \sqrt{2}$$

which simplifies to

$$\omega_{\text{opt}} = \frac{4}{2 + \sqrt{2}} = 1.172 \tag{2}$$

Now, starting with the initial values obtained, the boundary values, using Liebmann's over-relaxation iteration formula given by Eq. (1) and with the optimal value of ω given by Eq. (2), the successive iterations at the internal nodes along each row, starting from the bottom row can be computed. The computed values at the internal nodes for the first few iterations are depicted as in the Fig. 10.9:

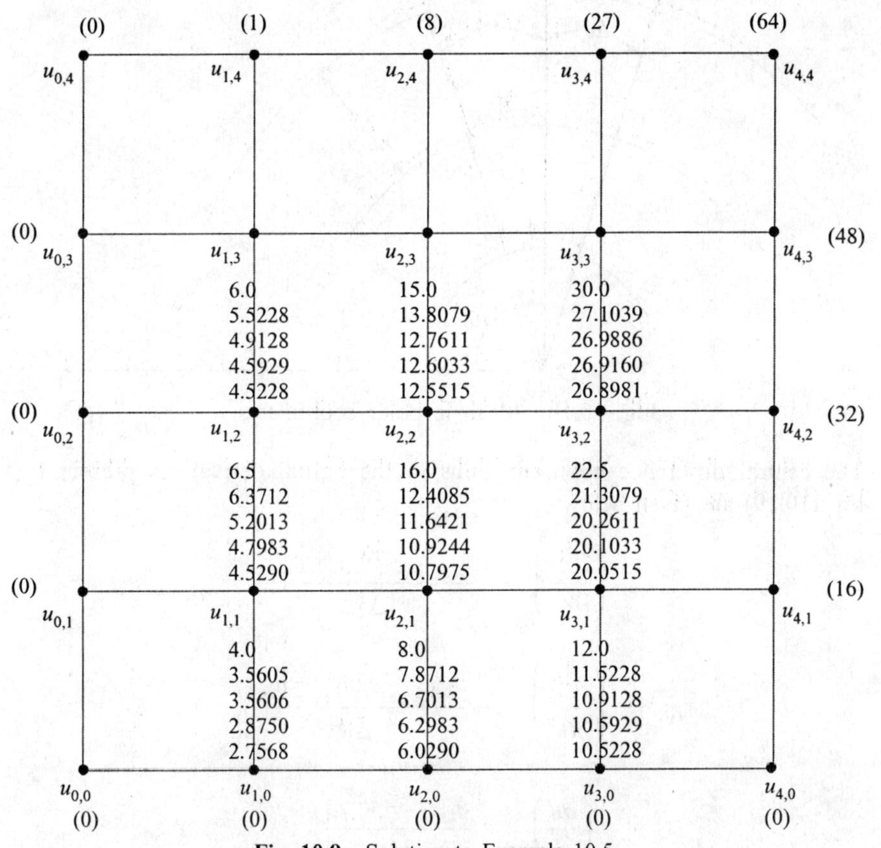

Fig. 10.9 Solution to Example 10.5.

It may be noted that the starting values at the internal nodes need not always be computed using diagonal and standard formulae. In fact, to start an iteration, one can initialize all the internal nodes to zero values and arrive at a convergent solution after few iterations. One can notice this fact from Examples 10.5, and 10.6.

10.4 LAPLACE EQUATION IN POLAR COORDINATES

In fluid mechanics, the flow of an incompressible inviscid and irrotational fluid involves the solution to Laplace equation in polar coordinates (r, θ), which is of the form

$$\nabla^2 u = \frac{\partial^2 u}{\partial r^2} + \frac{1}{r}\frac{\partial u}{\partial r} + \frac{1}{r^2}\frac{\partial^2 u}{\partial \theta^2} \tag{10.19}$$

using finite difference representation in polar coordinates as shown in Fig. 10.10.

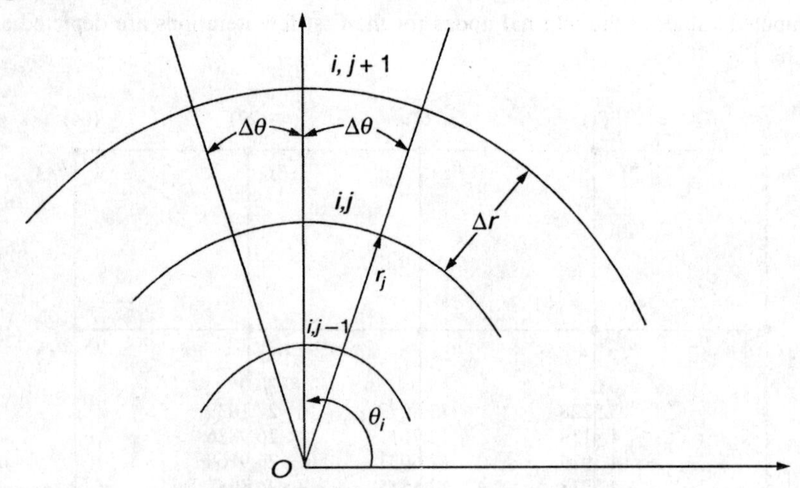

Fig. 10.10 Mesh in polar coordinates.

The central difference approximations to the partial derivatives present in Eq. (10.19) are given below:

$$\left(\frac{\partial^2 u}{\partial r^2}\right)_{i,j} = \frac{u_{i,j-1} - 2u_{i,j} + u_{i,j+1}}{\Delta r^2}$$

$$\left(\frac{\partial^2 u}{\partial \theta^2}\right)_{i,j} = \frac{u_{i-1,j} - 2u_{i,j} + u_{i+1,j}}{\Delta \theta^2}$$

$$\left(\frac{\partial u}{\partial r}\right)_{i,j} = \frac{u_{i,j-1} - u_{i,j+1}}{2\Delta r}$$

Thus, the finite difference representation of the Laplace equation (10.19) is given by

$$\frac{u_{i,j-1} - 2u_{i,j} + u_{i,j+1}}{\Delta r^2} + \frac{1}{j\Delta r}\,\frac{u_{i,j+1} - u_{i,j-1}}{2\Delta r} + \frac{1}{(j\Delta r)^2}\,\frac{u_{i-1,j} - 2u_{i,j} + u_{i+1,j}}{\Delta\theta^2} = 0 \tag{10.20}$$

Here, the mesh points are the points of intersection of r = constant (circles) and θ = constant (rays). On simplification, Eq. (10.20) can be rewritten as

$$\left(1-\frac{1}{2j}\right)u_{i,j-1} + \left(1+\frac{1}{2j}\right)u_{i,j+1} - 2\left[1+\frac{1}{(j\Delta\theta)^2}\right]u_{i,j} + \frac{1}{(j\Delta\theta)^2}\left(u_{i-1,j} + u_{i+1,j}\right) = 0 \tag{10.21}$$

when these equations are expanded for various values of i and j, we get a system of linear equations which can be easily solved for the unknown values at the internal nodes.

Similar approximations can be written down for Laplace equation in cylindrical and spherical polar coordinates, too.

EXERCISES

10.1 Solve the partial differential equation

$$u_{xx} + u_{yy} = -10(x^2 + y^2 + 10)$$

over a square region with sides $x = 0$, $y = 0$, $x = 3$, $y = 3$ taking the boundary condition $u = 0$ on all the boundary sides and mesh size

$$\Delta x = \Delta y = 1$$

10.2 In the theory of elasticity, the stress function u in the problem of torsion of a beam satisfies the Poisson's equation

$$u_{xx} + u_{yy} = -2, \qquad 0 \le x \le 1, \qquad 0 \le y \le 1$$

Find the stress function within the beam subject to the boundary conditions $u = 0$ on the sides $x = 0$, $x = 1$, $y = 0$, $y = 1$ using Gauss–Seidel iterative method and taking $\Delta x = \Delta y = 1/4$.

Note: Analytical solutions is (*see* Sankara Rao, 1995)

$$u(x, y) = x - x^2 - \frac{8}{\pi^3}\sum_{n=1}^{\infty}\left[\frac{\sin\,[(2n-1)\pi x]}{(2n-1)^3}\right]$$

$$\times\ \frac{\sinh\,[(2n-1)\pi y] + \sinh\,[(2n-1)\pi(1-y)]}{\sinh\,(2n-1)\pi}$$

10.3 Solve the elliptic partial differential equation $u_{xx} + u_{yy} = 0$ taking a unit square mesh, with boundary values as shown below, using Liebmann's method for the first four iterations after getting initial values at internal nodes using diagonal and standard formulae (see Fig. 10.11).

10.4 Solve the Poisson's equation $\nabla^2 u = 8x^2y^2$ for the square region with corners at (0, 0) and (1, 1) taking $\Delta x = \Delta y = 1/3$ and the values of u on the boundary are everywhere zero. Use Liebmann's iterative method and get the solution to the first five iterations.

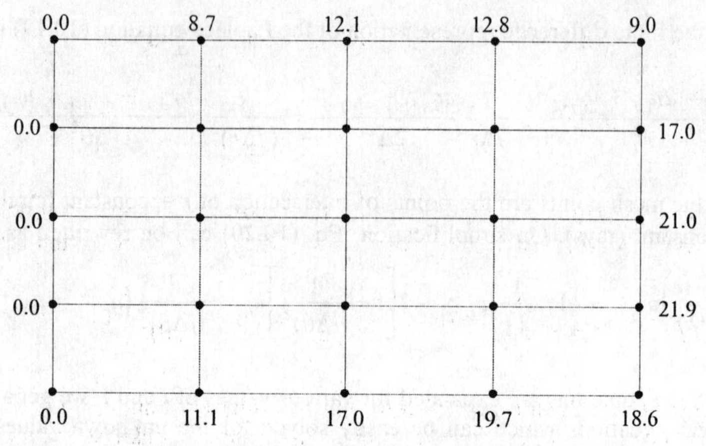

Fig. 10.11 Problem 10.3.

10.5 A rectangular plate is 8 units wide and 6 units high. The edges are held at temperatures shown in Fig. 10.12. Use Liebmann's method to find the solution to the governing Laplace equation $\nabla^2 u = 0$ after seven iterations at nodes spaced two units apart, assuming that heat flows only in x and y directions.

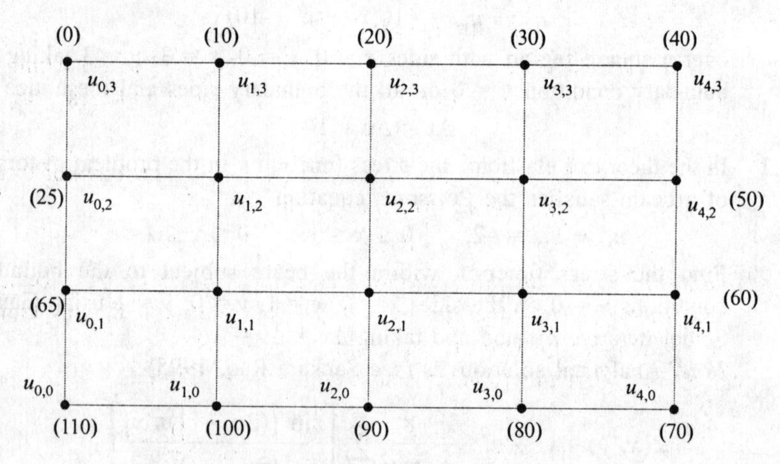

Fig. 10.12 Problem 10.5.

10.6 Consider the problem of torsion in a rectangular bar subject to twisting. The torsion function ϕ satisfies the Poisson's equation

$$\nabla^2 \phi + 2 = 0$$

with $\phi = 0$ on the boundary. Find ϕ over the cross-section of a rectangular bar bounded by $x = 0$, $y = 0$, $x = 3$, $y = 4$. Taking the mesh size of 1 unit in both the directions, solve the above Poisson's equation, using Liebmann's over-relaxation method and obtain the solution accurate to two decimal places.

Chapter 11

Hyperbolic Partial Differential Equations

11.1 INTRODUCTION

The hyperbolic partial differential equations play a dominant role in many branches of science and engineering. Wave equation is a typical example of hyperbolic equation, which model problems such as vibrating string fixed at both ends, Longitudinal vibrations in a bar, propagation of sound waves, electromagnetic waves and so on. They also occur in the field of computational fluid dynamics (CFD). It is well known that Euler's equations of gas dynamics are hyperbolic partial differential equations. However, as the subject, CFD is a highly developed area and is beyond the scope of an undergraduate programme, we shall consider only second order wave equation and discuss an explicit method for its numerical solution.

11.2 EXPLICIT FINITE DIFFERENCE METHOD

Consider one-dimensional second order wave equation

$$\frac{\partial^2 u}{\partial t^2} = c^2 \frac{\partial^2 u}{\partial x^2}, \quad 0 \le x \le l, \quad t \ge 0 \tag{11.1}$$

To solve this equation, using finite difference method, we need to specify the initial conditions and boundary conditions, which for example may be specified as follows:

$$\text{Initial conditions (IC): } u(x, 0) = f(x), \qquad \frac{\partial u}{\partial t}(x, 0) = 0 \tag{11.2}$$

$$\text{Boundary conditions (BC): } u(0, t) = u(l, t) = 0, \; t > 0 \tag{11.3}$$

The solutions obtained using FTCS schemes are in general unstable for hyperbolic partial differential equations. Hence, we choose central difference approximations to both the second order time and space derivatives, of Eq. (11.1), to get

$$\frac{u_i^{n+1} - 2u_i^n + u_i^{n-1}}{(\Delta t)^2} = c^2 \frac{u_{i-1}^n - 2u_i^n + u_{i+1}^n}{(\Delta x)^2}$$

where Δx is the mesh size in the x-direction and Δt is the time step. Let $r = \Delta t/\Delta x$, then the above equation reduces to

$$r^2c^2\left(u_{i-1}^n - 2u_i^n + u_{i+1}^n\right) = u_i^{n+1} - 2u_i^n + u_i^{n-1}$$

Equivalently,

$$u_i^{n+1} = 2(1 - r^2c^2)u_i^n + r^2c^2(u_{i-1}^n + u_{i+1}^n) - u_i^{n-1} \tag{11.4}$$

This formula involves three time-levels $(n - 1)$, n and $(n + 1)$. Let N is the number of divisions into which the interval $(0, l)$ has been divided, so that $\Delta x = l/N$. Now, the boundary conditions can be rewritten in the form

$$u_0^n = u_N^n = 0, \qquad n = 1, 2, 3, \ldots \tag{11.5}$$

while the initial conditions can be expressed for each node i as

$$u_i^0 = f(i\Delta x), \qquad i = 1, 2, \ldots, (N - 1) \tag{11.6}$$

Now using central difference approximation to the second initial condition, we get at the node (i, n)

$$\frac{u_i^{n-1} - u_i^{n+1}}{2\Delta t} = 0, \qquad \text{implying } u_i^{n+1} = u_i^{n-1} \tag{11.7}$$

For $n = 0$, Eq. (11.7) gives

$$u_i^1 = u_i^{-1} \tag{11.8}$$

while Eq. (11.4) becomes

$$u_i^1 = 2(1 - r^2c^2)\,u_i^0 + r^2c^2\left(u_{i-1}^0 + u_{i+1}^0\right) - u_i^{-1} \tag{11.9}$$

Eliminating u_i^{-1} from Eqs. (11.8) and (11.9), we obtain

$$u_i^1 = (1 - r^2c^2)u_i^0 + \frac{1}{2}r^2c^2\left(u_{i-1}^0 + u_{i+1}^0\right) \tag{11.10}$$

This formula can be used to compute u_i^1 $[i = 1, 2, \ldots, (N - 1)]$, that is, the value of u_i at the first time level, in terms of the initial conditions. Thus, knowing u_i^0 and u_i^1, we can compute u_i^2, using Eq. (11.4) by taking $n = 1$, for $i = 1, 2, \ldots, (N - 1)$. Knowing u_i^2, we can compute u_i^3 from Eq. (11.4) and so on. Here, Eq. (11.4) is an explicit finite difference scheme or formula, for finding the solution of the wave equation (11.1) numerically. For illustration, we consider the following example.

Example 11.1 Find the solution to the initial boundary value problem described by the wave equation

$$u_{tt} = u_{xx}, \qquad 0 \le x \le 1 \tag{1}$$

subject to the boundary conditions

$$u(0, t) = 0 = u(1, t), \qquad t > 0 \tag{2}$$

and the initial conditions

$$u(x, 0) = \sin(\pi x) \text{ and } u_t(x, 0) = 0, \qquad 0 \le x \le 1 \tag{3}$$

using an explicit finite difference scheme. Assume $\Delta x = 1/8$, $\Delta t = 1/8$. Compute u for eight time levels. Compare with the analytical solution

$$u(x, t) = \sin(\pi x) \cos(\pi t) \text{ at } t = \frac{1}{2}$$

Comment on the numerical solution at $t = 0$ and $t = 1$.

Solution In this example $c = 1$, $r = \Delta t/\Delta x = 1$. The boundary conditions can be put in finite difference notation as

$$u_0^n = u_N^n = 0, \qquad n = 0, 1, 2, 3, \ldots \tag{4}$$

where, $\Delta x = 1/N = 1/8$ and therefore, $N = 8$, $r = \Delta t/\Delta x = 1.0$, while the initial conditions are

$$u_i^0 = \sin \pi (i\Delta x), \qquad i = 0, 1, 2, \ldots, 8. \tag{5}$$

Using central difference approximation, the second initial condition becomes

$$\frac{u_i^{n+1} - u_i^{n-1}}{\Delta t} = 0 \qquad \text{implying } u_i^{n+1} = u_i^{n-1}$$

For $n = 0$, we get

$$u_i^1 = u_i^{-1} \tag{6}$$

The explicit finite difference scheme for Eq. (1) using Eq. (11.4) is given by

$$u_i^{n+1} = 2(1 - r^2)u_i^n + r^2(u_{i-1}^n + u_{i+1}^n) - u_i^{n-1} \tag{7}$$

For, $r = 1$, $n = 0$, using Eq. (6) we have

$$u_i^1 = \frac{u_{i-1}^0 + u_{i+1}^0}{2} \tag{8}$$

Using Eq. (8), we compute the initial values in the interval of integration (0, 1). The subsequent values of u for $t > 0$ can be computed from the following equation

$$u_i^{n+1} = u_{i-1}^n + u_{i+1}^n - u_i^{n-1} \tag{9}$$

which is obtained from Eq. (7) with $r = 1$. Thus, Eq. (5) is used to compute u at $t = 0$ ($n = 0$), while Eq. (8) is used to compute u at $t = \Delta t$ ($n = 1$). For subsequent values of u at $n = 2, 3, 4, \ldots$ we use Eq. (9). The resulting numerical values are tabulated as follows:

t	x						n
	0.0	1/8	2/8	3/8	4/8	5/8	
0.0	0.0	0.3827	0.7071	0.9239	1.0000	0.9239	0
1/8	0.0	0.3536	0.6533	0.8536	0.9239	0.8536	1
2/8	0.0	0.2706	0.5001	0.6533	0.7072	0.6533	2
3/8	0.0	0.1465	0.2706	0.3537	0.3827	0.3537	3
4/8	0.0	0.0000	0.0002	0.0000	0.0002	0.0000	4
	(0.0)	(0.0000)	(0.0000)	(0.0000)	(0.0000)	(0.0000)	
5/8	0.0	–0.1463	–0.2706	–0.3533	–0.3827	–0.3533	5
6/8	0.0	–0.2706	–0.4994	–0.6533	–0.7064	–0.6533	6
7/8	0.0	–0.3531	–0.6533	–0.8525	–0.9239	–0.8525	7
1.0	0.0	–0.3827	–0.7062	–0.9239	–0.9986	–0.9239	8
	(0.0000)	(–0.3827)	(–0.7071)	(–0.9239)	(–1.0000)	(–0.9239)	

The analytical solution to this problem is

$$u(x,t) = \sin(\pi x)\cos(\pi t)$$

(see Sankara Rao, 1995). The analytical values at $t = 1/2$, 1 are given in parenthesis, in the table. Obviously, there is a symmetry in the solution about $x = 0.5$. We observe from the table, that the numerical values agree with the analytical solution up to three decimal places only, as we have started computation by taking the initial and boundary conditions accurate to four decimal places.

After the completion of calculations for eight time steps, it can be seen from the u-values as given in the table, that the original initial values are reproduced, of course with a negative sign. This shows that, half a cycle has been completed in eight time steps.

Example 11.2 Solve the wave equation $u_{tt} = u_{xx}$, $0 \le x \le 1$, subject to the initial conditions $u(x, 0) = x^2$, $u_t(x, 0) = 1$ and the boundary conditions $u(0, t) = 0$, $u(1, t) = 1 + t$, $t > 0$ by taking $\Delta x = \Delta t = 0.2$ and compute the solution for the first four time steps.

Solution For given $\Delta x = \Delta t$ we have, $r = \Delta t/\Delta x = 1$. The explicit finite difference approximation to the given wave equation is

$$u_i^{n+1} = 2(1 - r^2)u_i^n + r^2(u_{i-1}^n + u_{i+1}^n) - u_i^{n-1}$$

Taking $r = 1$, the above equation reduces to

$$u_i^{n+1} = u_{i-1}^n + u_{i+1}^n - u_i^{n-1} \tag{1}$$

The derivative initial condition when approximated by central differences gives

$$\frac{u_i^{n+1} - u_i^{n-1}}{2\Delta t} = 1$$

That is,

$$u_i^{n+1} = u_i^{n-1} + 2\Delta t = u_i^{n-1} + 0.4$$

When $n = 0$, we obtain

$$u_i^1 = u_i^{-1} + 0.4 \tag{2}$$

while Eq. (1) gives

$$u_i^1 = u_{i-1}^0 + u_{i+1}^0 - u_i^{-1} \tag{3}$$

Eliminating u_i^{-1} between Eqs. (2) and (3), we have

$$u_i^1 = u_{i-1}^0 + u_{i+1}^0 - u_i^1 + 0.4 \quad \text{or} \quad u_i^1 = \frac{1}{2}(u_{i-1}^0 + u_{i+1}^0) + 0.2 \tag{4}$$

Thus, the initial condition, Eqs. (4) and (1) are enough to compute the solution to the given problem. The numerical values of u at different time steps are tabulated as below:

	x					
t	0.0	0.2	0.4	0.6	0.8	1.0
0.0	0.0	0.04	0.16	0.36	0.64	1.0
0.2	0.0	0.28	0.4	0.6	0.88	1.2
0.4	0.0	0.36	0.72	0.92	1.16	1.4
0.6	0.0	0.44	0.88	1.28	1.44	1.6
0.8	0.0	0.52	1.0	1.4	1.72	1.8

11.3 VON-NEUMANN STABILITY CONCEPT

The concept of stability of a finite difference scheme for a linear parabolic partial differential equation using matrix method has been developed in Section 9.5. Yet another important and most widely applied technique for stability analysis is due to Von-Neumann, developed during the World War II. Using this technique, the problem of stability analysis can be treated generally for linear equation (both parabolic and hyperbolic) with constant coefficients and with periodic boundary conditions. However, if the differential equation is non-linear, we shall make it linear by freezing the non-linear terms. In any case, linear stability analysis is a necessary condition for non-linear problems but not sufficient.

If $\bar{u}_i^n$ is an exact solution of the finite difference equation and u_i^n is its actual computed solution, then $\varepsilon_i^n = u_i^n - \bar{u}_i^n$, is the error at time level n and at the mesh point i. Recalling that errors are also solutions of the same discretized equation of a model equation, the basic principle of Von-Neumann method is that it expresses an initial line of errors in terms of finite Fourier series in space at each time level. Then, we determine whether the chosen computational algorithm is stable or not, based on the fact whether the separate

Fourier components of the error distribution decay or amplify as we progress from one time level to the next time level. To make this idea clear, let us consider a one-dimensional domain of length L, then the complex Fourier representation reflects the region $(0, L)$ onto the negative part $(-L, 0)$ and the fundamental frequency corresponds to the maximum wavelength $\lambda_{max} = 2L$. The corresponding wave number is $K = 2\pi/\lambda$, which attains its minimum value $K_{min} = \pi/L$; while the maximum wave number K_{max} of the finite spectrum on the interval $(-L, L)$ corresponds to the shortest resolvable wavelength on a mesh with spacing Δx. Obviously this shortest wavelength $\lambda_{min} = 2\Delta x$ (see Fig. 11.1), and therefore, $K_{max} = \pi/\Delta x$. Noting that $x_i = i\Delta x$, where $i = 0, 1, 2, ..., N$, we have $\Delta x = L/N$.

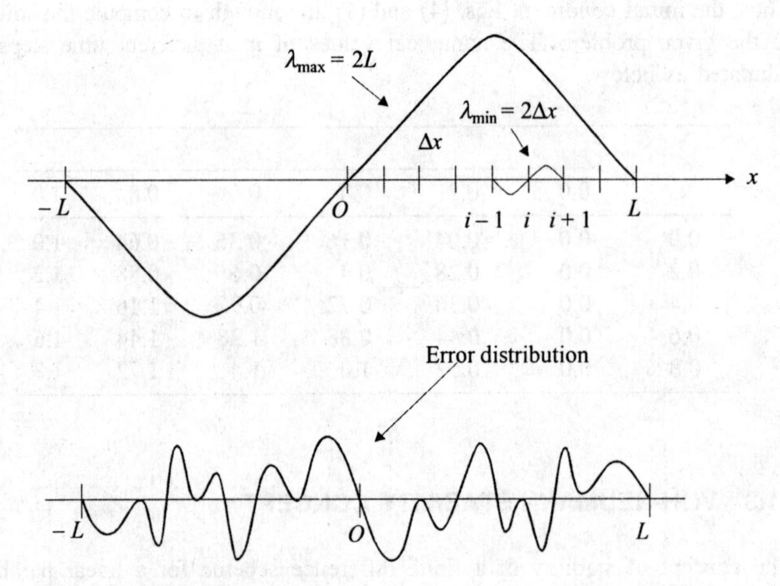

Fig. 11.1 Fourier representation of the error on the interval $(-L, L)$.

Now, any finite mesh function say ε_i^n or the solution u_i^n can be expressed into Fourier series in its complex exponential form as

$$\varepsilon_i^n \text{ (or } u_i^n) = V^n e^{In\pi x/L} = V^n e^{In\pi i\Delta x/(N\Delta x)} = V^n e^{IK_n i\Delta x} \tag{11.11}$$

where $I = \sqrt{(-1)}$ and $K_n = n\pi/(N\Delta x)$. Here, V^n is the amplitude function at time level n of a particular component whose wave number is K_n (wavelength $\lambda = 2\pi/K_n$). Let us also define the phase angle $\phi = K_n\Delta x$, so that Eq. (11.11) assumes the form

$$\varepsilon_i^n \text{ (or } u_i^n) = V^n e^{Ii\phi} \tag{11.12}$$

As long as the computational algorithm is linear, it is sufficient to study the propagation of the error due to just a single term. To demonstrate this principle, consider the wave equation

$$u_{tt} = c^2 u_{xx}, \qquad 0 \le x \le L, \quad t \ge 0$$

Applying the simple explicit finite difference scheme such as CTCS approximation, the above equation reduces to a form similar to Eq. (11.4). That is,

$$u_i^{n+1} = 2(1 - r^2c^2)u_i^n + r^2c^2(u_{i-1}^n + u_{i+1}^n) - u_i^{n-1}$$

where

$$r^2 = \frac{\Delta t^2}{\Delta x^2}$$

Inserting the form (11.12) for u_i^n into the above equation, we obtain

$$V^{n+1}e^{Ii\phi} = 2(1 - r^2c^2)\,V^n e^{Ii\phi} + r^2c^2\,[V^n e^{I(i-1)\phi} + V^n e^{I(i+1)\phi}] - V^{n-1}e^{Ii\phi} \quad (11.13)$$

For stability, the amplitude of any error harmonic should not grow with time. That is, $|V| \le 1$. After cancelling the common terms Eq. (11.13) becomes.

$$V = 2(1 - r^2c^2) + r^2c^2(e^{-I\phi} + e^{I\phi}) - V^{-1}$$

Further manipulations leads to

$$V + V^{-1} = 2 - 2r^2c^2\,(1 - \cos\phi)$$

$$= 2 - 4r^2c^2 \sin^2\frac{\phi}{2}$$

Therefore,

$$V^2 - 2\left(1 - 2r^2c^2 \sin^2\frac{\phi}{2}\right)V + 1 = 0 \quad (11.14)$$

which is a quadratic in V and hence will have two roots. For stability, we must have $|V| \le 1$. Also from Eq. (11.14) we can observe that the product of the two values of V is clearly unity. So, three cases arise.

Case 1: Both the roots are equal to unity. In that case the discriminant of the quadratic equation (11.14) is zero.

Case 2: One of the roots is greater than unity, in which case, stability of the chosen numerical scheme is not assured as the errors grow exponentially. Therefore, the discriminant of the quadratic equation (11.14) is greater than zero.

Case 3: Discriminant < 0 that is,

$$\left(1 - 2r^2c^2 \sin^2\frac{\phi}{2}\right)^2 - 1 \le 0$$

Thus, for stability

$$4r^4c^4 \sin^4\frac{\phi}{2} - 4r^2c^2 \sin^2\frac{\phi}{2} \le 0$$

That is

$$r^2c^2 \sin^2 \frac{\phi}{2} \le 1$$

This is always true if $r^2c^2 \le 1$. Which imply

$$c\frac{\Delta t}{\Delta x} \le 1 \tag{11.15}$$

Thus Eq. (11.15) is the required condition for the stability of the chosen numerical scheme. Von-Neumann technique can also be applied to study the problem of stability in the case of parabolic equation, which is discussed in the following example:

Example 11.3 Discuss the stability of FTCS scheme for two-dimensional heat equation $u_t = \alpha (u_{xx} + u_{yy})$

Solution Taking $\Delta x = \Delta y = h$, the FTCS scheme for the given heat equation yields

$$u_{i,j}^{n+1} - u_{i,j}^{n} = r\left(u_{i+1,j}^{n} + u_{i-1,j}^{n} + u_{i,j+1}^{n} + u_{i,j-1}^{n} - 4u_{i,j}^{n}\right) \tag{1}$$

where $r = \alpha\Delta t/h^2$. Inserting the Fourier component solution of Von-Neumann in the form

$$u_{i,j}^{n} = V^n e^{Ii\phi} e^{Ij\phi} \tag{2}$$

into Eq. (1), we obtain

$$V^{n+1} - V^n = rV^n\left(e^{I\phi} + e^{-I\phi} + e^{I\phi} + e^{-I\phi} - 4\right)$$

Let $G = V^{n+1}/V^n$, then the above equation can be rewritten as

$$G - 1 = r(2\cos\phi + 2\cos\phi - 4) = -8r\sin^2\frac{\phi}{2} \tag{3}$$

For stability of the chosen numerical algorithm, $|G| \le 1$ should be satisfied. That is

$$|1 - 8r\sin^2\frac{\phi}{2}| \le 1$$

We notice that

$$1 - 8r\sin^2\frac{\phi}{2} \le 1$$

for all values of r. Yet we have to show

$$-\left(1 - 8r\sin^2\frac{\phi}{2}\right) \le 1$$

That is,

$$1 - 8r \sin^2 \frac{\phi}{2} \geq -1$$

or

$$8r \sin^2 \frac{\phi}{2} \leq 2$$

Since the maximum value of $\sin \phi = 1$, we at once arrive at the condition that $r \leq 1/4$. In other words,

$$\alpha \frac{\Delta t}{h^2} \leq \frac{1}{4} \tag{4}$$

is the stability condition.

Example 11.4 Discuss the stability of Dufort and Frankel scheme for one-dimensional diffusion equation $T_t = \alpha T_{xx}$ given by Eq. (9.26); that is

$$(1 + 2r)\, T_i^{n+1} = 2r[T_{i-1}^n + T_{i+1}^n] + (1 - 2r)T_i^{n-1} \tag{1}$$

where $r = a\Delta t/\Delta x^2$, using Von-Neumann technique.

Solution Inserting the Von-Neumann Fourier component solution in the form

$$T_i^n = V^n e^{Ii\phi} \tag{2}$$

into Eq. (1), we get after cancellation of the common term

$$(1 + 2r)V = 2r\left(e^{-I\phi} + e^{I\phi}\right) + (1 - 2r)V^{-1}$$

which can be rewritten as

$$(1 + 2r)\, V^2 - (4r \cos \phi)\, V - (1 - 2r) = 0 \tag{3}$$

This being a quadratic will have two roots, given by

$$V_{1,2} = \frac{4r \cos\phi \pm \sqrt{16r^2 \cos^2\phi + 4(1 - 4r^2)}}{2(1 + 2r)}$$

On simplification, we get

$$V_{1,2} = \frac{2r \cos\phi \pm \sqrt{1 - 4r^2 \sin^2\phi}}{1 + 2r} \tag{4}$$

For stability, the condition $|V| < 1$ should be satisfied, The maximum value of $\cos \phi = 1$, when $\phi = 0$ in which case $\sin \phi = 0$. Therefore,

$$|V| = \frac{2r \pm 1}{2r + 1} \leq 1$$

is satisfied for all values of $r > 0$. Hence, Dufort and Frankel scheme is unconditionally stable for

$$r = \alpha \frac{\Delta t}{\Delta x^2} > 0$$

which is unusual for an explicit scheme.

EXERCISES

11.1 The function u satisfies the wave equation $u_{tt} = u_{xx}$, $0 \le x \le 1$. The boundary conditions are $u = 0$ at $x = 0$ and $x = 1$ for $t > 0$, and the initial conditions are

$$u(x, 0) = \sin\frac{x\pi}{8}, \qquad \frac{\partial u(x, 0)}{\partial t} = 0$$

Compute the solution for $x = 0\ (0.1)\ 0.5$ and $t = 0\ (0.1)\ 0.5$ and compare it with the analytical solution

$$u = \frac{1}{8}[\sin(\pi x)\cos(\pi t)] \qquad \text{at } t = 0.5$$

11.2 Solve the boundary value problem described by the hyperbolic partial differential equation

$$u_{tt} - c^2 u_{xx} = 0, \qquad 0 \le x \le l,\ t \ge 0$$

subject to the boundary conditions $u(0, t) = u(l, t) = 0$, $t \ge 0$, and the initial conditions

$$u(x, 0) = 10\sin\frac{\pi x}{l},\ 0 \le x \le l\ u_t(x, 0) = 0$$

by taking $l = 1$, $c = 1$, $\Delta x = 1/8$.

11.3 A string is tightly stretched between $x = 0$ and $x = L$ and is initially at rest. Each point of the string is given an initial velocity of

$$y_t(x, 0) = \mu\sin^3\frac{\pi x}{L}$$

Find numerically the displacement of the string with time $t = 0\ (0.5)\ 3.0$, assuming $y_{tt} = \alpha^2 y_{xx}$, $0 \le x \le L$ and taking $\alpha = 1$, $\mu = 1$, $\Delta x = 0.5$, $L = 3.0$.

11.4 Solve numerically $4u_{xx} = u_{tt}$, $0 \le x \le 4$, given that $u(0, t) = u(4, t) = 0$, and $u(x, 0) = x(4 - x)$, $u_t(x, 0) = 0$, taking the mesh size $\Delta x = 1$ and time step $\Delta t = 0.5$. Compute the solution for the first five time steps.

11.5 Find the numerical solution of the wave equation $u_{tt} = u_{xx}$, $0 \le x \le 1$, $t \ge 0$ subject to the initial conditions $u_t = 0$ at $t = 0$ and

$$u(x, 0) = \begin{cases} 5\,(x - 0.3), & 0.3 < x \le 0.5 \\ 5\,(0.7 - x), & 0.5 < x \le 0.7 \\ 0.0 & \text{for all other points,} \end{cases}$$

and boundary conditions

$$u(0, t) = u(1, t) = 0, \qquad t > 0$$

taking $\Delta t = \Delta x = 0.1$ for the first five-time steps.

11.6 Find the numerical solution to the wave equation for a vibrating string $u_{tt}(x, t) = 4u_{xx}(x, t)$, $0 \le x \le 1$, subject to the boundary conditions $u(0, t) = u(1, t) = 0$, for $t \ge 0$ and the initial conditions $u(x, 0) = \sin(\pi x) + \sin(2\pi x)$ and $u_t(x, 0) = 0$, $0 \le x \le 1$, using an explicit finite difference scheme. Assume $\Delta x = 0.1$, $\Delta t = 0.05$ and compute u for ten time steps.

Chapter 12

Boundary Value Problems

12.1 INTRODUCTION

An important class of problems known as boundary value problems occur in science and engineering, wherein the values of the function or its derivatives are given at more than one value of the independent variable, usually at the end points or boundaries of some domain. For example, in elasticity, computation of the deflection of a simply supported beam is well-known. Here, the deflections and its second derivatives are specified at the supports. Similarly, steady heat transfer problems with a distributed heat source fall in this class, where temperatures or temperature gradients are specified at two points.

It may be noted that every boundary value problem need not have a solution. For example, consider the boundary value problem denoted as BVP is given by

$$y'' + y = 0 \tag{12.1}$$

subject to the boundary conditions

$$y'(0) = 0, \qquad y(\pi/2) = 1$$

The general solution of the differential equation (12.1) is

$$y(x) = C_1 \cos x + C_2 \sin x$$

and the Wronskian

$$W(\cos x, \sin x) = \begin{vmatrix} \cos x & \sin x \\ -\sin x & \cos x \end{vmatrix} = 1$$

is non-vanishing, thus guaranteeing a unique solution to the given BVP. Now, the condition $y'(0) = 0$ implies $C_2 = 0$ and the remaining solution $y(x) = C_1 \cos x$ cannot satisfy $y(\pi/2) = 1$ for any C_1, because $\cos(\pi/2) = 0$. Hence, the above BVP has no solution.

However, most of the BVP's have a unique solution. For, let us consider the general second order linear differential equation

$$y'' + p_1(x)y' + p_0(x)y = 0 \tag{12.2}$$

with boundary conditions

$$y'(x_0) = a, \quad y(x_1) = b$$

Suppose, $y_1(x)$ and $y_2(x)$ are two linearly, independent solutions of the differential equation (12.2). Then, its general solution can be written as

$$y(x) = C_1 y_1(x) + C_2 y_2(x)$$

using the given boundary conditions, we have

$$C_1 y_1'(x_0) + C_2 y_2'(x_0) = a$$

$$C_1 y_1(x_1) + C_2 y_2(x_1) = b$$

These equations can be solved for unique C_1 and C_2 provided

$$\begin{vmatrix} y_1'(x_0) & y_2'(x_0) \\ y_1(x_1) & y_2(x_1) \end{vmatrix} \neq 0$$

This condition which is nothing but the Wronskian condition is usually satisfied, showing the existence of a unique solution to the BVP. But, if the points x_0 and x_1 are so chosen, such that $y_1'(x_0)y_2(x_1) - y_2'(x_0)y_1(x_1) = 0$, then the BVP is ill-posed. In that case, the BVP has no solution or has an infinite number of solutions.

In what follows, we shall discuss four diferent methods for solving BVP's. They are

1. Finite-difference method
2. The shooting method
3. Weighted residual methods, and finally
4. Cubic spline method.

12.2 FINITE-DIFFERENCE METHOD

In Chapter 9, we have presented various finite difference approximations such as forward difference, backward difference and central difference approximations to derivatives of various orders. By replacing derivatives in the given BVP by their corresponding finite difference approximations we arrive at a system of algebraic equations for the dependent variables at the node points whose solution under given boundary conditions gives us the numerical solution or approximate solution of the BVP. We shall illustrate this method through the following examples.

Example 12.1 The deflection y of a *fixed-fixed* beam of length 3 metres is governed by the differential equation

$$\frac{d^4 y}{dx^4} + 2y = \frac{x^2}{9} + \frac{2}{3}x + 4$$

and the boundary conditions are given by

$$y(0) = y'(0) = y(3) = y'(3) = 0$$

Find the deflection at the points $x = 1$ and $x = 2$.

Solution Taking $h = 1$, we denote $x_0 = 0$, $x_1 = 1$, $x_2 = 2$, $x_3 = 3$ with the corresponding y-values written as y_0, y_1, y_2, y_3. From the given data, we have $y_0 = y_3 = 0$, $y'_0 = y'_3 = 0$.

Now, replacing the fourth derivative by its finite difference approximation (See Example 9.1), the given differential equation becomes

$$\frac{1}{h^4}\left[y_{i+2} - 4y_{i+1} + 6y_i - 4y_{i-1} + y_{i-2}\right] + 2y_i$$

$$= \frac{x_i^2}{9} + \frac{2}{3}x_i + 4, \quad i = 1, 2$$

For $i = 1$,

$$y_3 - 4y_2 + 6y_1 - 4y_0 + y_{-1} + 2y_1 = \frac{1}{9} + \frac{2}{3} + 4$$

using the boundary condition $y_0 = y_3 = 0$, we get

$$-4y_2 + 8y_1 + y_{-1} = 43/9 \tag{1}$$

For $i = 2$

$$y_4 - 4y_3 + 6y_2 - 4y_1 + y_0 + 2y_2 = \frac{4}{9} + \frac{4}{3} + 4$$

Again using the boundary condition $y_0 = y_3 = 0$, we have

$$y_4 + 8y_2 - 4y_1 = 52/9 \tag{2}$$

Thus, we have only two equations (1) and (2) containing 4 unknowns. However, it may be observed that y_{-1} and y_4 are fictitious points. Using central difference approximations to the derivative boundary conditions at the left and right ends of the beam, we have

$$\frac{y_{-1} - y_1}{2h} = 0 \quad \text{and} \quad \frac{y_2 - y_4}{2h} = 0$$

Therefore,

$$y_{-1} = y_1, \quad y_4 = y_2 \tag{3}$$

Substituting (3) into Eqs. (1) and (2) we find

$$\left.\begin{aligned} 9y_1 - 4y_2 &= 43/9 \\ 4y_1 - 9y_2 &= -52/9 \end{aligned}\right\} \tag{4}$$

on solving, we get

$$y_1 = 1.01709, \qquad y_2 = 1.09402$$

Example 12.2 The deflection of a cantilever beam is governed by the differential equation

$$\frac{d^4 y}{dx^4} + 81y = 729x^2$$

The boundary conditions are given by

$$y(0) = y'(0) = 0, \qquad y''(1) = y'''(1) = 0$$

indicating zero deflection and slope at the fixed end. The last two boundary conditions prescribe zero bending moment and shear force at the free end. Find the deflection at the pivotal points of the beam assuming $h = 1/3$.

Solution Taking $h = 1/3$, the pivotal points are $x_0 = 0$, $x_1 = 1/3$, $x_2 = 2/3$, $x_3 = 1$ and the corresponding values of y are y_0, y_1, y_2 and y_3 respectively.

Replacing the fourth derivative in the governing equation by its finite difference equivalent, we get

$$\frac{1}{h^4}\left[y_{i+2} - 4y_{i+1} + 6y_i - 4y_{i-1}\ y_{i-2}\right]$$

$$+ 81y_i = 729\ x_i^2 \qquad i = 1, 2, 3 \tag{1}$$

For $i = 1$, $h = 1/3$, Eq. (1) gives

$$y_3 - 4y_2 + 7y_1 - 4y_0 + y_{-1} = 1$$

Using the boundary condition $y_0 = 0$, the above equation reduces to

$$y_3 - 4y_2 + 7y_1 + y_{-1} = 1 \tag{2}$$

For $i = 2$, $h = 1/3$, Eq. (1) gives

$$y_4 - 4y_3 + 7y_2 - 4y_1 = 4 \tag{3}$$

Similarly for $i = 3$, $h = 1/3$, we have

$$y_5 - 4y_4 + 6y_3 - 4y_2 + y_1 + y_3 = 9$$

or

$$y_5 - 4y_4 + 7y_3 - 4y_2 + y_1 = 9 \tag{4}$$

From Eqs. (2) to (4), we observe that y_{-1}, y_4 and y_5 are fictitious points and they can be eliminated using the derivative boundary conditions.

From Eqs. (9.9), (9.10) and (9.13), we have

$$y_i' = (y_{i+1} - y_{i-1})/2h$$

For $i = 0$, $y_0' = 0 = y_1 - y_{-1}$, implying

$$y_{-1} = y_1 \tag{5}$$

$$y_i'' = (y_{i+1} - 2y_i + y_{i-1})/h^2$$

For $i = 3$, $y''_3 = 0 = y_4 - 2y_3 + y_2$, implying

$$y_4 = 2y_3 - y_2 \tag{6}$$

$$y_i''' = (y_{i+2} - 2y_{i+1} + 2y_{i-1} - y_{i-2})/2h^3$$

For $i = 3$, using $y_3''' = 0$, we get

$$0 = y_5 - 2y_4 + 2y_2 - y_1, \text{ implying}$$

$$y_5 = 4y_3 - 4y_2 + y_1 \tag{7}$$

Now, using Eqs. (5), (6) and (7) in Eqs. (2), (3) and (4) we obtain

$$\left.\begin{aligned} 8y_1 - 4y_2 + y_3 &= 1 \\ 4y_1 - 6y_2 + 2y_3 &= -4 \\ 2y_1 - 4y_2 + 3y_3 &= 9 \end{aligned}\right\} \tag{8}$$

In Matrix notation

$$\begin{bmatrix} 8 & -4 & 1 \\ 4 & -6 & 2 \\ 2 & -4 & 3 \end{bmatrix} \begin{pmatrix} y_1 \\ y_2 \\ y_3 \end{pmatrix} = \begin{pmatrix} 1 \\ -4 \\ 9 \end{pmatrix} \tag{9}$$

Using crouts reduction scheme, Eq. (9) takes the form

$$\begin{bmatrix} 8 & 0 & 0 \\ 4 & -4 & 0 \\ 2 & -3 & 13/8 \end{bmatrix} \begin{bmatrix} 1 & -1/2 & 1/8 \\ 0 & 1 & -3/8 \\ 0 & 0 & 1 \end{bmatrix} \begin{pmatrix} y_1 \\ y_2 \\ y_3 \end{pmatrix} = \begin{pmatrix} 1 \\ -4 \\ 9 \end{pmatrix} \tag{10}$$

from which we find
$y_1 = 1.1539$, $y_2 = 3.9231$ and $y_3 = 7.4615$. Therefore, the deflections at the pivotal points are

$$y(1/3) = 1.1539, \quad y(2/3) = 3.9231 \quad \text{and} \quad y(1) = 7.4615.$$

Example 12.3 Solve the boundary value problem governed by the equation

$$\frac{d^2 y}{dx^2} + y = 0$$

with the boundary conditions

$$y(0) + y'(0) = 2 \quad \text{and} \quad y(\pi/2) + y'(\pi/2) = -1$$

by dividing the interval into four subintervals.

Solution Since the interval $(0, \pi/2)$ is divided into four subintervals, we take $h = \pi/8$. We denote the values of y by y_0, y_1, y_2, y_3 and y_4 corresponding to $x = 0, \pi/8, \pi/4, 3\pi/8, \pi/2$ respectively.

Now, replacing the second derivative in the governing equation by its finite difference equivalent, we have

$$(y_{i+1} - 2y_i + y_{i-1})/h^2 + y_i = 0$$

or

$$y_{i+1} + (h^2 - 2)y_i + y_{i-1} = 0 \tag{1}$$

For $i = 1$, 2 and 3, we get

$$\left.\begin{array}{l} y_2 - 1.8458\, y_1 + y_0 = 0 \\ y_3 - 1.8458\, y_2 + y_1 = 0 \\ y_4 - 1.8458\, y_3 + y_2 = 0 \end{array}\right\} \tag{2}$$

Using finite difference approximations to the boundary conditions, we have

$$y_i + (y_{i+1} - y_i)/h = 2$$

Setting $i = 0$, for left boundary, we get

$$y_0 + \frac{8}{\pi}(y_1 - y_0) = 2$$

or

$$\left(1 - \frac{8}{\pi}\right) y_0 = 2 - \frac{8}{\pi} y_1$$

or

$$-1.5465\, y_0 = 2 - 2.5465 y_1 \tag{3}$$

Similarly, we have for the second boundary

$$y_i + (y_i - y_{i-1})/h = -1$$

Setting $i = 4$, for the right side boundary yields

$$y_4 + \frac{8}{\pi}(y_4 - y_3) = -1$$

or

$$\left(1 + \frac{8}{\pi}\right) y_4 = -1 + \frac{8}{\pi} y_3$$

or

$$3.5465 y_4 = -1 + 2.5465 y_3 \tag{4}$$

With the help of Eqs. (3) and (4) we eliminate y_0 and y_4 in equation (2) and obtain

$$-0.1992 y_1 + y_2 = 1.2932 \tag{5}$$

$$y_1 - 1.8458 y_2 + y_3 = 0 \tag{6}$$

and

$$y_2 - 1.1278 y_3 = 0.2820 \tag{7}$$

Now, using Eqs. (5) and (7) in (6) we can show that

$$y_1 = 1.8425$$

then Eq. (5) gives

$$y_2 = 1.6602$$

Eq. (7) yields

$$y_3 = 1.2220$$

Finally, Eqs. (3) and (4) give

$$y_0 = 1.7407, \qquad y_4 = 0.5955$$

Hence, the required solution is tabulated below

x	0	$\pi/8$	$\pi/4$	$3\pi/8$	$\pi/2$
y	1.7407	1.8425	1.6602	1.2220	0.5955

12.3 SHOOTING METHOD

Shooting method is an iterative technique which uses one of the standard methods such as Runge-Kutta (R–K) method developed in Chapter 8 for solving initial value problems. For illustration, let us consider the boundary value problem described by the second order differential equation

$$y'' = f(x, y, y') \tag{12.3}$$

subject to the boundary conditions

$$y(0) = 0, \quad y(L) = y_L$$

where f is an arbitrary function and the data is prescribed at $x = 0$ and at $x = L$. The same differential equation describes an initial value problem, if the data is prescribed as

$$y(0) = 0, \quad y'(0) = y_p$$

To solve the above BVP, we reduce it to two first order differential equations by substitution, such as

$$u = y, \qquad v = y' \tag{12.4}$$

Thus, Eq. (12.3) becomes

$$\left.\begin{aligned} u' &= v \\ v' &= f(x, u, v) \end{aligned}\right\} \tag{12.5}$$

while the boundary conditions can be written as

$$u(0) = y_0, \quad u(L) = y_L$$

Now, we can solve this system as an initial value problem, provided $u'(0) = v(0)$ were given in addition to $u(0)$. Since $v(0)$ is not specified, we use a "guess", for $v(0)$ or $y'(0)$ and integrate both the equations (12.5) upto $x = L$. At this point, we compare $u(L)$ with y_L. In general they do not agree. Now, we make another "guess", for $v(0)$ or $y'(0)$ and the process is repeated. Suppose, the first two guesses are denoted as $y'_1(0)$ and $y'_2(0)$ and we have the solutions of Eq. (12.3) or equivalently that of Eq. (12.5) at $x = L$, denoted by $y_1(L)$ and $y_2(L)$. Using the idea of linear interpolation, we form the straight line between the points $[y_1(L), y_1'(0)]$ and $[y_1(L), y_2'(0)]$. This straight line is a crude approximation to the actual curve of $y(L)$ vs $y'(0)$ as shown in Fig. 12.1.

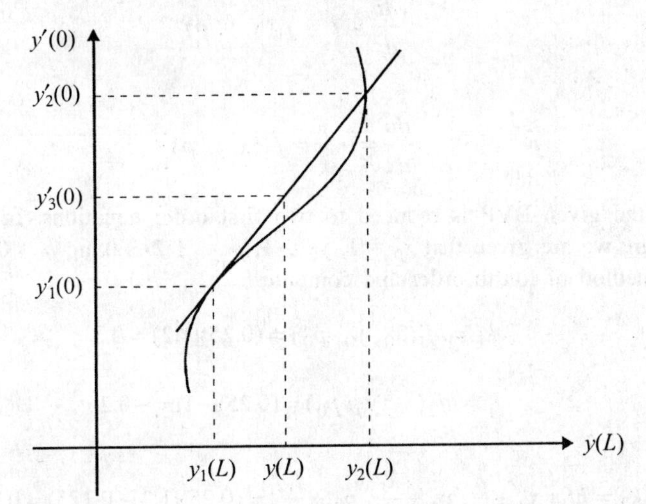

Fig. 12.1 Functional relationship between $y(L)$ and $y'(0)$.

The equation of the line between the above two points is given by

$$y'(0) = y_2'(0) + m[y(L) - y_2(L)] \tag{12.6}$$

where m the slope of the line defined by

$$m = \frac{y_2'(0) - y_1'(0)}{y_2(L) - y_1(L)}$$

The next "guess" for $y'(0)$ can be taken as the point of intersection of the vertical line from y_L with the straight line (12.6), thus we have

$$y_3'(0) = y_2'(0) + m[y(L) - y_2(L)]$$

which suggests an iterative formula

$$y_{n+1}'(0) = y_n'(0) + m_{n-1}[y(L) - y_n(L)], \qquad \text{for } n = 1, 2, 3 \ldots \tag{12.7}$$

where

$$m_{n-1} = \frac{y_n'(0) - y_{n-1}'(0)}{y_{n(L)} - y_{n-1}(L)} \tag{12.8}$$

These iterations are carried out until $y(L)$ is sufficiently close to y_L. For illustration, we consider the following example.

Example 12.4 Solve the boundary value problem

$$x\frac{d^2y}{dx^2} + y = 0, \; y(1) = 1, \quad y(1.25) = 1.3513$$

Using shooting method. Take $h = 0.25$ and assume the initial guesses for $y'(1)$ as 1.2 and 1.5 respectively.

Solution Let

$$\frac{dy}{dx} = p = f_1(x, y, p) \tag{1}$$

then

$$\frac{dp}{dx} = -\frac{y}{x} = f_2(x, y, p) \tag{2}$$

Thus, the given BVP is reduced to two first order equations. In the present problem, we are given that $x_0 = 1$, $y_0 = 1$, $p_0 = 1.2$. Taking $h = 0.25$, we use R–K method of fourth order and compute.

$$k_1 = hf_1(x_0, y_0, p_0) = (0.25)(1.2) = 0.3$$

$$l_1 = hf_2(x_0, y_0, p_0) = (0.25)(-1) = -0.25$$

$$k_2 = hf_1\left(x_0 + \frac{h}{2}, y_0 + \frac{k_1}{2}, p_0 + \frac{l_1}{2}\right) = (0.25)(1.2 - 0.125) = 0.2688$$

$$l_2 = hf_2\left(x_0 + \frac{h}{2}, y_0 + \frac{k_1}{2}, p_0 + \frac{l_1}{2}\right) = (0.25)(-1)\left[\frac{1 + 0.15}{1 + 0.125}\right] = -0.2556$$

$$k_3 = hf_1\left(x_0 + \frac{h}{2}, y_0 + \frac{k_2}{2}, p_0 + \frac{l_2}{2}\right) = (0.25)(1.2 - 0.1278) = 0.2681$$

$$l_3 = hf_2\left(x_0 + \frac{h}{2}, y_0 + \frac{k_2}{2}, p_0 + \frac{l_2}{2}\right) = (0.25)(-1)\left[\frac{1 + 0.1344}{1 + 0.125}\right] = -0.2521$$

$$k_4 = hf_1(x_0 + h, y_0 + k_3, p_0 + l_3) = (0.25)(1.2 - 0.2521) = 0.2370$$

$$l_4 = hf_2(x_0 + h, y_0 + k_3, p_0 + l_3) = (0.25)(-1)\left[\frac{1 + 0.2681}{1 + 0.25}\right] = -0.2536$$

Now, we find

$$y_1(1.25) = y_0 + \frac{1}{6}[k_1 + 2k_2 + 2k_3 + k_4] = 1.2685 \tag{3}$$

and

$$y'(1.25) = p_1 = p_0 + \frac{1}{6}[l_1 + 2l_2 + 2l_3 + l_4] = 0.7468$$

Taking the second guess, $y_2'(1) = 1.5$, we solve the system (1) and (2) again by using $x_0 = 1$, $y_0 = 1$, $p_0 = 1.5$ and $h = 0.25$. Thus, R–K fourth order numerical scheme yields

$$k_1 = 0.375, \qquad l_1 = -0.25$$
$$k_2 = 0.3438, \qquad l_2 = -0.2639$$
$$k_3 = 0.3420, \qquad l_3 = -0.2604$$
$$k_4 = 0.3099, \qquad l_4 = -0.2684$$

thus giving

$$\begin{aligned} y_2(1.25) &= y_0 + \frac{1}{6}[k_1 + 2k_2 + 2k_3 + k_4] \\ &= 1 + \frac{1}{6}[0.375 + 0.6876 + 0.6840 + 0.3099] \\ &= 1.3428 \end{aligned} \tag{4}$$

To start the iteration given by Eq. (12.7), using Eqs. (3) and (4), we compute the slope of the straight line as

$$m = \frac{y_2'(1) - y_1'(1)}{y_2(1.25) - y_1(1.25)} = \frac{(1.5 - 1.2)}{(1.3428 - 1.2685)} = 4.0377$$

and thereby getting

$$y_3'(1) = 1.5 + 4.0377[1.3513 - 1.3428] = 1.5343 \tag{5}$$

Thus, we observe that by taking $y_1'(1) = 1.2$ and $y_2'(1) = 1.5$, we have obtained $y_1(1.25) = 1.2685$ and $y_2(1.25) = 1.3428$, which is still not close to the given value of $y(1.25) = 1.3513$. Therefore, we take the new value $y_3'(1) = 1.5343$ and repeat the use of fourth order R–K method to compute

$$k_1 = 0.3836 \qquad l_1 = -0.25$$
$$k_2 = 0.3523 \qquad l_2 = -0.2648$$
$$k_3 = 0.3505 \qquad l_3 = -0.2614$$
$$k_4 = 0.3182 \qquad l_4 = 1.0804$$

thus getting

$$\begin{aligned} y_3(1.25) &= y_0 + \frac{1}{6}[k_1 + 2k_2 + 2k_3 + k_4] \\ &= 1 + 0.3512 = 1.3512 = y(1.25) \end{aligned} \tag{6}$$

which is found to be accurate to three decimal places of the given value of $y(1.25)$.

It may be noted that by taking $y'(1) = 1.5343$, we can determine the solution of the given BVP, even in the extended interval $1 \le x \le 2$ by posing it as an initial value problem such as

$$\frac{dy}{dx} = p, \quad \frac{dp}{dx} = -\frac{y}{x}$$

with the initial conditions $y(1) = 1, y'(1) = 1.5343$

12.4 WEIGHTED RESIDUAL METHODS

The principal idea of weighted residual methods (WRMs) is to find an approximate solution to differential equations. Let us consider a general differential equation of the form

$$L(u) = f \tag{12.9}$$

which is valid in a particular domain Ω, subject to the boundary conditions such as

$$B_j u = g_j \tag{12.10}$$

on the boundary Γ of the above domain. In order to find an approximate solution to the above boundary value problem, we introduce a set of functions of the form

$$\hat{u} = u_0 + \sum_{j=1}^{n} C_j \phi_j \tag{12.11}$$

where u_0 is chosen to satisfy the boundary conditions, exactly if possible. The coefficients C_j are unknowns which are to be determined by solving a system of equations and ϕ_j are the linearly-independent functions also called trial functions, supposed to be known. Since (12.11) is an approximate function, we observe that, when substituted in Eq. (12.9), it will not satisfy exactly. Thus, we get the error or equation residual denoted by R, which is a continuous function of spatial coordinates in general and can be written as

$$L(\hat{u}) - f = R \tag{12.12}$$

In one spatial dimension, the approximating functions might be polynomials or trignometric functions, such as

$$\phi_j(x) = x^{j-1} \quad \text{or} \quad \phi_j(x) = \sin j\pi x$$

Now, let us introduce a set of weighting functions W_j, $j = 1, 2, \ldots$ and construct an inner product (R, W_j) defined by

$$(R, W_j) = \iiint_{\Omega} W_j \, R d\Omega = 0, \; j = 1, 2, .., n \tag{12.13}$$

In other words, we are forcing the residual or error of the approximate differential equation (12.12) to become zero in an average sense. By letting $j = 1, 2, ..., n$ a system of equations for the Cj's is generated.

There are many ways of choosing the weighting functions W_j, also called test functions. Their choices lead to various methods. They are (1) Galerkin method, (2) Least squares method, (3) Method of Moments and (4) Collocation method.

In the Galerkin method, we choose the weighting function W_j from the same family as the trial function (basis function). Thus, we have

$$(R, \phi_j) = 0, \quad j = 1, 2, \ldots, n \tag{12.14}$$

In the Least square method, the weighting function is nothing but the residual itself. Here, we minimize the inner product with respect to every coefficient C_j, thus getting

$$\frac{\partial}{\partial C_j}(R, R) = 0 \tag{12.15}$$

In the method of moments, the weighting functions are chosen from any set of linearly, independent functions such as 1, x, x^2, x^3, . . . for one-dimensional problems, so that

$$(R, x^j) = 0, \quad j = 1, 2, \ldots \tag{12.16}$$

In the Collocation method, we choose a set of points in the domain as collocation points, the weighting function being the Dirac delta function (See Sankara Rao, 1995), that is

$$W_j = \delta(x - x_j) \tag{12.17}$$

such that

$$(R, \delta(x - x_j)) = R(x_i) = 0, \tag{12.18}$$

implying that the residual is zero at n specified collocation points x_j.

For illustration of WRMs, here follows an example

Example 12.5 Using Galerkin method, solve the following BVP described by the differential equation

$$\frac{d^2u}{dx^2} + u + x = 0, \quad 0 < x < 1 \tag{1}$$

subject to the boundary conditions

$$u(0) = 0, \quad \left.\frac{du}{dx}\right|_{x=1} = 0 \tag{2}$$

Solution Let us consider the polynomial trial function in the form

$$u(x) = u_0 + \sum_{j=1}^{3} C_j\phi_j \, . \quad \text{That is,}$$

$$u(x) = u_0 + C_1x + C_2x^2 + C_3x^3 \tag{3}$$

Satisfaction of the given boundary conditions yield

$$u_0 = 0, \quad C_1 = -2C_2 - 3C_3$$

Thus, we choose the approximating function as

$$\hat{u}(x) = C_2(x^2 - 2x) + C_3(x^3 - 3x) \tag{4}$$

Now, substituting Eq. (4) in the differential equation, we get the residual $R(x)$ as

$$R(x) = 2C_2 + 6C_3x + C_2(x^2 - 2x) + C_3(x^3 - 3x) + x$$

which simplifies to

$$R(x) = 2C_2 + x + C_2 (x^2 - 2x) + C_3(x^3 + 3x) \tag{5}$$

Since, there are two unknowns C_2 and C_3, we take the weighting function as

$$W_1 = x^2 - 2x, \qquad W_2 = x^3 + 3x$$

and evaluate two integrals, such as

$$\int_0^1 (x^2 - 2x)R(x)\,dx = 0$$

and

$$\int_0^1 (x^3 + 3x)R(x)\,dx = 0$$

The first integral gives

$$\int_0^1 (x^2 - 2x)[2C_2 + x + C_2(x^2 - 2x) + C_3(x^3 + 3x)]\,dx = 0$$

or

$$\frac{2}{3}C_2 - 2C_2 + \frac{1}{4} - \frac{2}{3} + C_2\left(\frac{1}{5} + \frac{4}{3} - 1\right) + C_3\left(\frac{1}{6} + \frac{3}{4} - \frac{2}{5} - 2\right) = 0$$

or

$$48C_2 + 89C_3 = -25 \tag{6}$$

The second integral gives

$$\int_0^1 (x^3 + 3x)[2C_2 + x + C_2(x^2 - 2x) + C_3(x^3 + 3x)]\,dx = 0$$

on evaluation, we have

$$847C_2 + 1824C_3 = -504 \tag{7}$$

Solving Eqs. (6) and (7) we find

$$C_2 = -0.0611, \qquad C_3 = -0.2479$$

Therefore,

$$C_1 = 0.8659$$

Hence, the required solution is

$$u = 0.8659x - 0.0611x^2 - 0.2479x^3 \tag{8}$$

For example

$$u(0.5) = 0.3867 \tag{9}$$

while the exact solution is

$$u = \left(\frac{\sin x}{\cos 1} - x\right)\Bigg|_{x=0.5} = 0.3873 \tag{10}$$

12.5 CUBIC SPLINE METHOD

In Section 6.8, we have introduced the concept of a cubic spline for interpolation, while in Section 6.8.1, we gave a detailed account of the construction of a cubic spline from its properties. In this section, an alternate attempt is made to construct a cubic spline which is more suitable to solve BVPs. Of course, we have preserved the notation as such.

Definition 12.1 A cubic spline is a piecewise cubic polynomial, that is twice continuously differentiable. Given the data points (x_i, y_i), $i = 0, 1, 2, ..., n$ to determine a function $S(x)$ such that

(i) $S(x_i) = y_i$, $i = 0, 1, 2, ..., n$
(ii) $S(x)$ is cubic in each subinterval (x_i, x_{i+1}) for $i = 0, 1, 2, ..., (n - 1)$ and
(iii) $S(x) \in c^2\ [x_0, x_n]$ are satisfied.

Starting from this basic definition, we shall proceed to construct the cubic spline. For, let us assume that the knots x_i were arbitrarily-placed such that $x_{i+1} - x_i = h_i$. Since $S(x)$ is cubic, it is easy to see that $S''(x)$ is linear and hence in the interval (x_i, x_{i+1}), we can write it as

$$S''(x) = \frac{1}{h_i}\left[(x_{i+1} - x)\, S''(x_i) + (x - x_i) S''(x_{i+1})\right] \tag{12.19}$$

Following the usual notation

$$S''(x_i) = M_i \tag{12.20}$$

and integrating Eq. (12.19) twice with respect to x, we get

$$S(x) = \frac{1}{h_i}\left[\frac{(x_{i+1} - x)^3}{6} M_i + \frac{(x - x_i)^3}{6} M_{i+1}\right]$$

$$+\ C_i\ (x_{i+1} - x) + d_i\ (x - x_i) \tag{12.21}$$

Where C_i and d_i are constants of integration.

Noting that $S(x_i) = y_i$, $S(x_{i+1}) = y_{i+1}$, we find from Eq. (12.21) that

$$C_i = \frac{1}{h_i}\left[y_i - \frac{h_i^2}{6} M_i\right]$$

and (12.22)

$$d_i = \frac{1}{h_i}\left[y_{i+1} - \frac{h_i^2}{6} M_{i+1}\right]$$

Thus, in the interval (x_i, x_{i+1}), we have a cubic spline in the form

$$S(x) = \frac{1}{h_i}\left[\frac{(x_{i+1}-x)^3}{6}M_i + \frac{(x-x_i)^3}{6}M_{i+1}\right]$$

$$+\frac{(x_{i+1}-x)}{h_i}\left(y_i - \frac{h_i^2}{6}M_i\right) + \frac{(x-x_i)}{h_i}\left(y_{i+1} - \frac{h_{i+1}^2}{6}M_{i+1}\right) \tag{12.23}$$

Now, to utilize the third condition, that is, the continuity of first and second derivative of $S(x)$ we observe

$$S'(x) = \frac{1}{h_i}\left[\frac{-(x_{i+1}-x)^2}{2}M_i + \frac{(x-x_i)^2}{2}M_{i+1}\right.$$

$$\left. -\left(y_i - \frac{h_i^2}{6}M_i\right) + \left(y_{i+1} - \frac{h_i^2}{6}M_{i+1}\right)\right] \tag{12.24}$$

To get the right hand derivative we set $x = x_i$ in Eq. (12.24) and obtain

$$S'(x_i\,+) = \frac{-h_i}{6}(2M_i + M_{i+1}) + \frac{1}{h_i}(y_{i+1} - y_i) \tag{12.25}$$

Similarly in the interval (x_{i-1}, x_i), we will have

$$S'(x_i\,-) = \frac{h_{i-1}}{6}(2M_i + M_{i-1}) + \frac{1}{h_{i-1}}(y_i - y_{i-1}) \tag{12.26}$$

For continuity of $S(x)$ we must have

$$S'(x_i\,+) = S'(x_i\,-)$$

which gives the following recurrence relation

$$h_{i-1}M_{i-1} + 2\,(h_{i-1} + h_i)\,M_i + h_iM_{i+1}$$

$$= 6\left[\frac{y_{i+1}-y_i}{h_i} - \frac{y_i - y_{i-1}}{h_{i-1}}\right],\ i = 1, 2, \dots (n-1) \tag{12.27}$$

It may be noted that this equation is identical to Eq. (6.83).

In some applications, it is conveient to work with $S'(x_i)$ rather than with $S''(x_i)$. In such a case, we set $S'(x_i) = m_i$ and get from Eq. (12.24) that

$$S'(x_{i+1}) = m_{i+1} = \frac{h_i}{3}M_{i+1} + \frac{h_i}{6}M_i + \frac{y_{i+1}-y_i}{h_i}$$

on the same interval (x_i, x_{i+1}), we also get

$$S'(x_i) = m_i = -\frac{h_i}{3}M_i - \frac{h_i}{6}M_{i+1} + \frac{y_{i+1} - y_i}{h_i}$$

From these two relations, we at once have

$$m_{i+1} - m_i = \frac{h_i}{2}(M_i + M_{i+1}) \tag{12.28}$$

Similarly, we can also derive the following relation between m_i's as

$$\frac{m_{i-1}}{h_{i-1}} + 2\left(\frac{1}{h_{i-1}} + \frac{1}{h_i}\right)m_i + \frac{m_{i+1}}{h_i} = \frac{3(y_i - y_{i-1})}{h_{i-1}^2} + \frac{3(y_{i+1} - y_i)}{h_i^2} \tag{12.29}$$

Very often, we will be interested in equally-spaced knots, that is $x_i = x_o + ih$, $i = 1, 2, ..., n$. In that case Eqs. (12.27) and (12.29) assume simple forms as

$$M_{i-1} + 4M_i + M_{i+1} = \frac{6}{h^2}(y_{i-1} - 2y_i + y_{i+1}) \tag{12.30}$$

and

$$m_{i-1} + 4m_i + m_{i+1} = \frac{3}{h}(y_{i+1} - y_{i-1}) \quad \text{for } i = 1, 2, ..., (n-1) \tag{12.31}$$

Eqs. (12.30) and (12.31), each constitute $(n - 1)$ equations in $(n + 1)$ unknowns. Obviously, we require two more relations in order to have a unique interpolating spline, so that the systems (12.30) and (12.31) are complete. There are several ways of choosing these relations from the end point conditions, so that different types of splines are determined. The cubic splines thus constructed find several applications. For instance, they can be used to solve boundary value problems such as

$$y'' + p(x)y' + q(x)y = r(x), \; x_0 \le x \le x_n \tag{12.32}$$

subject to the boundary conditions

$$y(x_0) = a, \quad y(x_n) = b$$

Now, suppose we seek a cubic spline method of solution; for simplicity let us assume equally-spaced knots, then Eqs. (12.25) and (12.26) become

$$S'(x_i +) = -\frac{h}{3}M_i - \frac{h}{6}M_{i+1} + \frac{1}{h}(y_{i+1} - y_i), \quad i = 0,1,2,\ldots,(n-1) \tag{12.33}$$

and

$$S'(x_i -) = \frac{h}{3}M_i + \frac{h}{6}M_{i-1} + \frac{1}{h}(y_i - y_{i-1}), \quad i = 1,2,\ldots,n \tag{12.34}$$

From the given differential equation (12.32), we have on using Eq. (12.33) that

$$M_i = -p_i y_i' - q_i y_i + r_i$$

$$= -p_i\left[-\frac{h}{3}M_i - \frac{h}{6}M_{i+1} + \frac{1}{h}(y_{i+1} - y_i)\right] - q_i y_i + r_i$$

or

$$\left(1-\frac{h}{3}p_i\right)M_i - \frac{h}{6}p_i M_{i+1} = r_i - q_i y_i - \frac{p_i}{h}(y_{i+1}-y_i),$$

$$i = 0, 1, 2, \ldots, (n-1) \qquad (12.35)$$

Similarly, using Eq. (12.34), Eq. (12.32) gives

$$\frac{h}{6}p_i M_{i-1} + \left(1+\frac{h}{3}p_i\right)M_i = r_i - q_i y_i - \frac{p_i}{h}(y_i - y_{i-1}),$$

$$i = 1, 2, \ldots, (n) \qquad (12.36)$$

Thus, Eqs. (12.35) and (12.36) together with the boundary conditions give us $(2n + 2)$ equations whose solution readily yields $(2n + 2)$ unknowns $y_0, y_1, \ldots, y_n$ and $M_o, M_1, \ldots, M_n$. For illustration, let us consider the following example.

Example 12.6 Solve the boundary value problem governed by the differential equation

$$y'' + \frac{4x}{1+x^2}y' + \frac{2}{1+x^2}y = 0$$

subject to the boundary conditions $y(0) = 1$, $y(2) = 0.2$ using cubic spline method.

Solution Introducing the notation

$$y''(x_i) = S''(x_i) = M_i$$

the given differential equation can be written as

$$M_i = -\frac{4x_i}{1+x_i^2}y_i' - \frac{2}{1+x_i^2}y_i \qquad (1)$$

Suppose, we divide the interval [0, 2] into two equal subintervals and then using Eqs. (12.33) and (12.34) to replace y'_i, we find

$$M_i = -\frac{4x_i}{1+x_i^2}\left[-\frac{h}{3}M_i - \frac{h}{6}M_{i+1} + \frac{1}{h}(y_{i+1}-y_i)\right] - \frac{2}{1+x_i^2}y_i \quad \text{for } i = 0,1 \quad (2)$$

and

$$M_i = -\frac{4x_i}{1+x_i^2}\left[\frac{h}{3}M_i + \frac{h}{6}M_{i-1} + \frac{1}{h}(y_i - y_{i-1})\right] - \frac{2}{1+x_i^2}y_i \quad \text{for } i = 1,2 \quad (3)$$

Taking $h = 1$, $x_0 = 0$, $x_1 = 1$, $x_2 = 2$, Eqs. (2) and (3) reduce to

$$M_0 = -2y_0 \tag{4}$$

$$M_1 = -2\left[-\frac{1}{3}M_1 - \frac{1}{6}M_2 + (y_2 - y_1)\right] - y_1$$

or

$$\frac{1}{3}M_1 - \frac{1}{3}M_2 = y_1 - 2y_2 \tag{5}$$

$$M_1 = -2\left[\frac{1}{3}M_1 + \frac{1}{6}M_0 + (y_1 - y_0)\right] - y_1$$

or

$$\frac{5}{3}M_1 + \frac{1}{3}M_0 = -3y_1 + 2y_0 \tag{6}$$

$$M_2 = -\frac{8}{5}\left[+\frac{1}{3}M_2 + \frac{1}{6}M_1 + (y_2 - y_1)\right] - \frac{2}{5}y_2$$

or

$$\frac{23}{15}M_2 + \frac{4}{15}M_1 = -2y_2 + \frac{8}{5}y_1 \tag{7}$$

Now, applying the boundary conditions

$y_0 = 1$ and $y_2 = 0.2$, Eqs. (4) to (7) simplifies to

$$\left.\begin{aligned} &M_0 = -2 \\ &\frac{1}{3}M_1 - \frac{1}{3}M_2 - y_1 = -\frac{2}{5} \\ &\frac{5}{3}M_1 + 3y_1 = \frac{8}{3} \\ &\frac{4}{15}M_1 + \frac{23}{15}M_2 - \frac{8}{5}y_1 = -\frac{2}{5} \end{aligned}\right\} \tag{8}$$

whose solution is found to be

$$M_0 = -2, \qquad M_1 = 184/295, \qquad M_2 = 58/295$$

$$y_1 = 96/177 = 0.5423 \tag{9}$$

It may be noted that the exact solution of the given BVP is

$$y(x) = 1/(1 + x^2) \tag{10}$$

from which we find

$$y_1 = y(1) = 0.5 \tag{11}$$

To study the effect of reducing h, let us divide the interval [0, 2] into four equal parts, so that $h = 0.5$ on repeating the above steps, we obtain

$$M_0 = -2, \quad M_1 = -0.3834, \quad M_2 = 0.5576, \quad M_3 = 0.3738, \quad M_4 = 0.1873$$

and

$$\left.\begin{aligned} y(0.5) &= 0.8234 \\ y(1.0) &= 0.5228 \\ y(1.5) &= 0.3147 \end{aligned}\right\} \qquad (12)$$

observe that, as the value of h is reduced, the numerical solution approaches the exact solution and the error is reduced.

EXERCISES

12.1 The deflection of a beam on simple supports is governed by the equation

$$\frac{d^4 y}{dx^4} + 81y = 81x^2$$

with the boundary conditions

$$y(0) = y''(0) = 0, \quad y(1) = y''(1) = 0.$$

compute the deflections at $x = 1/3$ and $x = 2/3$.

12.2 Solve the boundary value problem

$$x\frac{d^2 y}{dx^2} + y = 0$$

subject to the boundary conditions

$$y(1) = 1, \quad y(2) = 2, \quad \text{taking } h = 0.25$$

12.3 Solve the boundary value problem described as

$$y'' = x + y$$

subject to the boundary conditions

$$y(0) = y(1) = 0$$

by dividing the interval into four subintervals

12.4 Solve the boundary value problem described by the equation

$$\frac{d^4 y}{dx^4} - \frac{d^3 y}{dx^3} + y = x^3$$

with the boundary conditions

$$y(0) = 2, \quad y'(0) = -1, \quad y(1) = -2, \quad y'(1) = 1$$

taking $h = 0.2$

12.5 Solve the boundary value problem

$$y'' = x + (1 - x/5)y,$$
$$y(1) = 2, \quad y(3) = -1$$

with $h = 0.2$, assuming the initial guesses as

$$y_1'(1) = -1.5, \quad y_2'(1) = -3.0$$

by writing a computer program

12.6 Solve the BVP described by the differential equation

$$\frac{d^2y}{dx^2} + y = 3x^2$$

subject to $y(0) = 0$, $y(2) = 3.5$ using Galerkin method.

12.7 Using Galerkin technique, solve the BVP governed by the differential equation

$$y'' = 3x + 1$$

with the boundary conditions: $y(0) = 0$, $y(1) = 0$ taking a quadratic in x as the approximating function. Compare it with the analytical solution

$$y = \frac{x^3}{2} + \frac{x^2}{2} - x$$

12.8 A laminar boundary layer on a flat plate is self similar and is governed by the differential equation

$$f''' + ff'' = 0$$

where $f = f(\eta)$ and η is a similarity variable f and its derivatives are proportional to certain fluid mechanical quantities: $f \propto \psi$, the stream function, $f' = u/U$, where u is a local fluid velocity and U is the free stream fluid velocity and $f'' \propto \tau$, the shear stress. Condition of "no slip" at the wall and free stream conditions at large distances from the wall are to be satisfied. That is,

$$f'(0) = f(0) = 0, \quad f'(\infty) = 1$$

write a computer program to solve the above BVP, using shooting method. Take $\infty = 10$, $h = \Delta\eta = 0.01$ (Refer, Boundary layer theory', Schlichting, 1960).

12.9 Using cubic spline method, solve the BVP governed by

$$\frac{d^2y}{dx^2} + y = 3x^2,$$
$$y(0) = 0, \quad y(2) = 3.5$$

Taking $h = 1$.

Chapter 13

Approximation of Functions

13.1 INTRODUCTION

One of the basic problems in numerical analysis is the approximation of a real continuous function $f(x)$ by an approximating function say by a polynomial or a ratio of polynomials with a minimum error. Once an approximating function is chosen, we must know the 'goodness' of the approximation. In order to measure the 'goodness' of an approximation, the concept of, norm, its choice and minimization are some of the questions need to be answered. This topic, infact, is quite essential for developing computer algorithms to compute standard functions, such as, sines, cosines, logarithms, exponential etc... which are repeatedly used in many scientific calculations to a well-defined accuracy. One way to approximate a function by a polynomial is to use a truncated Taylor series expansion and to evaluate the first few terms. For obvious reasons, Taylor series method is usually not appropriate to compute a standard function and most modern algorithms do not use it. In these modern times, we are approximating functions by polynomials such as Chebyshev polynomials.

Suppose, we have a function $f(x)$ which is known to be continuous in some given interval and is approximated by some polynomial $P_n(x)$ of degree, n, or less; then various measures of closeness of approximation of $f(x)$ by $P_n(x)$ is expressed through the concept of 'Norm'.

The L_p–norm of a function $f(x)$ denoted by $\| f \|_p$ or $L_p(f)$ is defined as

$$\| f \|_p = L_p(f) = \left\{ \int_0^1 f(x)^p \, dx \right\}^{1/p}, \quad p \geq 1 \tag{13.1}$$

By a best L_p–approximation $P_n(x)$ to $f(x)$, we mean that

$$\int_0^1 |f(x) - P_n(x)|^p \, dx \tag{13.2}$$

is minimized. However, for most of the practical applications, we use only L_1 and L_2 norms, which are known as mean error norm and mean square error norm respectively. An approximation in which the L_2–norm is minimized is called a least–squares approximation. In addition to L_1 and L_2 norms, there is another widely used and important norm called L_∞–norm, also called Chebyshev norm defined by

$$\| f(x)\|_\infty = \max_{a\le x\le b} |f(x)| \tag{13.3}$$

An approximation $P_n(x)$ to $f(x)$ in which the maximum error

$$\|f(x) - P_n(x)\|_\infty = \max_{a\le x\le b} |f(x) - P_n(x)| \tag{13.4}$$

is minimized, is called mini-max approximation to $f(x)$.

We shall state below two basic theorems without proof for the approximation of functions using L_∞–norm.

Theorem 13.1 (Existence of approximating polynomial): If $f(x)$ is a continuous function in $[a, b]$ and given any $\in > 0$, then there exists a polynomial $P_n(x)$ of degree n, or less such that for $x \in [a, b]$

$$|f(x) - P_n(x)| \le \in \tag{13.5}$$

This is known as Weierstrass approximation theorem.

Theorem 13.2 (Uniqueness of approximating polynomial): If $f(x)$ is a continuous function in $[a, b]$, then for a given n there exists a unique polynomial $F_n(x)$ of degree n or less such that

$$E_n = \max_{a\le x\le b} |f(x) - F_n(x)| \le \max_{a\le x\le b} |f(x) - P_n(x)| \tag{13.6}$$

for any polynomial $P_n(x)$ of degree n or less (See Young and Gregory, 1972).

13.2 LEAST–SQUARES APPROXIMATION

In a least-squares approximation, the given continuous function $f(x)$ is approximated in the interval $[a, b]$ by a polynomial $P_n(x)$ of degree n or less such that the L_2–norm

$$\int_0^1 |f(x) - P_n(x)|^2 \, dx \tag{13.7}$$

is minimized. For a discrete set of points $[x_0, x_1, x_2, ..., x_n]$ we minimize

$$\sum_{i=0}^{n} [f(x_i) - P_n(x_i)]^2 \tag{13.8}$$

Let

$$P_n(x) = \sum_{i=0}^{n} c_i x^i, \tag{13.9}$$

Then, we choose c_i such that $\| f(x) - P_n(x) \|^2$ is minimized. In other words, we find the coefficients c_i so that the function

$$I = \sum_{i=0}^{n} [f(x_i) - P_n(x_i)]^2 \tag{13.10}$$

is minimum. For simplicity of illustration, let us consider the case, when $n = 2$, then Eqs. (13.9) and (13.10) gives.

$$I = \sum_{i=0}^{2} \left[f(x_i) - (c_0 + c_1 x_i + c_2 x_i^2) \right]^2 \tag{13.11}$$

Differentiating partially with respect to c_i and equating to zero, we get for $i = 0, 1, 2$

$$\frac{\partial I}{\partial c_0} = 2\left[f(x_0) - (c_o + c_1 x_0 + c_2 x_0^2) \right] + 2\left[f(x_1) - (c_0 + c_1 x_1 + c_2 x_1^2) \right]$$

$$+ 2\left[f(x_2) - (c_0 + c_1 x_2 + c_2 x_2^2) \right] = 0$$

$$\frac{\partial I}{\partial c_1} = 2\left[f(x_0) - (c_o + c_1 x_0 + c_2 x_0^2) \right](-x_0) + 2\left[f(x_1) - (c_0 + c_1 x_1 + c_2 x_1^2) \right]$$

$$(-x_1) + 2\left[f(x_2) - (c_0 + c_1 x_2 + c_2 x_2^2) \right](-x_2) = 0$$

and

$$\frac{\partial I}{\partial c_2} = 2\left[f(x_0) - (c_o + c_1 x_0 + c_2 x_0^2) \right](-x_0^2) + 2\left[f(x_1) - (c_0 + c_1 x_1 + c_2 x_1^2) \right]$$

$$(-x_1^2) + 2\left[f(x_2) - (c_0 + c_1 x_2 + c_2 x_2^2) \right](-x_2^2) = 0$$

on simplification the above equations become

$$3c_0 + c_1(x_0 + x_1 + x_2) + c_2(x_0^2 + x_1^2 + x_2^2) = f(x_0) + f(x_1) + f(x_2),$$

$$(x_0 + x_1 + x_2)\, c_0 + (x_0^2 + x_1^2 + x_2^2) c_1$$

$$+ (x_0^3 + x_1^3 + x_2^3)\, c_2 = x_o f(x_0) + x_1 f(x_1) + x_2 f(x_2),$$

and

$$(x_0^2 + x_1^2 + x_2^2)\, c_0 + (x_0^3 + x_1^3 + x_2^3)\, c_1 + (x_0^4 + x_1^4 + x_2^4) c_2 =$$

$$x_0^2 f(x_0) + x_1^2 f(x_1) + x_2^2 f(x_2).$$

These equations are called normal equations. In compact matrix notation, they can be recast as

$$\begin{bmatrix} 3 & x_0+x_1+x_2 & x_0^2+x_1^2+x_2^2 \\ x_0+x_1+x_2 & x_0^2+x_1^2+x_2^2 & x_0^3+x_1^3+x_2^3 \\ x_0^2+x_1^2+x_2^2 & x_0^3+x_1^3+x_2^3 & x_0^4+x_1^4+x_2^4 \end{bmatrix}\begin{pmatrix} c_0 \\ c_1 \\ c_2 \end{pmatrix}$$

$$= \begin{pmatrix} f(x_0)+f(x_1)+f(x_2) \\ x_0 f(x_0)+x_1 f(x_1)+x_2 f(x_2) \\ x_0^2 f(x_0)+x_1^2 f(x_1)+x_2^2 f(x_2) \end{pmatrix} \tag{13.12}$$

Equivalently

$$\begin{bmatrix} 1 & 1 & 1 \\ x_0 & x_1 & x_2 \\ x_0^2 & x_1^2 & x_2^2 \end{bmatrix}\begin{bmatrix} 1 & x_0 & x_0^2 \\ 1 & x_1 & x_1^2 \\ 1 & x_2 & x_2^2 \end{bmatrix}\begin{pmatrix} c_0 \\ c_1 \\ c_2 \end{pmatrix} = \begin{bmatrix} 1 & 1 & 1 \\ x_0 & x_1 & x_2 \\ x_0^2 & x_1^2 & x_2^2 \end{bmatrix}\begin{pmatrix} f(x_0) \\ f(x_1) \\ f(x_2) \end{pmatrix} \tag{13.13}$$

In general, the above equation can also be written as

$$[A][A]^T (C) = [A](f)$$

where

$$[A] = \begin{bmatrix} 1 & 1 & \cdots & 1 \\ x_0 & x_1 & \cdots & x_n \\ x_0^2 & x_1^2 & \cdots & x_n^2 \\ \vdots & \vdots & & \vdots \\ x_0^n & x_1^n & \cdots & x_n^n \end{bmatrix}, \quad C = \begin{pmatrix} c_0 \\ c_1 \\ \vdots \\ c_n \end{pmatrix}, \quad f = \begin{pmatrix} f(x_0) \\ f(x_1) \\ \vdots \\ f(x_n) \end{pmatrix}$$

For illustration, we consider the following example.

Example 13.1 Find the least squares approximation to the function $f(x) = e^x$, defined in [0, 1] as

$$f(0.0)=1.0000, \quad f(0.5)=1.6487, \quad f(1.0)=2.7183.$$

Solution Let $c_o + c_1 x + c_2 x^2$ be $P_n(x)$, the polynomial approximation to $f(x) = e^x$, then we have from the normal equations (13.13).

$$\begin{bmatrix} 1 & x_0 & x_0^2 \\ 1 & x_1 & x_1^2 \\ 1 & x_2 & x_2^2 \end{bmatrix}\begin{pmatrix} c_0 \\ c_1 \\ c_2 \end{pmatrix} = \begin{pmatrix} 1.0000 \\ 1.6487 \\ 2.7183 \end{pmatrix}$$

or

$$\begin{pmatrix} c_0 \\ c_1 \\ c_2 \end{pmatrix} = \begin{bmatrix} 1 & x_0 & x_0^2 \\ 1 & x_1 & x_1^2 \\ 1 & x_2 & x_2^2 \end{bmatrix}^{-1} \begin{pmatrix} 1.0000 \\ 1.6487 \\ 2.7183 \end{pmatrix}$$

$$= \begin{bmatrix} 1 & 0 & 0 \\ 1 & 0.5 & 0.25 \\ 1 & 1 & 1 \end{bmatrix}^{-1} \begin{pmatrix} 1.0000 \\ 1.6487 \\ 2.7183 \end{pmatrix} = \begin{bmatrix} 1 & 0 & 0 \\ -3 & 4 & -1 \\ 2 & -4 & 2 \end{bmatrix} \begin{pmatrix} 1.0000 \\ 1.6487 \\ 2.7183 \end{pmatrix}$$

$$= \begin{pmatrix} 1.0000 \\ 0.8765 \\ 0.8418 \end{pmatrix}$$

Hence, the required polynomial approximation is

$$e^x = 1.0000 + 0.8765x + 0.8418x^2$$

13.3 CHEBYSHEV POLYNOMIAL APPROXIMATION

The Chebyshev polynomials are widely used for approximating functions in the range [–1, 1]. It is an established fact, that the best approximation to a function $f(x)$ in the sense of both L_∞–norm and L_2–norm is obtained by choosing a linear combination of Chebyshev polynomials.

Chebyshev polynomials are defined as

$$T_n(x) = \cos n\theta \tag{13.14}$$

where $\theta = \cos^{-1}x$ or $\cos\theta = x$, $-1 \le x \le 1$.

Recalling the trignometric identities such as

$$\cos 2\theta = 2\cos^2\theta - 1,$$

$$\cos 3\theta = 4\cos^3\theta - 3\cos\theta,$$

and

$$\cos(n+1)\theta + \cos(n-1)\theta = 2\cos\theta\cos n\theta,$$

following the above definition, we can write

$$T_2(x) = 2x^2 - 1,$$

$$T_3(x) = 4x^3 - 3x,$$

and

$$T_{n+1}(x) + T_{n-1}(x) = 2xT_n(x)$$

Thus, we have the following recurrence relation

$$\left.\begin{aligned} &T_0(x) = 1, \qquad T_1(x) = x \\ &T_{n+1}(x) = 2xT_n(x) - T_{n-1}(x) \end{aligned}\right\} \tag{13.15}$$

from which the first few Chebyshev polynomials can be listed as

$$
\begin{aligned}
T_0(x) &= 1 \\
T_1(x) &= x \\
T_2(x) &= 2x^2 - 1 \\
T_3(x) &= 4x^3 - 3x \\
T_4(x) &= 8x^4 - 8x^2 + 1 \\
T_5(x) &= 16x^5 - 20x^3 + 5x \qquad (13.16)\\
T_6(x) &= 32x^6 - 48x^4 + 18x^2 - 1 \\
T_7(x) &= 64x^7 - 112x^5 + 56x^3 - 7x \\
T_8(x) &= 128x^8 - 256x^6 + 160x^4 - 32x^2 + 1 \\
T_9(x) &= 256x^9 - 576x^7 + 432x^5 - 120x^3 + 9x \\
T_{10}(x) &= 512x^{10} - 1280x^8 + 1120x^6 - 400x^4 + 50x^2 - 1
\end{aligned}
$$

It may be observed from the above list that the coefficient of x^n in $T_n(x)$ is 2^{n-1}. The first four Chebyshev polynomials are graphically shown in Fig. 13.1.

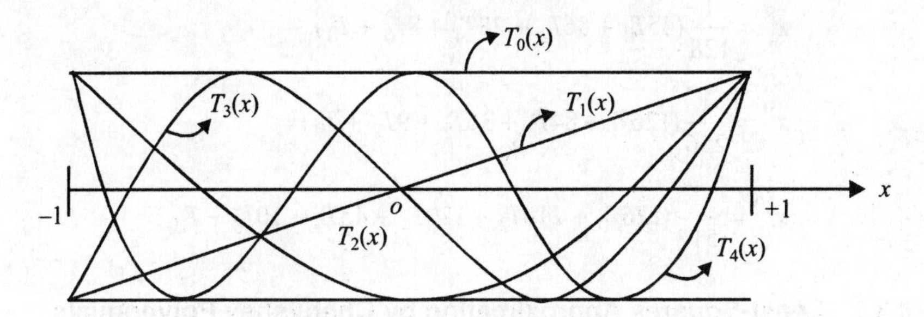

Fig. 13.1 First four Chebyshev polynomials.

Since $T_n(x) = \cos n\theta$ and $|\cos n\theta| = 1$ for $n\theta = 0, \pi, 2\pi, \ldots$ Further, since $x = \cos\theta$ and $-1 \le x \le 1$, we find $0 \le \theta \le \pi$. Also observe that $T_n(x) = 0$ for $n\theta = (2i + 1)\pi/2$ or $\theta = (2i + 1)\pi/2n$. That is, $T_n(x) = 0$ for

$$x_i = \cos\left[\frac{(2i+1)\pi}{2n}\right], i = 0,1,\ldots,(n-1) \qquad (13.17)$$

and therefore, $T_n(x)$ has only n roots (zeros). It may also be noted that $T_n(x)$ attains its maximum magnitude, unity, at $(n + 1)$ arguments including the end points, that is at

$$x = \cos(i\pi/n),\ i = 0, 1, \ldots, n \qquad (13.18)$$

In many applications, it may be necessary to express the powers of x as a linear combination of Chebyshev polynomials. They can be obtained by rearranging Eqs. (13.16) as

$$
\begin{aligned}
1 &= T_0 \\
x &= T_1 \\
x^2 &= \frac{1}{2}(T_0 + T_2) \\
x^3 &= \frac{1}{4}(3T_1 + T_3) \\
x^4 &= \frac{1}{8}(3T_0 + 4T_2 + T_4) \\
x^5 &= \frac{1}{16}(10T_1 + 5T_3 + T_5) \\
x^6 &= \frac{1}{32}(10T_0 + 15T_2 + 6T_4 + T_6) \\
x^7 &= \frac{1}{64}(35T_1 + 21T_3 + 7T_5 + T_7) \\
x^8 &= \frac{1}{128}(35T_0 + 56T_2 + 28T_4 + 8T_6 + T_8) \\
x^9 &= \frac{1}{256}(126T_1 + 84T_3 + 36T_5 + 9T_7 + T_9) \\
x^{10} &= \frac{1}{512}(126T_0 + 210T_2 + 120T_4 + 45T_6 + 10T_8 + T_{10})
\end{aligned}
\tag{13.19}
$$

13.3.1 Least-Squares Approximation by Chebyshev Polynomials

It can be easily verified that Chebyshev polynomials form an orthogonal set. That is,

$$
\int_{-1}^{1} \frac{T_n(x)T_m(x)}{\sqrt{1-x^2}}\,dx = \begin{cases} 0, & n \neq m \\ \pi, & n = m = 0 \\ \pi/2, & n = m \neq 0 \end{cases} \tag{13.20}
$$

Suppose, $f(x)$ is a continuous function defined in the interval $[-1, 1]$ and if we change the variable x to θ, using $x = \cos\theta$, we get

$$
f(x) = f(\cos\theta) = g(\theta), \qquad 0 \le \theta \le \pi
$$

observe that $g(\theta)$ is a periodic function, which on expressing in terms of Fourier cosine series, result in

$$
g(\theta) = \frac{a_0}{2} + \sum_{n=1}^{\infty} a_n \cos n\theta
$$

where

$$a_n = \frac{2}{\pi}\int_0^{\pi} g(\theta)\cos n\theta\, d\theta$$

Thus, $f(x)$ can be written as

$$f(x) = \frac{a_0}{2} + \sum_{n=1}^{\infty} a_n T_n(x) \tag{13.21}$$

where

$$a_n = \frac{2}{\pi}\int_{-1}^{1} \frac{f(x)T_n(x)}{\sqrt{1-x^2}}\, dx \tag{13.22}$$

This is called the Chebyshev series of the function $f(x)$. Hence, the Chebyshev polynomial approximation $P_n(x)$ to $f(x)$ can be obtained by truncating at some point. Thus,

$$P_n(x) = \frac{c_0}{2} + \sum_{k=1}^{n} c_k T_k(x) \tag{13.23}$$

where,

$$c_k = \frac{2}{\pi}\int_{-1}^{1} \frac{f(x)T_k(x)}{\sqrt{1-x^2}}\, dx, \; k = 0, 1, \ldots, n \tag{13.24}$$

For illustration, we consider the following example.

Example 13.2 Show that the chebyshev series for $\cos^{-1}x$ is

$$\cos^{-1} x = \frac{\pi}{2} - \frac{4}{\pi}T_1(x) - \frac{4}{9\pi}T_3(x) - \frac{4}{25\pi}T_5(x) - \frac{4}{49\pi}T_7(x)$$

Solution Let $\cos^{-1}x = P_7(x) = \dfrac{c_o}{2} + c_1T_1(x) + c_2T_2(x) + \ldots + c_7T_7(x)$

Then, the coefficients of $P_7(x)$ can be obtained using Eq. (13.24) as

$$c_0 = \frac{2}{\pi} \times \frac{1}{2}\int_{-1}^{1} \frac{\cos^{-1}(x)(1)}{\sqrt{1-x^2}}\, dx = -\frac{1}{\pi}\Big[(\cos^{-1} x)^2/2\Big]_{-1}^{1} = \frac{\pi}{2}$$

Similarly

$$\begin{aligned} c_k &= \frac{2}{\pi}\int_{-1}^{1} \frac{\cos^{-1}(x)T_k(x)}{\sqrt{1-x^2}}\, dx \\ &= \frac{2}{\pi}\int_{-1}^{1} \frac{\cos^{-1}(x)\cos k\theta}{\sqrt{1-x^2}}\, dx \\ &= \frac{2}{\pi}\int_{-1}^{1} \frac{\cos^{-1}(x)\cos(k\cos^{-1} x)}{\sqrt{1-x^2}}\, dx \end{aligned}$$

Substituting $\cos^{-1}x = t$, we have

$$c_k = \frac{2}{\pi}\int_0^{\pi} t\cos kt dt = \left[t\frac{\sin kt}{k} + \frac{\cos kt}{k^2}\right]_0^{\pi}$$

or

$$c_k = \frac{2}{\pi k^2}\left[(-1)^k - 1\right]$$

which gives

$$c_1 = -\frac{4}{\pi}, \quad c_2 = 0, \quad c_3 = -\frac{4}{9\pi}, \quad c_4 = 0$$

$$c_5 = -\frac{4}{25\pi}, \quad c_6 = 0 \quad c_7 = -\frac{4}{49\pi}.$$

Hence, the required Chebyshev polynomial is

$$P_7(x) = \frac{\pi}{2} - \frac{4}{\pi}T_1(x) - \frac{4}{9\pi}T_3(x) - \frac{4}{25\pi}T_5(x) - \frac{4}{49\pi}T_7(x)$$

13.4 ECONOMIZED POWER SERIES

At first, we shall discuss the important concept that Chebyshev polynomials can be used to derive a polynomial approximation to a function that is significantly efficient than the Maclaurin's series expansion in the sense of accuracy. For illustration, consider the Maclaurin series expansion for e^{-x} as

$$e^{-x} = 1 - x + \frac{x^2}{2} - \frac{x^3}{6} + \frac{x^4}{24} - \frac{x^5}{120} + \frac{x^6}{720} - \ldots \tag{13.25}$$

Similarly, we can obtain Chebyshev series expansion of e^{-x} by evaluating the coefficient integrals given in Eq. (13.24), which of course is bit involved, and therefore, we resort to alternate procedure of replacing the powers of x in Eq. (13.25) by their equivalent chebyshev polynomials, using the set of Eqs. (13.19), thus getting

$$e^{-x} = T_0 - T_1 + \frac{1}{4}(T_0 + T_2) - \frac{1}{24}(3T_1 + T_3) + \frac{1}{192}(3T_0 + 4T_2 + T_4)$$

$$-\frac{1}{1920}(10T_1 + 5T_3 + T_5) + \frac{1}{32\times 720}(10T_0 + 15T_2 + 6T_4 + T_6) - \ldots$$

or

$$e^{-x} = 1.26606T_0 - 1.13021T_1 + 0.27148T_2 - 0.04427T_3$$

$$+ 0.00546875T_4 - 0.0005208T_5 + 0.0000434T_6 - \ldots$$

Since the maximum magnitude of $T_n(x)$ is unity, the last three terms can be truncated to get the first four terms of the chebyshev series. For illustration of the fact, that Chebyshev polynomial expansion is more accurate, than that of Maclaurin series expansion to approximate e^{-x}, we revert back to expansion in powers of x, using Eqs. (13.19), thus getting the first four terms of the above equation as

$$e^{-x} = 1.26606 - 1.13021x + 0.27148(2x^2 - 1) - 0.04427(4x^3 - 3x)$$

or

$$e^{-x} = 0.99458 - 0.9974x + 0.54296x^2 - 0.17708x^3.$$

Thus, both the chebyshev series and Maclaurin series for the function e^{-x}, retaining terms upto x^3 only, can be written as

$$\left.\begin{aligned} e^{-x} &= 0.99458 - 0.9974x + 0.54296x^2 - 0.17708x^3 \\ \text{and} \quad e^{-x} &= 1 - x + 0.5x^2 - 0.16667x^3 \end{aligned}\right\} \quad (13.26)$$

Now, for the purpose of comparison, we tabulate below both chebyshev and Maclaurin expansions as given in Eq. (13.26) for $-1 \le x \le 1$ with error as shown in Table 13.1.

Table 13.1 Comparison of Chebyshev Series for e^{-x} with Its Maclaurin Series as in Eq. (13.26)

x	e^{-x}	Chebyshev	Error	Maclaurin	Error
–1.0	2.71828	2.71236	0.00592	2.66667	0.05161
–0.8	2.22554	2.23066	–0.00512	2.20534	0.0202
–0.6	1.82212	1.82674	–0.00462	1.81600	0.00612
–0.4	1.49183	1.49175	0.00008	1.49067	0.00116
–0.2	1.22140	1.21720	0.0042	1.22133	0.00007
0.0	1.00000	0.99458	0.00542	1.00000	0.00000
0.2	0.81873	0.81540	0.00333	0.81867	0.00006
0.4	0.67032	0.67116	–0.00084	0.66933	0.00099
0.6	0.54881	0.55182	–0.00301	0.54400	0.00481
0.8	0.44933	0.45349	–0.00416	0.43467	0.01466
1.0	0.36788	0.36300	0.00488	0.33333	0.03455

Observe Table 13.1 and Fig. 13.2, where we have compared the error of the Chebyshev expansion for e^{-x} with that of Maclaurin series by retaining terms upto x^3. We notice that, in Chebyshev expansion the errors are small and uniformly distributed throughout the interval including the origin. While in Maclaurin expansion, the error is almost negligible near and at the origin and the error builds up at the ends of the interval.

Now, let us see how Chebyshev polynomials help us to construct an economized power series when compared with the Maclaurin series for a

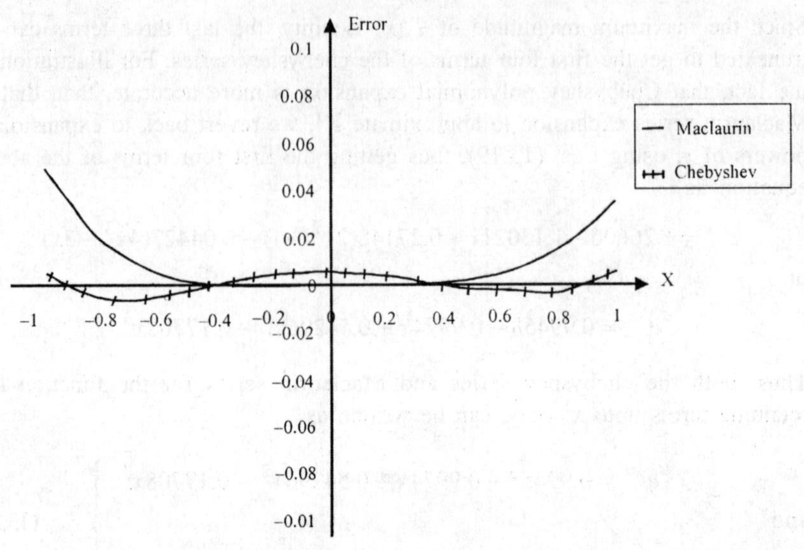

Fig. 13.2 Error estimates for Chebyshev and Maclaurin series for e^{-x}.

function. For illustration, consider the expansion of the function e^{-x}, that is,

$$e^{-x} = 1 - x + \frac{x^2}{2} - \frac{x^3}{6} + \frac{x^4}{24} - \frac{x^5}{120} + \frac{x^6}{720} - \cdots \tag{13.25}$$

Suppose, we wish to truncate the series to approximate e^{-x} in the interval [0, 1] with a precision of 0.001, then we have to retain terms upto x^6 [as $1/720 = 0.00139$] in Maclaurin series. Suppose, we subtract $T_6/(720 \times 32)$ from the truncated series of (13.25), we observe with the help of Eq. (13.16) that x^6 term gets cancelled. Also since the maximum value of T_6 is unity in the interval [0, 1], this will alter the sum of the truncated series by an amount

$$\frac{1}{(720 \times 32)} \approx 0.0000434$$

which is very small when compared with our required precision of 0.001. The resulting series would be

$$e^{-x} = 1 - x + \frac{x^2}{2} - \frac{x^3}{6} + \frac{x^4}{24} - \frac{x^5}{120} + \frac{x^6}{720} - \frac{1}{(720 \times 32)}(32x^6 - 48x^4 + 18x^2 - 1)$$

on simplification, we arrive at

$$\begin{aligned} e^{-x} &= 1.000043 - x + 0.449219x^2 - 0.166667x^3 \\ &\quad + 0.043750x^4 - 0.008333x^5 \end{aligned} \tag{13.27}$$

This is the economized power series in the sense that it is a fifth degree polynomial that approximates e^{-x} to the prescribed precision. Now, let us compare the errors in both the series

$$\left.\begin{aligned} e^{-x} &= 1 - x + 0.5x^2 - 0.166667x^3 + 0.041667x^4 \\ &\quad - 0.008333x^5 + 0.001389x^6 \\ \text{and} \\ e^{-x} &= 1.000043 - x + 0.499219x^2 - 0.166667x^3 \\ &\quad + 0.043750x^4 - 0.008333x^5 \end{aligned}\right\} \tag{13.28}$$

as depicted in Table 13.2.

Table 13.2 Comparison of Maclaurin and Economized Power Series

x	e^{-x}	Maclaurin sixth degree	Error	Economized fifth degree	Error
0.0	1.00000	1.00000	0.00000	1.00004	–0.00004
0.2	0.81873	0.81873	0.00000	0.81875	–0.00002
0.4	0.67032	0.67032	0.00000	0.67029	0.00003
0.6	0.54881	0.54882	–0.00001	0.54878	0.00003
0.8	0.44933	0.44937	–0.00004	0.44940	–0.00007
1.0	0.36788	0.36805	–0.00017	0.36801	–0.00013

From this table, we observe that the maximum magnitude of the error in Maclaurin sixth degree series is 0.00017, while in economized fifth degree series it is found to be 0.00013. Hence, we have succeeded in economizing the power series in that, we are getting more or less same precision with less number of terms.

13.5 PADE APPROXIMATION

One of the most popular domains in the theory of, approximation of functions, is the Pade approximation. In Section 13.4, we have seen that the approximation of a function using Chebyshev polynomials is more efficient in the interval [–1, 1] than by its Maclaurin series representation. However, Pade approximation is found to be much more efficient in terms of computer time and storage requirements for the constants involved. Usually in the Pade approximation of a function, we represent a function as the quotient of two polynomials such that

$$f(x) \approx R_N(x) = \frac{P_m(x)}{Q_n(x)} = \frac{a_0 + a_1x + a_2x^2 + \cdots + a_mx^m}{1 + b_1x + b_2x^2 + \cdots + b_nx^n} \tag{13.29}$$

where $N = m + n$. The Pade approximation and Maclaurin expansion of a function are so related in that the coefficients are determined such that both $f(x)$ and $R_N(x)$ and their first N derivatives agree at $x = 0$.

That is,

$$f^{(k)}(0) = R_N^{(k)}(0), \qquad k = 0, 1, 2, \ldots, N \tag{13.30}$$

Suppose that the Maclaurin series for $f(x)$ be written as

$$f(x) = c_0 + c_1x + c_2x^2 + \cdots + c_Nx^N. \tag{13.31}$$

Then, we can write

$$f(x) - R_N(x) = c_0 + c_1x + c_2x^2 + \cdots + c_Nx^N - \frac{a_0 + a_1x + a_2x^2 + \cdots + a_mx^m}{1 + b_1x + b_2x^2 + \cdots + b_nx^n}$$

that is,

$$f(x) - R_N(x) = \frac{(c_0 + c_1x + c_2x^2 + \cdots + c_Nx^N)(1 + b_1x + b_2x^2 + \cdots + b_nx^n) - (a_0 + a_1x + a_2x^2 + \cdots + a_mx^m)}{(1 + b_1x + b_2x^2 + \cdots + b_nx^n)} \tag{13.32}$$

Since c_r are constant coefficients in Maclaurin series expansion, they can be determined from the relation

$$c_r = \frac{f^{(r)}(0)}{r!} \tag{13.33}$$

Now, using the idea expressed in Eq. (13.30) for $k = 0$ we get

$$c_0 - a_0 = 0 \tag{13.34}$$

Similarly, for the first N derivatives of $f(x)$ and $R_N(x)$ to be equal at $x = 0$, we require that the coefficients of the powers of x upto and including x^N must vanish in the numerator of Eq. (13.32) which gives N more equations such as

$$\left.\begin{aligned}
&b_1c_0 + c_1 - a_1 = 0\\
&b_2c_0 + b_1c_1 + c_2 - a_2 = 0\\
&b_3c_0 + b_2c_1 + b_1c_2 + c_3 - a_3 = 0\\
&\vdots\\
&b_nc_{m-n} + b_{n-1}c_{m-n+1} + \cdots + c_m - a_m = 0\\
&b_nc_{m-n+1} + b_{n-1}c_{m-n+2} + \cdots + c_{m+1} = 0\\
&b_nc_{m-n+2} + b_{n-1}c_{m-n+3} + \cdots + c_{m+2} = 0\\
&\vdots\\
&b_nc_{N-n} + b_{n-1}c_{N-n+1} + \cdots + c_N = 0
\end{aligned}\right\} \tag{13.35}$$

It may be observed that in each Eq. (13.35), the sum of the subscripts of each term is the same which is of course equal to the power of the x-term in the numerator. Thus, Eqs. (13.34) and (13.35) together, when solved give us the required coefficients in Pade approximation. For illustration of the concept of Pade approximation, we consider the following couple of examples.

Example 13.3 Find the Pade approximation for the function e^x with numerator and denominator each of polynomial of degree three. Find the maximum error in the interval [0, 1].

Solution Suppose $f(x) = e^x$ and let $R_N(x)$ is a rational function which is an approximation to $f(x)$ in which the number of constants are $N+1 = (m + n + 1)$. In the present problem, we are given $m = n = 3$, and therefore, $N + 1 = 7$ or $N = 6$. Thus, we form Maclaurin series for e^x as

$$e^x = 1 + \frac{x}{1} + \frac{x^2}{2} + \frac{x^3}{6} + \frac{x^4}{24} + \frac{x^5}{120} + \frac{x^6}{720} \tag{1}$$

Then,

$$f(x) - R_6(x) = \left[1 + \frac{x}{1} + \frac{x^2}{2} + \frac{x^3}{6} + \frac{x^4}{24} + \frac{x^5}{120} + \frac{x^6}{720}\right] - \frac{a_0 + a_1x + a_2x^2 + a_3x^3}{1 + b_1x + b_2x^2 + b_3x^3}$$

$$= \frac{\left\{\left[1 + \frac{x}{1} + \frac{x^2}{2} + \frac{x^3}{6} + \frac{x^4}{24} + \frac{x^5}{120} + \frac{x^6}{720}\right](1 + b_1x + b_2x^2 + b_3x^3) - (a_0 + a_1x + a_2x^2 + a_3x^3)\right\}}{(1 + b_1x + b_2x^2 + b_3x^3)} \tag{2}$$

Equating the coefficients of x^0 to x^6 in the numerator of Eq. (2) to zero, we have

$$\left.\begin{aligned}
a_0 &= 1\\
a_1 &= 1 + b_1\\
a_2 &= b_2 + b_1 + (1/2)\\
a_3 &= b_3 + b_2 + \frac{b_1}{2} + \frac{1}{6}\\
b_3 + \frac{b_2}{2} &+ \frac{b_1}{6} + \frac{1}{24} = 0\\
\frac{b_3}{2} + \frac{b_2}{6} &+ \frac{b_1}{24} + \frac{1}{120} = 0\\
\frac{b_3}{6} + \frac{b_2}{24} &+ \frac{b_1}{120} + \frac{1}{720} = 0
\end{aligned}\right\} \tag{3}$$

Solving the last three equations of (3) we get

$$b_1 = -\frac{1}{2}, \quad b_2 = \frac{1}{10}, \quad b_3 = -\frac{1}{120}$$

using these values, the first four equations of (3) yield

$$a_1 = \frac{1}{2}, \quad a_2 = \frac{1}{10}, \quad a_3 = \frac{1}{120}, \quad a_0 = 1$$

Hence, the required Pade approximation to e^x is

$$e^x = \frac{\left(1+\frac{x}{2}+\frac{x^2}{10}+\frac{x^3}{120}\right)}{\left(1-\frac{x}{2}+\frac{x^2}{10}-\frac{x^3}{120}\right)} \tag{4}$$

The following table gives the comparison of errors in both Pade and Maclaurin approximations to e^x

x	True value of e^{-x}	Pade approximation	Error	Maclaurin approximation	Error
0.2	1.221403	1.22140	3×10^{-6}	1.22140	3×10^{-6}
0.4	1.491825	1.49182	-5×10^{-6}	1.49183	-5×10^{-6}
0.6	1.822119	1.82212	-1×10^{-6}	1.82212	-1×10^{-6}
0.8	2.225541	2.22555	-9×10^{-6}	2.22554	1×10^{-6}
1.0	2.718282	2.71831	-2.8×10^{-5}	2.71825	3.2×10^{-5}

From the above table, it is clear that enough terms are available even in Maclaurin series to get five decimal accuracy when x takes values from 0.2 through 0.8. However, when $x = 1.0$, which is the limit for convergence of the Maclaurin servies, the magnitude of the error is more when compared with Pade approximation. In this example, the maximum error due to Pade approximation is 2.8×10^{-5} and Pade approximation is more accurate.

Example 13.4 Find the Pade approximation for the function log $(1 + x)$ with numerator and denomenator each of polynomial of degree three.

Solution Let $f(x) = \log(1 + x)$ and $R_6(x)$ is its Pade approximation. Then

$$f(x) - R_6(x) = \log(1+x) - \frac{a_0 + a_1x + a_2x^2 + a_3x^3}{1 + b_1x + b_2x^2 + b_3x^3}$$

$$\left(x - \frac{x^2}{2} + \frac{x^3}{3} - \frac{x^4}{4} + \frac{x^5}{5} - \frac{x^6}{6}\right)(1 + b_1x + b_2x^2 + b_3x^3)$$

$$= \frac{-(a_0 + a_1x + a_2x^2 + a_3x^3)}{(1 + b_1x + b_2x^2 + b_3x^3)} \tag{1}$$

Equating the coefficients of x^0 through x^6 in the numerator of Eq. (1) to zero, we get

$$\left.\begin{aligned} a_0 &= 0 \\ a_1 &= 1 \\ a_2 &= -\frac{1}{2} + b_1 \end{aligned}\right\}$$

$$\left.\begin{aligned}
a_3 &= \frac{1}{3} - \frac{b_1}{2} + b_2 \\
\frac{b_1}{3} &- \frac{b_2}{2} + b_3 - \frac{1}{4} = 0 \\
-\frac{b_1}{4} &+ \frac{b_2}{3} - \frac{b_3}{2} + \frac{1}{5} = 0 \\
\frac{b_1}{5} &- \frac{b_2}{4} + \frac{b_3}{3} - \frac{1}{6} = 0
\end{aligned}\right\} \tag{2}$$

The last three equations of (2) can be written in matrix form as

$$\begin{bmatrix} \frac{1}{3} & -\frac{1}{2} & 1 \\ -\frac{1}{4} & \frac{1}{3} & -\frac{1}{2} \\ \frac{1}{5} & -\frac{1}{4} & \frac{1}{3} \end{bmatrix} \begin{pmatrix} b_1 \\ b_2 \\ b_3 \end{pmatrix} = \begin{pmatrix} \frac{1}{4} \\ -\frac{1}{5} \\ \frac{1}{6} \end{pmatrix} \tag{3}$$

whose solution is found to be

$$b_1 = \frac{3}{2}, \quad b_2 = \frac{3}{5}, \quad b_3 = \frac{1}{20} \tag{4}$$

Substituting these values in the first four equations of (2) we obtain

$$a_0 = 0, \qquad a_1 = 1, \qquad a_2 = 1, \qquad a_3 = \frac{11}{60}$$

Hence, the required Pade approximation to log $(1 + x)$ is

$$\log(1+x) = \frac{\left(x + x^2 + \frac{11}{60}x^3\right)}{\left(1 + \frac{3}{2}x + \frac{3}{5}x^2 + \frac{1}{20}x^3\right)}$$

13.6 FOURIER SERIES APPROXIMATION

One way of approximating functions is through polynomials. Another way of representing functions is through Fourier series such as

$$f(x) = \frac{a_0}{2} + \sum_{n=1}^{\infty} (a_n \cos nx + b_n \sin nx) \tag{13.36}$$

where $f(x)$ is defined in the interval $(c, c + 2\pi)$.

This series is in the form of a sum of sine and cosine terms with proper coefficients a_0, a_n, b_n ($n = 1, 2, 3, ...$) which are called Fourier coefficients of $f(x)$. These fourier coefficients can be determined by recalling the orthogonality property of sines and cosiness, for integer values of m and n. That is

$$\int_c^{c+2\pi} \sin nx \, dx = 0 \tag{13.37}$$

$$\int_c^{c+2\pi} \cos nx \, dx = \begin{cases} 0, & n \neq 0 \\ 2\pi, & n = 0 \end{cases} \tag{13.38}$$

$$\int_c^{c+2\pi} \sin mx \, \cos nx \, dx = \frac{1}{2}\int_c^{c+2\pi} [\sin(m+n)x + \sin(m-n)x] \, dx$$

$$= -\frac{1}{2}\left[\frac{\cos(m+n)x}{(m+n)} + \frac{\cos(m-n)x}{(m-n)}\right]_c^{c+2\pi}$$

$$= 0, \quad m \neq n \tag{13.39}$$

$$\int_c^{c+2\pi} \cos mx \, \cos nx \, dx = \frac{1}{2}\int_c^{c+2\pi} [\cos(m+n)x + \cos(m-n)x] \, dx$$

$$= \frac{1}{2}\left[\frac{\sin(m+n)x}{(m+n)} + \frac{\sin(m-n)x}{(m-n)}\right]_c^{c+2\pi}$$

$$= 0, \quad m \neq n \tag{13.40}$$

$$\int_c^{c+2\pi} \sin mx \sin nx \, dx = \frac{1}{2}\int_c^{c+2\pi} [\cos(m-n)x - \cos(m+n)x] \, dx$$

$$= \frac{1}{2}\left[\frac{\sin(m-n)x}{(m-n)} - \frac{\sin(m+n)x}{(m+n)}\right]_c^{c+2\pi}$$

$$= 0, \quad m \neq n \tag{13.41}$$

For $m = n$ and $n \neq 0$

$$\int_c^{c+2\pi} \sin^2 nx \, dx = \left[\frac{x}{2} - \frac{\sin 2nx}{4n}\right]_c^{c+2\pi} = \pi \tag{13.42}$$

$$\int_c^{c+2\pi} \cos^2 nx \, dx = \left[\frac{x}{2} + \frac{\sin 2nx}{4n}\right]_c^{c+2\pi} = \pi \tag{13.43}$$

$$\int_c^{c+2\pi} \sin nx \cos nx \, dx = \frac{1}{2}\int_c^{c+2\pi} \sin 2nx \, dx = 0 \tag{13.44}$$

Suppose, the function $f(x)$ as defined in Eq. (13.36) is periodic and of period 2π and that the series on its right hand side is uniformly convergent and can be integrated term by term in the interval $(c, c + 2\pi)$, then we can determine the values of a_0, a_n, and b_n by integrating Eq. (13.36). That is,

$$\int_c^{c+2\pi} f(x)dx = \frac{a_0}{2}\int_c^{c+2\pi} dx + \int_c^{c+2\pi}\left(\sum_{n=1}^{\infty} a_n \cos nx\right)dx$$

$$+\int_c^{c+2\pi}\left(\sum_{n=1}^{\infty} b_n \sin nx\right)dx$$

$$= \frac{a_0}{2}(c + 2\pi - c) + 0 + 0 = a_0\pi$$

Therefore,

$$a_0 = \frac{1}{\pi}\int_c^{c+2\pi} f(x)\,dx \tag{13.45}$$

To determine a_n, we multiply both sides of Eq. (13.36) by cos nx and integrate with respect to x between the limits c to $(c + 2\pi)$, thus having

$$\int_c^{c+2\pi} f(x)\cos nx\,dx = \frac{a_0}{2}\int_c^{c+2\pi}\cos nx\,dx$$

$$+\int_c^{c+2\pi}\left(\sum_{n=1}^{\infty} a_n \cos nx\right)\cos nx\,dx$$

$$+\int_c^{c+2\pi}\left(\sum_{n=1}^{\infty} b_n \sin nx\right)\cos nx\,dx = 0 + a_n\pi + 0$$

which yields

$$a_n = \frac{1}{\pi}\int_c^{c+2\pi} f(x)\cos nx\,dx \tag{13.46}$$

Similarly, to find b_n, we multiply both sides of Eq. (13.36) by sin nx and integrate with respect to x between the limits c to $c + 2\pi$, thus having.

$$\int_c^{c+2\pi} f(x)\sin nx\,dx = \frac{a_0}{2}\int_c^{c+2\pi}\sin nx\,dx$$

$$+\int_c^{c+2\pi}\left(\sum_{n=1}^{\infty} a_n \cos nx\right)\sin nx\,dx$$

$$+\int_c^{c+2\pi}\left(\sum_{n=1}^{\infty} b_n \sin nx\right)\sin nx\,dx = 0 + 0 + b_n\pi$$

which gives

$$b_n = \frac{1}{\pi}\int_c^{c+2\pi} f(x)\sin nx\, dx \tag{13.47}$$

13.6.1 For Periods other than 2π

Suppose, the period of $f(x)$ is not 2π but is of $2l$ that is, $f(x)$ is defined in the interval $c \le x \le (c + 2l)$, then, we can change this interval into an interval of length 2π using the transformation

$$\frac{x}{l} = \frac{z}{\pi} \quad \text{or} \quad z = \frac{\pi x}{l} \tag{13.48}$$

Thus, the function $f(x)$ of period $2l$ is transformed into $F(z)$ of period 2π in the interval $\left(\frac{\pi c}{l}, \frac{\pi c}{l} + 2\pi\right)$ which can be expressed into Fourier series as

$$F(z) = \frac{a_0}{2} + \sum_{n=1}^{\infty} a_n \cos nz + \sum_{n=1}^{\infty} b_n \sin nz \tag{13.49}$$

or

$$F(z) = F\left(\frac{\pi x}{l}\right) = f(x) = \frac{a_0}{2} + \sum_{n=1}^{\infty} a_n \cos\left(\frac{n\pi x}{l}\right) + \sum_{n=1}^{\infty} b_n \sin\left(\frac{n\pi x}{l}\right) \tag{13.50}$$

where

$$\left.\begin{aligned} a_0 &= \frac{1}{l}\int_c^{c+2l} f(x)dx, \\ a_n &= \frac{1}{l}\int_c^{c+2l} f(x)\cos\left(\frac{n\pi x}{l}\right)dx, \\ b_n &= \frac{1}{l}\int_c^{c+2l} f(x)\sin\left(\frac{n\pi x}{l}\right)dx. \end{aligned}\right\} \tag{13.51}$$

As a special case, if $c = -l$, then the interval change to $(-l, l)$ or $-l \le x \le l$ and the above Fourier coefficients become

$$\left.\begin{aligned} a_0 &= \frac{1}{l}\int_{-l}^{l} f(x)\, dx, \\ a_n &= \frac{1}{l}\int_{-l}^{l} f(x)\cos\left(\frac{n\pi x}{l}\right)dx, \\ b_n &= \frac{1}{l}\int_{-l}^{l} f(x)\sin\left(\frac{n\pi x}{l}\right)dx. \end{aligned}\right\} \tag{13.52}$$

If $f(x)$ is an even function, then the Fourier cofficients reduces to

$$a_0 = \frac{2}{l}\int_0^l f(x)\,dx,\; a_n = \frac{2}{l}\int_0^l f(x)\cos\left(\frac{n\pi x}{l}\right)dx$$

$$b_n = 0 \tag{13.53}$$

Similarly, if $f(x)$ is an odd function, then the Fourier coefficients become

$$a_0 = 0,\; a_n = 0,\; b_n = \frac{2}{l}\int_0^l f(x)\sin\left(\frac{n\pi x}{l}\right)dx \tag{13.54}$$

Example 13.5 Find the Fourier coefficients for $f(x) = |x|$ in the interval $(-\pi, \pi)$ and hence find its Fourier series.

Solution Let

$$f(x) = \frac{a_0}{2} + \sum_{n=1}^{\infty} a_n \cos nx + \sum_{n=1}^{\infty} b_n \sin nx \tag{1}$$

Then, observe that $f(-x) = |-x| = |x| = f(x)$. Therefore, the given function is an even function and hence

$$b_n = 0 \tag{2}$$

and

$$a_0 = \frac{2}{\pi}\int_0^{\pi} f(x)\,dx = \frac{2}{\pi}\int_0^{\pi}|x|dx = \frac{2}{\pi}\int_0^{\pi} x\,dx$$

or

$$\frac{a_0}{2} = \frac{\pi}{2} \tag{3}$$

$$a_n = \frac{2}{\pi}\int_0^{\pi} f(x)\cos\frac{n\pi x}{\pi}dx = \frac{2}{\pi}\int_0^{\pi} x\cos nx\,dx$$

$$= \frac{2}{\pi}\left[x\left(\frac{\sin nx}{n}\right) - (1)\left(\frac{\cos nx}{n^2}\right)\right]_0^{\pi}$$

$$= \frac{2}{\pi}\left[\frac{\cos n\pi}{n^2} - \frac{1}{n^2}\right] = \frac{2}{n^2\pi}\left[(-1)^n - 1\right]$$

$$= \begin{cases} 0, & \text{for } n \text{ even} \\ 4/n^2\pi, & \text{for } n \text{ odd} \end{cases} \tag{4}$$

Substituting (2), (3) and (4) in (1) we obtain the required Fourier series as

$$|x| = \frac{\pi}{2} - \frac{4}{\pi}\left[\cos x + \frac{\cos 3x}{3^2} + \frac{\cos 5x}{5^2} + \cdots\right]$$

Example 13.6 Find the Fourier coefficients and hence the Fourier expansion of the function

$$f(x) = x^2 - 2$$

in the interval (–2, 2).

Solution Let

$$f(x) = \frac{a_0}{2} + \sum_{n=1}^{\infty} a_n \cos\frac{n\pi x}{2} + \sum_{n=1}^{\infty} b_n \sin\frac{n\pi x}{2} \tag{1}$$

We observe that $f(x) = x^2 - 2$ is an even function and therefore,

$$b_n = 0 \tag{2}$$

Now,

$$a_0 = \frac{1}{2}\int_{-2}^{2} f(x)dx = \int_0^2 (x^2 - 2)\,dx = -\frac{4}{3}$$

Therefore,

$$\frac{a_0}{2} = -\frac{2}{3} \tag{3}$$

and

$$a_n = \frac{2}{2}\int_0^2 (x^2 - 2)\cos\left(\frac{n\pi x}{2}\right)dx$$

$$= \left[(x^2 - 2)\frac{\sin\left(\frac{n\pi x}{2}\right)}{\left(\frac{n\pi}{2}\right)} - 2x\left\{-\frac{\cos\left(\frac{n\pi x}{2}\right)}{\left(\frac{n^2\pi^2}{4}\right)}\right\} + 2\left\{-\frac{\sin\left(\frac{n\pi x}{2}\right)}{\left(\frac{n^3\pi^3}{8}\right)}\right\}\right]_0^2$$

$$= \frac{4\times 4}{n^2\pi^2}\cos n\pi = (-1)^n \frac{16}{(n^2\pi^2)} \tag{4}$$

Substituting (2), (3) and (4) in (1) we get the Fourier expansion for the given function as

$$x^2 - 2 = -\frac{2}{3} - \frac{16}{\pi^2}\left[\cos\frac{\pi x}{2} - \frac{1}{2^2}\cos \pi x + \frac{1}{3^2}\cos\frac{3\pi x}{2} - \cdots\right]$$

Example 13.7 Find the Fourier series expansion for

$$f(x) = \begin{cases} -\pi, & -\pi \le x \le 0 \\ x, & 0 \le x \le \pi \end{cases}$$

Solution Let

$$f(x) = \frac{a_0}{2} + \sum_{n-1}^{\infty} a_n \cos nx + \sum_{n=1}^{\infty} b_n \sin nx \tag{1}$$

Then,

$$a_0 = \frac{1}{\pi}\int_{-\pi}^{\pi} f(x)\, dx = \frac{1}{\pi}\left[-\int_{-\pi}^{0} \pi\, dx + \int_{0}^{\pi} x\, dx\right]$$

$$= -[x]_{-\pi}^{0} + \frac{1}{\pi}\left[\frac{x^2}{2}\right]_0^{\pi} = -\pi + \frac{\pi}{2} = \frac{\pi}{2} \tag{2}$$

and

$$a_n = \frac{1}{\pi}\int_{-\pi}^{\pi} f(x)\cos nx\, dx = \frac{1}{\pi}\left\{\int_{-\pi}^{0} (-\pi)\cos nx\, dx + \int_{0}^{\pi} (x)\cos nx\, dx\right\}$$

$$= \frac{1}{\pi}\left[(-\pi)\left(\frac{\sin(nx)}{n}\right)_{-\pi}^{0}\right] + \frac{1}{\pi}\left[(x)\left(\frac{\sin nx}{n}\right) - \left(-\frac{\cos nx}{n^2}\right)\right]_0^{\pi}$$

$$= 0 + \frac{1}{\pi}\left[\frac{\cos n\pi}{n^2} - \frac{1}{n^2}\right] = \frac{1}{\pi n^2}\left[(-1)^n - 1\right]$$

Therefore,

$$a_n = \frac{[(-1)^n - 1]}{(\pi n^2)} \tag{3}$$

Similarly

$$b_n = \frac{1}{\pi}\int_{-\pi}^{\pi} f(x)\sin nx\, dx$$

$$= \frac{1}{\pi}\left\{\int_{-\pi}^{0} (-\pi)\sin nx\, dx + \int_{0}^{\pi} (x)\sin nx\, dx\right\}$$

$$= \frac{1}{\pi}\left\{\left[(-\pi)\left(\frac{-\cos nx}{n}\right)\right]_{-\pi}^{0} + \left[(x)\left(-\frac{\cos nx}{n}\right) - \left(-\frac{\sin nx}{n^2}\right)\right]_0^{\pi}\right\} \tag{4}$$

or

$$b_n = \frac{1}{n} - \frac{\cos n\pi}{n} - \frac{\cos n\pi}{n} = \frac{1}{n}[1 - 2\cos n\pi] \tag{4}$$

Now, substituting (2), (3) and (4) in (1) we have the required Fourier expansion as

$$f(x) = -\frac{\pi}{4} + \sum_{n=1}^{\infty} \frac{1}{\pi n^2}\left[(-1)^n - 1\right] \cos nx$$

$$+ \sum_{n=1}^{\infty} \frac{1}{n}\left[1 - 2(-1)^n\right] \sin nx$$

13.7 HARMONIC ANALYSIS

When a function $f(x)$ is specified by some analytical expression, its Fourier coefficients are obtained by integration as explained in Section 13.6. However, if the function is given as a table of values in an interval at some equi-spaced points, then its Fourier coefficients can be computed through formal numerical integration, using either trapezoidal rule or by Simpson's rule, which is known as Harmonic analysis or discrete Fourier transform. Here, follows a couple of examples for illustration.

Example 13.8 The variation of a periodic current is given in the following table over a period P

t(sec)	0	$P/6$	$P/3$	$P/2$	$2P/3$	$5P/6$	P
A (amp.)	1.98	1.30	1.05	1.30	–0.88	–0.25	1.98

Show by harmonic analysis, that there is a direct current part of 0.75 amp., in the variable current and also determine the amplitude of the first harmonic.

Solution In the present problem, the length of the interval is P. Therefore, $2l = P$ or $l = P/2$. Let

$$A = \frac{a_0}{2} + \sum_{n=1}^{\infty} a_n \cos\left(\frac{n\pi t}{P/2}\right) + \sum_{n=1}^{\infty} b_n \sin\left(\frac{n\pi t}{P/2}\right) \quad (1)$$

The term $a_1 \cos\left(\frac{2\pi t}{P}\right) + b_1 \sin\left(\frac{2\pi t}{P}\right)$ is said to be the fundamental or first harmonic. The term $a_2 \cos\left(\frac{4\pi t}{P}\right) + b_2 \sin\left(\frac{4\pi t}{P}\right)$ is called second harmonic and so on. The values of t, A, $\cos\left(\frac{2\pi t}{P}\right)$ and $\sin\left(\frac{2\pi t}{P}\right)$ are tabulated as shown below:

t	A	$\cos\left(\frac{2\pi t}{P}\right)$	$\sin\left(\frac{2\pi t}{P}\right)$
0	1.98	1.0	0.0
$P/6$	1.30	0.5	0.866
$P/3$	1.05	–0.5	0.866
$P/2$	1.30	–1.0	0.0
$2P/3$	–0.88	–0.5	–0.866
$5P/6$	–0.25	0.5	–0.866
P	1.98	1.0	0.0

The interval (0, P) is subdivided into six subintervals, and therefore, $h = P/6$. Using Trapezoidal rule.

$$\int_a^b f(x)\,dx = \frac{h}{2}[f_0 + 2(f_1 + f_2 + \cdots + f_{n-1}) + f_n]$$

The Fourier components are computed as follows

$$a_0 = \frac{1}{P/2}\int_0^P A\,dt \approx \frac{1}{P/2}\cdot\frac{P}{12}[1.98 + 2(1.3 + 1.05 + 1.3 - 0.88 - 0.25) + 1.98]$$

$$= \frac{1}{6}[1.98 + 5.04 + 1.98] = \frac{1}{6}[9] = 1.5$$

$$a_1 = \frac{1}{P/2}\int_0^P A\cos\left(\frac{2\pi t}{P}\right)dt = \frac{1}{6}[(1.98)(1)$$

$$+ 2(0.5\times 1.30 - 0.5\times 1.05 - 1.0\times 1.3 + 0.88\times 0.5$$

$$- 0.25\times 0.5) + (1.98)(1)]$$

$$= \frac{1}{6}[2\times 1.98 - 1.72] = 0.3733$$

$$b_1 = \frac{1}{P/2}\int_0^P A\sin\left(\frac{2\pi t}{P}\right)dt = \frac{1}{6}[(1.98)(0) + 2(1.3\times 0.866$$

$$+ 1.05\times 0.866 + 1.3\times 0 + 0.88\times 0.866 + 0.25\times 0.866)$$

$$+ (1.98)(0)] = 6.02736/6 = 1.0046$$

Hence,

$$A = 0.75 + 0.3733\cos\left(\frac{2\pi t}{P}\right) + 1.0046\sin\left(\frac{2\pi t}{P}\right) + \cdots$$

Thus, the direct current = $\frac{a_0}{2}$ = 0.75 amp.

The amplitude of the first harmonic is $= \sqrt{a_1^2 + b_1^2}$

$$= \sqrt{(0.3733)^2 + (1.0046)^2} = 1.0717$$

Example 13.9 An experiment showed the displacement y, of a part of a mechanism as given in the table with corresponding angular movement $x°$ of the crank. The values represent periodic function in the interval [0°, 360°]. Express y as a Fourier series neglecting the harmonics above the second.

$x°$	0	30	60	90	120	150	180	210	240	270	300	330
y	1.80	1.10	0.30	0.16	1.50	1.30	2.16	1.25	1.30	1.52	1.76	2.00

Solution In the present problem, the length of the interval is (0°, 360°) or $(0, 2\pi)$. Therefore, period $2\pi = 2l$ or $l = \pi$. Let

$$y = \frac{a_0}{2} + \sum_{n=1}^{\infty} (a_n \cos nx + b_n \sin nx) \tag{1}$$

The values of x, $y \cos x$, $\sin x$, $\cos 2x$, $\sin 2x$ are given in the following table.

y	$x°$	$\cos x$	$\sin x$	$\cos 2x$	$\sin 2x$
1.80	0	1.0	0.0	1.0	0.0
1.10	30	0.8660	0.5	0.5	0.8660
0.30	60	0.5	0.8660	–0.5	0.8660
0.16	90	0.0	1.00	–1.0	0.0
1.50	120	–0.5	0.8660	–0.5	–0.8660
1.30	150	–0.8660	0.5	0.5	–0.8660
2.16	180	–1.0	0.0	1.0	0.0
1.25	210	–0.8660	–0.5	0.5	0.8660
1.30	240	–0.5	–0.8660	–0.5	0.8660
1.52	270	0.0	–1.0	–1.0	0.0
1.76	300	0.5	–0.8660	–0.5	–0.8660
2.00	330	0.8660	–0.5	0.5	–0.8660
1.80	360	1.0	0.0	1.0	0.0

In this problem, $h = 2\pi/12$, $\pi/6$. Now using Trapezoidal rule, that is,

$$\int_a^b f(x)\,dx = \frac{h}{2}[f_0 + 2(f_1 + f_2 + \cdots + f_{n-1}) + f_n]$$

the Fourier coefficients are computed as follows

$$a_0 = \frac{1}{\pi}\int_0^{2\pi} y\,dx = \frac{1}{\pi} \cdot \frac{\pi}{12}[1.80 + 2\,(1.10 + 0.30$$

$$+ 0.16 + 1.5 + 1.3 + 2.16 + 1.25 + 1.3 + 1.52$$

$$+ 1.76 + 2.0) + 1.80] = 2.6917$$

Similarly

$$a_1 = \frac{1}{\pi}\int_0^{2\pi} y \cos x \, dx = \frac{1}{\pi} \cdot \frac{\pi}{12}[(1.80)(1) + 2\{(1.10)(0.866)$$

$$+ (0.3)(0.5) + (0.16)(0) + (1.5)(-0.5) + (1.3)(-0.866)$$

$$+ (2.16)(-1) + (1.25)(-0.866) + (1.3)(-0.5) + (1.52)(0)$$

$$+ (1.76)(0.5) + (2.0)(0.866)\} + (1.8)(1)]$$

$$= -0.0423$$

$$b_1 = \frac{1}{\pi}\int_0^{2\pi} y \sin x \, dx = \frac{1}{\pi} \cdot \frac{\pi}{12}[(1.80)(0) + 2\{(1.1)(0.5)$$

$$+ (0.3)(0.866) + (0.16)(1) + (1.5)(0.866) + (1.3)(0.5)$$

$$+ (2.16)(0) + (1.25)(-0.5) + (1.3)(-0.866) + (1.52)(-1.0)$$

$$+ (1.76)(-0.866) + (2.0)(-0.5)\} + (1.8)(0)]$$

$$= -0.4839$$

$$a_2 = \frac{1}{\pi}\int_0^{2\pi} y \cos 2x = \frac{1}{\pi} \cdot \frac{\pi}{12}[(1.80)(1.0) + 2\{(1.10)(0.5)$$

$$+ (0.3)(-0.5) + (0.16)(-1.0) + (1.5)(-0.5) + (1.3)(0.5)$$

$$+ (2.16)(1) + (1.25)(0.5) + (1.3)(-0.5) + (1.52)(-1.0)$$

$$+ (1.76)(-0.5) + (2.0)(0.5)\} + (1.8)(1)]$$

$$= 0.4458$$

and

$$b_2 = -0.3767$$

Hence, the required Fourier series is given by

$$y = 1.3459 + (-0.0423 \cos x - 0.4839 \sin x)$$

$$+ (0.4458 \cos 2x - 0.3767 \sin 2x)$$

13.8 THE FAST FOURIER TRANSFORM

The fast Fourier transform denoted by FFT is a mathematical technique which has a wide variety of applications in areas such as conventional radar communications, signal processing, sonics and accoustics, biomedical engineering, applied mechanics and electromagnetics etc... The basic idea of the Fourier transform is that it allows one to have a look at a function or waveform in both time and frequency domains by decomposing into a sum of sinusoids of different frequencies. Many applications involving continuous Fourier transform depend on the use of a digital computer for its implementation, which naturally leads to the use of discrete Fourier transform, and therefore, to that of FFT. In the classical approach, we determine the Fourier coefficients, as explained in Section 13.7, in that, we observe that, sines and cosines have to be evaluated several times for angles around the origin as shown in Fig. 13.3.

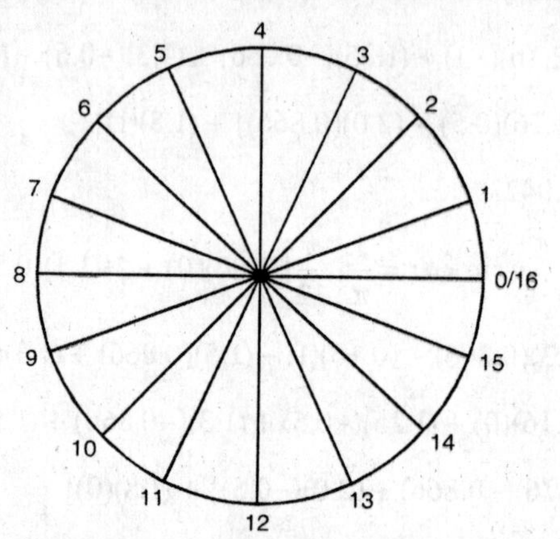

Fig. 13.3 Angles used for sixteen points.

In order to compute cos nx and sin nx, we move round the circle and use every n th value and hence easy to see that, we repeat previous values and require excessive computer time for large n. In 1965, it was Cooley and Tukey who took advantage of this fact of recomputation and published their mathematical algorithm which had become known as 'Fast Fourier Transform, FFT.

In developing FFT, instead of using the usual Fourier series for a periodic function $f(x)$ of period 2π, such as

$$f(x) = \frac{a_0}{2} + \sum_{n=1}^{\infty} [a_n \cos nx + b_n \sin nx] \tag{13.55}$$

where

$$a_n = \frac{1}{\pi} \int_{-\pi}^{\pi} f(x) \cos nx \, dx, \; n = 0, 1, 2, \ldots \tag{13.56}$$

and

$$b_n = \frac{1}{\pi} \int_{-\pi}^{\pi} f(x) \sin nx \, dx, \; n = 1, 2, 3, \ldots \tag{13.57}$$

its equivalent exponential form is often used in the literature, which is developed as follows. Applying an Euler's indentity and using i for $\sqrt{-1}$, we write

$$e^{inx} = \cos nx + i \sin nx$$

and therefore, we have

$$\cos nx = [e^{inx} + e^{-inx}]/2 \tag{13.58}$$

and

$$\sin nx = \left[e^{inx} - e^{-inx}\right]/2i \tag{13.59}$$

Thus, Eq. (13.55) can be written as

$$f(x) = \frac{a_0}{2} + \frac{1}{2}\sum_{n=1}^{\infty}(a_n - ib_n)e^{inx}$$

$$+\frac{1}{2}\sum_{n=1}^{\infty}(a_n + ib_n)e^{-inx} \tag{13.60}$$

This expression is further simplified by introducing negative values of n in Eqs. (13.56) and (13.57). That is,

$$a_{-n} = \frac{1}{\pi}\int_{-\pi}^{\pi} f(x)\cos(-nx)\,dx$$

$$= \frac{1}{\pi}\int_{-\pi}^{\pi} f(x)\cos nx\,dx = a_n,\ n = 1, 2 \ldots \tag{13.61}$$

and

$$b_{-n} = \frac{1}{\pi}\int_{-\pi}^{\pi} f(x)\sin(-nx)\,dx$$

$$= -\frac{1}{\pi}\int_{-\pi}^{\pi} f(x)\sin nx\,dx = -b_n,\ n = 1, 2, \ldots \tag{13.62}$$

Therefore, we find

$$\sum_{n=1}^{\infty} a_n\, e^{-inx} = \sum_{n=-1}^{-\infty} a_n e^{inx} \tag{13.63}$$

and

$$\sum_{n=1}^{\infty} i\, b_n e^{-inx} = -\sum_{n=-1}^{-\infty} i\, b_n\, e^{inx} \tag{13.64}$$

Substituting Eqs. (13.63) and (13.64) in Eq. (13.60), we get

$$f(x) = \frac{a_0}{2} + \frac{1}{2}\sum_{n=-\infty}^{\infty}(a_n - ib_n)e^{+inx}$$

or

$$f(x) = \sum_{n=-\infty}^{\infty} \alpha_n\, e^{inx} \tag{13.65}$$

where,

$$\alpha_n = \frac{1}{2}(a_n - ib_n), \quad n = 0, \pm 1, \pm 2, \ldots$$

Now, observe that Eq. (13.65) is the Fourier series expressed in the exponential form. If we combine Eqs. (13.56), (13.57), (13.61) and (13.62) we obtain

$$\alpha_n = \frac{1}{2\pi}\int_{-\pi}^{\pi} f(x)\, i^{-in\,x}\, dx\,,$$

$$n = 0, \pm 1, \pm 2 + \ldots \tag{13.66}$$

Thus, the final expression of the Fourier series in the exponential form is given by Eq. (13.65) and its complex coefficients in the form of Eq. (13.66). Here, $|\alpha_n|$ is called the power spectrum of f and these show the frequencies in $f(x)$. Hence, if $f(x)$ is known in the time domain, we can find f by computing α_n's which is an important aspect of the wave analysis.

In case the values of $f(x)$ is specified at equidistant points in the interval $(0, 2\pi)$ say at $x_j = 2\pi j/N, j = 0, 1, 2, \ldots, (N-1)$ and if $f(x)$ is periodic, and therefore $f_N = f_0$, $f_{N+1} = f_1$ and so on, then, we use numerical integration and carry out harmonic analysis as explained in section 13.7. Very often, we have a sampled continuous signal, in which case, it is better to use discrete Fourier transform. This is generally defined in the literature on FFT as

$$X(n) = \sum_{k=0}^{N-1} x_0(k)\, e^{-i2\pi nk/N}\,, n = 0, 1, 2, \ldots, (N-1) \tag{13.67}$$

Here, $x_0(k)$ are the N values of the signal samples in the time domain, while $X(n)$ represents the coefficients of N frequency terms. It may be observed that Eq. (13.67) describe N linear equations for finding the unknown $X(n)$ which require the multiplication of an N component vector by an $N \times N$ matrix. In order to present the FFT algorithm as described by Cooley and Tukey (1965), let us take N = 4 and introduce the notation

$$W = e^{-i2\pi/N} \tag{13.68}$$

thereby, Eq. (13.67) can be expanded as

$$X(0) = W^0 x_0(0) + W^0 x_0(1) + W^0 x_0(2) + W^0 x_0(3)$$

$$X(1) = W^0 x_0(0) + W^1 x_0(1) + W^2 x_0(2) + W^3 x_0(3)$$

$$X(2) = W^0 x_0(0) + W^2 x_0(1) + W^4 x_0(2) + W^6 x_0(3)$$

$$X(3) = W^0 x_0(0) + W^3 x_0(1) + W^6 x_0(2) + W^9 x_0(3)$$

In compact matrix notation, we may write the above set of equations as

$$\begin{pmatrix} x(0) \\ x(1) \\ x(2) \\ x(3) \end{pmatrix} = \begin{bmatrix} W^0 & W^0 & W^0 & W^0 \\ W^0 & W^1 & W^2 & W^3 \\ W^0 & W^2 & W^4 & W^6 \\ W^0 & W^3 & W^6 & W^9 \end{bmatrix} \begin{pmatrix} x_0(0) \\ x_0(1) \\ x_0(2) \\ x_0(3) \end{pmatrix} \tag{13.69}$$

If we examine Eq. (13.69), it reveals that, W and possibly $x_0(k)$ being complex, we require N^2 complex multiplications and $N(N-1)$ complex additions to carry out matrix computation on its right hand side. The success of FFT algorithm is attributed to the fact that the number of multiplications and additions are greatly reduced. Though, there are many variations in FFT algorithm, we shall present below, the one due to Cooley and Tukey. Using the relationship

$$W^{nk} = W^{nk \bmod \mathrm{N}}$$

we rewrite Eq. (13.69) as

$$\begin{pmatrix} x(0) \\ x(1) \\ x(2) \\ x(3) \end{pmatrix} = \begin{bmatrix} 1 & 1 & 1 & 1 \\ 1 & W^1 & W^2 & W^3 \\ 1 & W^2 & W^0 & W^2 \\ 1 & W^3 & W^2 & W^1 \end{bmatrix} \begin{pmatrix} x_0(0) \\ x_0(1) \\ x_0(2) \\ x_0(3) \end{pmatrix} \tag{13.70}$$

For clarification, we may recall that nk mod (N) means the remainder on division of nk by N. Thus, for example,

$$W^9 = W^{9 \bmod (4)} = W^1 \text{ and so on.}$$

Following the method of factorization due to Cooley and Tukey, we can factor Eq. (13.70) into an equivalent form as

$$\begin{pmatrix} X(0) \\ X(2) \\ X(1) \\ X(3) \end{pmatrix} = \begin{bmatrix} 1 & W^0 & 0 & 0 \\ 1 & W^2 & 0 & 0 \\ 0 & 0 & 1 & W^1 \\ 0 & 0 & 1 & W^3 \end{bmatrix} \begin{bmatrix} 1 & 0 & W^0 & 0 \\ 0 & 1 & 0 & W^0 \\ 1 & 0 & W^2 & 0 \\ 0 & 1 & 0 & W^2 \end{bmatrix} \begin{pmatrix} x_0(0) \\ x_0(1) \\ x_0(2) \\ x_0(3) \end{pmatrix} \tag{13.71}$$

Actual multiplication of the two matrices in Eq. (13.71) yields the square matrix in Eq. (13.70) with the only exception that the 1st and 2nd rows are interchanged. Suppose, we denote the interchanged column vector by $\bar{X}(n)$, then

$$\bar{X}(n) = \begin{pmatrix} X(0) \\ X(2) \\ X(1) \\ X(3) \end{pmatrix} \tag{13.72}$$

This particular type of factorization is the crucial point for the efficiency of FFT algorithm, which is seen in the following steps: In the first step, we use the factored form (13.71) and the vector $\bar{X}(n)$ is obtained in two stages. Of course, the number of stages depend on N. At the end of first stage, we denote the result as

$$\begin{bmatrix} 1 & 0 & W^0 & 0 \\ 0 & 1 & 0 & W^0 \\ 1 & 0 & W^2 & 0 \\ 0 & 1 & 0 & W^2 \end{bmatrix} \begin{pmatrix} x_0(0) \\ x_0(1) \\ x_0(2) \\ x_0(3) \end{pmatrix} = \begin{pmatrix} x_1(0) \\ x_1(1) \\ x_1(2) \\ x_1(3) \end{pmatrix} \tag{13.73}$$

Here, to compute $x_1(0)$, we need one complex multiplication and one complex addition, which is seen from

$$x_1(0) = x_0(0) + W^0 x_0(2)$$

Similarly, the element $x_1(1)$ is determined by one complex multiplication and addition. Noting that $W^2 = -W^\circ$, we find

$$x_1(2) = x_0(0) + W^2 x_0(2) = x_0(0) - W^0 x_0(2)$$

observe that the complex multiplication $W^0 x_0(2)$ has already been carried out in determining $x_1(0)$, therefore, in computing $x_1(2)$, we require only one complex addition. Adopting similar reasoning, we note that only one complex addition is required to compute $x_1(3)$. Thus, to compute $x_1(k)$, we require only two complex multiplications and four complex additions. In the second stage, we compute

$$\begin{bmatrix} 1 & W^0 & 0 & 0 \\ 1 & W^2 & 0 & 0 \\ 0 & 0 & 1 & W^1 \\ 0 & 0 & 1 & W^3 \end{bmatrix} \begin{pmatrix} x_1(0) \\ x_1(1) \\ x_1(2) \\ x_1(3) \end{pmatrix} = \begin{pmatrix} x_2(0) \\ x_2(1) \\ x_2(2) \\ x_2(3) \end{pmatrix} = \begin{pmatrix} X(0) \\ X(2) \\ X(1) \\ X(3) \end{pmatrix} \tag{13.74}$$

which also require two complex multiplications and four complex additions. Thus, computation of $\overline{X}(n)$ by means of Eq. (13.71) requires in all, four complex multiplications and eight complex additions.

However, observe that computation of $X(n)$ by direct multiplication using Eq. (13.70), requires 16 complex multiplications and 12 complex additions. See! how much of saving in computation time? Just by the matrix factorization process introduced by Cooley and Tukey. At this point, we are in a position to see the efficiency of the FFT algorithm. When $N = 2^r$, the FFT algorithm factorizes an $N \times N$ matrix into r matrices each of $N \times N$ having a special property of minimizing the number of complex multiplications and additions. In the above example, we have considered for $N = 4$, the number of complex multiplications are $(Nr)/2 = (4 \times 2)/2 = 4$ and the complex additions are $(Nr) = 4 \times 2 = 8$, if FFT algorithm is followed. However, if we follow the direct method, we note that the number of complex multiplications involved are $N^2 = 4^2 = 16$, while the number of complex additions required are $N(N-1) = 4 \times 3 = 12$. If we assume that the computing time is proportional to the number of multiplications involved, then the ratio of direct to FFT computing time in general is

$$\frac{N^2}{(Nr)/2} = \frac{2N}{r} \tag{13.75}$$

For $N = 4$, it is 4/1, while for large N say $N = 1024 = 2^{10}$, the reduction in computing time will be $(2 \times 1024)/10 \geq 200/1$. If $N = 1024$, the number of stages are $r = 10$.

In the final step, that is, to get $X(n)$ from $\bar{X}(n)$ we proceed as follows. We rewrite $\bar{X}(n)$ by replacing its arguments with their equivalent binary forms and the bits are then reversed or flipped, thus getting $X(n)$. For illustration, consider the case when $N = 4$.

$$\bar{X}(n) = \begin{pmatrix} X(0) \\ X(2) \\ X(1) \\ X(3) \end{pmatrix} \text{ becomes } \begin{pmatrix} X(00) \\ X(10) \\ X(01) \\ X(11) \end{pmatrix} \text{ flips to } \begin{pmatrix} X(00) \\ X(01) \\ X(10) \\ X(11) \end{pmatrix}$$

$$= \begin{pmatrix} X(0) \\ X(1) \\ X(2) \\ X(3) \end{pmatrix} = X(n) \tag{13.76}$$

For N greater then 4, it is cumbersome to explain the matrix factorization process as in Eq. (13.71). However, it is convenient to represent through directed flow graphs, so that it can be extended easily for large N. Eventually a computer program can be written with little effort. In general there will be r stages where $N = 2^r$. Thus, for $N = 4$, there will be 2 stages, for $N = 8$, the number of stages are 3, while for $N = 16$, there will be 4 stages and so on. For more details, the interested reader may refer Brigham (1988).

EXERCISES

13.1 Show that the Chebyshev series for $\sin^{-1}_x$ is given by

$$\sin^{-1} x = \frac{4}{\pi}\left[T_1 + \frac{1}{9}T_3 + \frac{1}{25}T_5 + \frac{1}{49}T_7 + \cdots\right]$$

by evaluating the coefficient integrals.

13.2 Obtain the Chebyshev series expansion for

$$\log(1 + x) = x - \frac{x^2}{2} + \frac{x^3}{6} - \frac{x^4}{4} + \frac{x^5}{5}$$

13.3 Find the Chebyshev series expansion for

$$e^x = 1 + x + \frac{x^2}{2} + \frac{x^3}{6} + \frac{x^4}{24}$$

13.4 Find the first few terms of the Chebyshev series for the Maclaurin series of sin x given by

$$\sin x = x - \frac{x^3}{6} + \frac{x^5}{120}$$

Economize the result to give a third degree polynomial as

$$\sin x = \frac{169}{192}T_1 - \frac{5}{128}T_3 = 0.9974x - 0.1562x^3$$

Compare the errors in both the series.

13.5 The function $\tan^{-1}x$ can be expressed in terms of Maclaurin series as

$$\tan^{-1} x = x - \frac{x^3}{3} + \frac{x^5}{5} - \frac{x^7}{7} + \frac{x^9}{9} - \cdots$$

Economize this to give a third degree polynomial.

13.6 What transformation change any finite range $a \le y \le b$ bijectively maps onto the range $-1 \le x \le 1$.

13.7 Find the Pade approximation for the function $\tan^{-1}x$ with numerator and denominator having polynomials of degree five and four respectively.

13.8 Find the Pade approximation for the function e^x with numerator and denominator having cubic and linear polynomials respectively.

13.9 Find the Pade approximation R_{22} for the function log $(1 + x)$.

13.10 Find the Pade approximation R_{54} for the function tan x.

Appendix

Thomas Algorithm for Tridiagonal Systems

For tridiagonal system the LU decomposition method leads to an efficient algorithm, known as Thomas algorithm. Consider a system of the form

$$a_k x_{k-1} + b_k x_k + c_k x_{k+1} = f_k,\ k = 1, ..., n \tag{A.1}$$

with

$$a_1 = c_n = 0 \tag{A.2}$$

the following algorithm is obtained.

Step 1:

$$\left.\begin{aligned} &\beta_1 = b_1,\ \beta_k = b_k - a_k \frac{C_{k-1}}{\beta_{k-1}},\ k = 2, \ldots, n \\ &\gamma_1 = \frac{f_1}{\beta_1},\ \gamma_k = \frac{(-a_k \gamma_{k-1} + f_k)}{\beta_k},\ k = 2, \ldots, n \end{aligned}\right\} \tag{A.3}$$

Step 2:

$$x_n = \gamma_n$$

$$x_k = \gamma_k - x_{k+1} \frac{c_k}{\beta_k},\quad k = (n-1), \ldots, 1 \tag{A.4}$$

This requires in all, $5n$ operations.

It has been established that the above algorithm will always converge if the tridiagonal system is diagonally-dominant. That is, if

$$\left.\begin{aligned} &|b_k| \geq |a_k| + |c_k|,\quad k = 2, \ldots, (n-1) \\ &|b_1| > |c_1| \text{ and } |b_n| > |a_n| \end{aligned}\right\} \tag{A.5}$$

Bibliography

1. Ahlberg, J.H., Nilson, E.N. and Walsh, J.L., *The Theory of Splines and Their Applications,* Academic Press, New York, 1967.

2. Ames, W.F., *Numerical Methods for Partial Differential Equations,* Thomas Nelson, London, 1969.

3. Brigham, E. Oron, *The Fast Fourier Transform and Its Applications*, Prentice Hall, Inc. Englewood Cliffs, New Jersey, 1988.

4. Cooley, J.W. and Tukey, J.W., *An Algorithm for the Machine Calculations of Complex Fourier Series*, Mathematics of Computation, Vol. 19, No. 90, pp. 297–301, 1965.

5. Collatz, L., *Functional Analysis and Numerical Mathematics,* Academic Press, New York, 1966.

6. Conte, S.D. and Carl de Boor, *Elementary Numerical Analysis—An Algorithmic Approach*, 3rd international edition, Tata McGraw-Hill, New Delhi, 1981.

7. Crank, J. and Nicolson, P., 'A practical method for numerical evaluation of solutions of partial differential equations of the heat conduction type', *Proc. Camb. Phil. Soc.*, Vol. 43, pp. 50–67, 1947.

8. Davis, P. and Rabinowitz, P., *Numerical Integration*, Blaisdell, New York, 1967.

9. Dorn, W.S., Daniel D. and Mccracken, *Numerical Methods with FORTRAN IV Case Studies,* John Wiley, New York, 1972.

10. Faddeeva, D.K. and Faddeeva, V.N., *Computational Methods of Linear Algebra,* translated by Williams, R.C., Freeman, San Francisco, 1963.

11. Froberg, C.E., *Introduction to Numerical Analysis*, Addison–Wesley, Reading, Massachusetts, 1965.

12. Gear, C.W., *Numerical Initial Value Problems in Ordinary Differential Equations*, Prentice Hall, Englewood Cliffs, New Jersey, 1971.

13. Gerald, C.F. and Wheatley, P.O., *Applied Numerical Analysis*, Addison–Wesley, Reading, Massachusetts, 1994.

14. Goodwin, E.T., *Modern Computing Methods,* National Physical Laboratory, Teddington, Middlesex, 1961.

15. Hildebrand, F.B., *Introduction to Numerical Analysis*, Tata McGraw-Hill, New Delhi, 1982.

16. Jain, P.C. and Sankara Rao, K., 'Numerical solution of unsteady viscous incompressible fluid flow past a circular cylinder', *Phys. Fluids*, Supplement II, Vol. 12, pp. II 57–II 64, 1969.

17. Kopal, Z., *Numerical Analysis*, John Wiley, New York, 1955.

18. Krishnamurthy, E.V. and Sen, S.K., *Numerical Algorithms*, East–West Publishers, New Delhi, 1986.

19. Mitchell, A.R. and Griffiths, G.F., *The Finite Difference Method in Partial Differential Equations*, John Wiley, New York, 1980.

20. Peaceman, D.W. and Rachford, H.H., 'The numerical solution of parabolic and elliptic differential equations', *J. Soc. Indust. Appl. Maths.,* Vol. 3, pp. 28–41, 1955.

21. Ralston, A. and Wilf, H.S., *Mathematical Methods for Digital Computers,* John Wiley, New York, 1965.

22. Richtmyer, R.D., *Difference Methods for Initial Value Problems,* Interscience, New York, 1957.

23. Sankara Rao, K., *Introduction to Partial Differential Equations,* Prentice-Hall of India, New Delhi, 1995.

24. Scarborough, J.B., *Numerical Mathematical Analysis,* The John Hopkins University Press, Baltimore, 1966.

25. Smith, G.D., *Numerical Solution of Partial Differential Equations*, Oxford University Press, London, 1965.

26. Spath, H., 'Interpolation by certain quintic splines', *Computer Journal,* Vol. 12, No. 3, pp. 292–93, 1969.

27. Stroud, A.H. and Secrest, D., *Gaussian Quadrature Formulas,* Prentice Hall, Englewood Cliffs, New Jersey, 1966.

28. Wilkinson, J.H., *The Algebraic Eigenvalue Problem,* Oxford University Press, London, 1965.

29. Young, D.M. and Gregory, R.T., *A Survey of Numerical Mathematics,* Vols. 1 and 2, Addison Wesley, Reading, Massachusetts, 1972.

Answers to Exercises

CHAPTER 2

2.1 Given $f(x) = x^3 - 3x - 5$, we note $f(2) = -3$, $f(3) = 13$. Hence, a real root lies between $x_1 = 2$ and $x_2 = 3$. Various approximations are given below:

n	x_{n+1}	$f(x_{n+1})$
2	2.5	3.125
3	2.25	−0.3594
4	2.375	1.2715
5	2.3125	0.4290
6	2.28125	0.0281

2.2 Given $f(x) = x^3 + x - 3$, we note $f(1) = -1$, $f(2) = 7$. Hence, a real root lies between $x_1 = 1$ and $x_2 = 2$. Various approximations are given below:

n	x_{n+1}	$f(x_{n+1})$
2	1.125	−0.45117
3	1.17798	−0.1874
4	1.1994	−0.07519
5	1.2079	−0.02972
6	1.21125	−0.011693
7	1.21257	-4.5705×10^{-3}
8	1.21308	-1.78507×10^{-3}
9	1.21328	-7.13168×10^{-4}
10	1.21336	-2.794298×10^{-4}

Therefore, the required root is 1.2134

2.3 Given $f(x) = x^6 - x^4 - x^3 - 1$, we note that $f(1.4) = -0.056064$, $f(1.5) = 1.953125$. Hence, the real root lies between $x_1 = 1.4$ and $x_2 = 1.5$. Various approximations to the root are given below:

n	x_{n+1}	$f(x_{n+1})$
2	1.40279	−0.012735
3	1.40342	-2.86077×10^{-3}
4	1.40356	-6.609198×10^{-4}

Hence, the required, real root is 1.4036.

2.4 Given $f(x) = x^3 - \sin x + 1$. We note that $f(-1) = 0.84147$, $f(-2) = -6.0907$. Since $f(-1)$ and $f(-2)$ are of opposite signs, the real root lies between -2 and -1. Let $x_1 = -1$, $x_2 = -2$, using Regula-Falsi method, various approximations to the root are given below:

n	x_{n+1}	$f(x_{n+1})$
2	−1.12139	0.49055
3	−1.18688	0.25526
4	−1.21959	0.12495

Thus, the required root after three successive approximations is −1.2196.

2.6 Given $f(x) = \cos x - xe^x$, we observe $f(0) = 1$, $f(1) = -2.177979$. Hence, the root lies between $x_1 = 0$ and $x_2 = 1$. Using Regula-Falsi method, various approximations to the root are given below:

n	x_{n+1}	$f(x_{n+1})$
2	0.31467	0.51987
3	0.44673	0.20355
4	0.49402	0.70802×10^{-1}
5	0.50995	0.23608×10^{-1}
6	0.51520	0.77601×10^{-2}
7	0.51692	0.25389×10^{-2}
8	0.51749	0.82936×10^{-3}

2.7 Rewrite the equation in the form $x = (\cos x + 3)/2 = \phi(x)$. We observe $f(0) = -4$, $f(\pi/2) = \pi - 3 > 0$, showing that a root lies in the interval $(0, \pi/2)$. Also, $|\phi'(x)| = \sin(x/2) < 1$ for all x in $(0, \pi/2)$. We start with $x_0 = \pi/2$. Then the successive iterations are

$$x_1 = \phi(x_0) = 1.5, \qquad f(x_1) = 0.07074$$

$$x_2 = \frac{1}{2}\left(\cos\frac{1.5 \times 180}{\pi} + 3\right) = 1.5353686, \qquad f(x_2) = -0.0353169$$

$$x_3 = 1.5177, \qquad f(x_3) = 0.01766$$

$$x_4 = 1.526, \qquad f(x_4) = -7.21865 \times 10^{-3}$$

$$x_5 = 1.522, \qquad f(x_5) = 4.77696 \times 10^{-3}$$

$$x_6 = 1.524, \qquad f(x_6) = -1.22075 \times 10^{-3}$$

Hence, the root is 1.524.

2.8 $f(x) = x \log_{10} x - 4.77$

$f'(x) = \log_{10} x + \log_{10} e = \log_{10} x + 0.4343$

Since, $e = 2.71828$, $\log_{10} e = 0.4343$.

Newton's–Raphson formula is

$$x_{n+1} = x_n - \frac{f(x_n)}{f'(x_n)} = x_n - \frac{x_n \log_{10} x_n - 4.77}{\log_{10} x_n + 0.4343}$$

Note that $f(6) = -0.1011, f(7) = 1.1457$. Therefore, the required root lies in the interval (6, 7). Taking $x_0 = 7$ as the initial approximation, the successive approximations are

$$x_1 = 7 - \frac{1.145686}{1.2794} = 6.1045, \qquad f(x_1) = 0.026$$

$$x_2 = 6.1045 - \frac{0.026}{1.21995} = 6.0832, \qquad f(x_2) = 3.23613 \times 10^{-5}$$

$$x_3 = 6.0832 - \frac{3.23613 \times 10^{-5}}{1.21843} = 6.0832, \qquad f(x_3) = 3.23613 \times 10^{-5}$$

Hence, the required root is 6.0832.

2.11 $f(x) = x^3 - 3x - 5, f'(x) = 3x^2 - 5, f''(x) = 6x, f(2) = -3, f(3) = 13$. Taking $x_0 = 3$, the successive iterations by Newton's method are

n	x_n	$f(x_n)$
1	2.45833	2.48164
2	2.29431	0.19399
3	2.27915	1.59433×10^{-3}

2.12 $f(x) = x^4 - x - 10, f'(x) = 4x^3 - 1, f''(x) = 12x^2, f(1) = -10, f(2) = 4$. Taking $x_0 = 2$, the successive iterations by Newton's method are

n	x_n	$f(x_n)$
1	1.87096	0.38268
2	1.85577	4.6201×10^{-3}
3	1.85558	-6.2104×10^{-5}

2.13 0.5175

2.14 $x_{n+1} = \dfrac{1}{p}\left[\dfrac{(p-1)x_n^p + N}{x_n^{p-1}}\right]$

2.15 Taking $x_0 = 2.5$, the first approximation is 2.30666

2.16 Let $f(x) = x^3 - 3x - 5$, then $f(1) = -7, f(2) = -3, f(3) = 13$. Muller's method can be conveniently started by taking

$$x_{i-2} = 1, \quad x_{i-1} = 2 \quad x_i = 3$$

$$f_{i-2} = -7, \quad f_{i-1} = -3 \quad f_i = 13$$

The first approximation is

$$x_{i+1} = x_i + \lambda h_i = 3 - 0.74043 = 2.25957$$

2.17 Muller's method can be conveniently started by taking

$$x_{i-2} = 1.8, \qquad x_{i-1} = 2.0 \qquad x_i = 2.2$$
$$f_{i-2} = 0.07385, \qquad f_{i-1} = -0.09070 \qquad f_i = -0.29150$$

The first approximation is given by

$$x_{i+1} = x_i + \lambda h_i = 2.2 - (1.52373)\,(0.2) = 1.89525$$

2.18 $i = 1$, the new polynomial is $x^3 - 10x^2 + 17x - 1 = 0$

$i = 2$, the new polynomial is $x^3 - 66x^2 + 269x - 1 = 0$

$i = 3$, the new polynomial is $x^3 - 3818x^2 + 72229x - 1 = 0$

The corresponding approximations of the roots are

$$\sqrt{\frac{1}{17}} = 0.2425, \qquad \sqrt{\frac{17}{10}} = 1.3038, \qquad \sqrt{10} = 3.1623$$

$$\sqrt[4]{\frac{1}{269}} = 0.2469, \qquad \sqrt[4]{\frac{269}{66}} = 1.4208, \qquad \sqrt[4]{66} = 2.8503$$

$$\sqrt[8]{\frac{1}{72229}} = 0.2469, \qquad \sqrt[8]{\frac{72229}{3818}} = 1.4441, \qquad \sqrt[8]{\frac{3818}{1}} = 2.8037$$

Using the given polynomial, the roots are found to be -0.2469, 1.4441 and 2.8037.

2.19 Given $f(x) = x \sin x - 1$, we observe the $f(0) = -1.0$, $f(2) = 0.81859$. Using the method of false position, various approximations to the root are given as

n	x_{n+1}	$f(x_{n+1})$
2	1.09975	–0.02002
3	1.12124	0.00983
4	1.11416	0.563×10^{-5}

2.20 $(x^2 - 1.5x + 4.3)(x^2 - 4.2x + 16)$

2.21 $(x^2 + 5x + 2)(2x^2 - 3x + 7)$

2.22 $x = 0.5086, \quad y = 0.8411$

Exact solution is, $x = 0.5$, $y = 0.866$

2.23 $x = 0.5005$, $y = 0.9993$

Exact solution is $x = 0.5$, $y = 1.0$.

CHAPTER 3

3.1 (i) The upper-triangular form is

$$x_1 + \frac{x_2}{2} + \frac{x_3}{3} = 1$$

$$\frac{x_2}{12} + \frac{x_3}{12} = -\frac{1}{2}$$

$$\frac{x_3}{180} = \frac{1}{6}$$

The solution is $x_1 = 9$, $x_2 = -36$, $x_3 = 30$.

(ii) The upper-triangular form is

$$x_1 + \frac{x_2}{4} + \frac{x_3}{4} = 1$$

$$x_2 - \frac{3}{5}x_3 = \frac{4}{5}$$

$$-4x_3 = 2$$

Hence, its solution is

$$x_1 = 1, \quad x_2 = \frac{1}{2}, \quad x_3 = -\frac{1}{2}$$

(iii) The solution of the given system is given by $x = 5$, $y = 4$, $z = -7$, $w = 1$.

3.2 (i) $x_1 = 22/9$, $x_2 = 3$, $x_3 = 11/9$

(ii) $x = 0.4879$, $y = -8.6943$, $z = -6.1637$

3.3 (i) $x_1 = 1$, $x_2 = 1$, $x_3 = 1$

(ii) $x = 3$, $y = 1$, $z = 3$

3.4 (i) $[L] = \begin{bmatrix} 1 & 0 & 0 \\ 2 & -3 & 0 \\ 3 & -2 & -\frac{14}{3} \end{bmatrix}, \quad [U] = \begin{bmatrix} 1 & 1 & 1 \\ 0 & 1 & -\frac{1}{3} \\ 0 & 0 & 1 \end{bmatrix}$

and the required solution is $x = 1$, $y = -2$, $z = 4$.

(ii) $[L] = \begin{bmatrix} 6 & 0 & 0 \\ -1 & \frac{35}{6} & 0 \\ 0 & -1 & \frac{204}{35} \end{bmatrix}, \quad [U] = \begin{bmatrix} 1 & -\frac{1}{6} & 0 \\ 0 & 1 & -\frac{6}{35} \\ 0 & 0 & 1 \end{bmatrix}$

The required solution is $x_1 = 11/17$, $x_2 = 15/17$, $x_3 = 11/17$.

3.5 $[L] = \begin{bmatrix} 4 & 0 & 0 \\ 2 & 2.5 & 0 \\ 3 & 1.25 & 0.5 \end{bmatrix}, \quad [U] = \begin{bmatrix} 1 & -0.75 & 0.5 \\ 0 & 1 & 2.4 \\ 0 & 0 & 1 \end{bmatrix}$

Then the solution to the given system is $x_1 = -5.2$, $x_2 = -8.6$, $x_3 = 3$.

3.6 $x = 1.1196, y = 0.8699, z = 0.1415.$

3.8 Starting from the initial zero vector, the first seven iterations due to Jacobi are tabulated as below:

Iteration number, r	x_1	x_2	x_3	x_4
1	0.5	0.5	0.25	0.25
2	0.6875	0.6875	0.4375	0.4375
3	0.7813	0.7813	0.5313	0.5313
4	0.8281	0.8281	0.5781	0.5781
5	0.8516	0.8516	0.6016	0.6016
6	0.8633	0.8633	0.6133	0.6133
7	0.8691	0.8691	0.6206	0.6206

3.9

Iteration number, r	x_1	x_2	x_3
1	3.5	2.25	1.625
2	4.625	3.625	2.3125
3	5.3125	4.3125	2.6563
4	5.6563	4.6563	2.8282
5	5.8282	4.8282	2.9141

3.10

Iteration number, r	x	y	z
1	0.85	−1.0275	1.0109
2	1.00247	−0.999825	0.99978
3	0.999969	−1.00000	1.00000

3.11 $x = 2.6, y = 0.3432, z = 0.741$

3.12 $x_1 = 1.0, x_2 = 1.0, x_3 = 1.0$

3.13 $A^{-1} = \begin{bmatrix} \frac{1}{2} & -\frac{1}{2} & \frac{1}{2} \\ -4 & 3 & -1 \\ \frac{5}{2} & -\frac{3}{2} & \frac{1}{2} \end{bmatrix}$

3.14 $A^{-1} = \begin{bmatrix} 1 & -\frac{1}{5} & -\frac{2}{5} \\ 1 & -\frac{1}{5} & -\frac{7}{5} \\ -1 & \frac{2}{5} & \frac{4}{5} \end{bmatrix}$

3.15 (i) $A^{-1} = \begin{bmatrix} 3 & 1 & \frac{3}{2} \\ -\frac{5}{4} & -\frac{1}{4} & -\frac{3}{4} \\ -\frac{1}{4} & -\frac{1}{4} & -\frac{1}{4} \end{bmatrix}$

(ii) $A^{-1} = \begin{bmatrix} 2 & -1 & 0 \\ -1 & -3 & 2 \\ 0 & 2 & -1 \end{bmatrix}$

CHAPTER 4

4.1 The largest eigenvalue is $\lambda = 6.941$ and the corresponding eigenvector is

$$X = \begin{pmatrix} 0.298 \\ 0.063 \\ 1.0 \end{pmatrix}$$

4.2 The largest eigenvalue is $\lambda = 20.185$ and the corresponding eigenvector is

$$X = \begin{pmatrix} 0.061 \\ 1.0 \\ 0.026 \end{pmatrix}$$

4.3 At the end of fifth iteration, the dominant eigenvalue is $\lambda = 16.447$ and the corresponding eigenvector is

$$X = \begin{pmatrix} 0.243 \\ -0.085 \\ 1.0 \end{pmatrix}$$

4.4 At the end of sixth iteration, the largest eigenvalue is found to be $\lambda = 5.1564$ and the corresponding eigenvector is

$$X = \begin{pmatrix} 1.0 \\ 0.6968 \\ 0.3491 \end{pmatrix}$$

4.5 At the end of two rotations, the eigenvalues are found to be 1.408, 0.123 and 0.0028 and the corresponding eigenvectors are given by matrix

$$S = \begin{bmatrix} 1.0 & 1.0 & 1.0 \\ 0.556 & -0.965 & -5.591 \\ 0.391 & -1.186 & 5.395 \end{bmatrix}$$

4.6 The eigenvalues are 11.0990, 3.4142, 0.9009, 0.5858. The eigenvectors are

$$\begin{bmatrix} 0.8198 & 1.0000 & 1.0000 & -0.4142 \\ 1.0000 & 0.4142 & -0.8198 & 1.0000 \\ 1.0000 & -0.4142 & -0.8198 & -1.0000 \\ 0.8198 & -1.0000 & 1.0000 & 0.4142 \end{bmatrix}$$

4.7 (i) At the end of third rotation, the eigenvalues of the given matrix are found to be 4.999, –1, –1 and the eigenvectors are the column vectors of

$$\begin{bmatrix} 0.577 & -0.707 & -0.408 \\ 0.577 & 0.707 & -0.408 \\ 0.577 & 0 & 0.817 \end{bmatrix}$$

(ii) At the end of three rotations, the eigenvalues are found to be 6.3723, 2.0, 0.6277 and the corresponding eigenvectors are given by the matrix

$$S = \begin{bmatrix} 0.8431 & 1.0000 & -0.5931 \\ 1.0000 & 0.0000 & 1.0000 \\ 0.8431 & -1.0000 & -0.5931 \end{bmatrix}$$

4.8 (i) The dominant eigenvalue = 3.987 and the corresponding eigenvector is $[0.6677 \ \ 1]^T$.
(ii) The dominant eigenvalue = 5.3719 and the corresponding eigenveetor is $[0.4574 \ \ 1]^T$.

CHAPTER 5

5.1 The required fit is $y = 0.1235x^2 + 0.3213$

5.2 The required least square fit is $y = 9.1x - 3.0$, $E = 70.7$

5.3 $Q = -0.1341p^2 + 3.5836p + 6.2149$

5.4 $y = 0.5012x^{1.9977}$

5.5 $y = 1.201e^{0.00967}$

5.6 $y = 3.1875x + 13.0313$

5.7 $y = 1.0000 + 0.8766x + 0.8417x^2$

5.8 $y = 107.72 + 1.796x + 0.0035x^2$

5.9 $y = 0.70x + 2.60$

5.10 $y = 0.875x^2 - 1.7x + 2.125$

5.11 $y = 0.3634e^{0.7475x}$

CHAPTER 6

6.1 $\Delta^2 y_1 = y_3 - 2y_2 + y_1, \quad \Delta^4 y_0 = y_4 - 4y_3 + 6y_2 - 4y_1 + y_0$

6.2 Working backward from the second difference of the function to the first differences and then to the function values, we get

Δy_n row is 0 1 5 18 36 60
and then
y_n row is 0 0 1 6 24 60 120

6.6 Using linear interpolation formula

$$y = y_1 + \frac{y_2 - y_1}{x_2 - x_1}(x - x_1) \quad \text{and} \quad x_1 = 2, \quad x_2 = 4$$

$f(3)$ value is 325. Actual value is 125.

6.7 3.8988 kgf/cm^2

6.8 $y_x = \frac{1}{3}(2x^4 - 24x^3 + 100x^2 - 168x + 93)$

6.9 $f(x) = x^2 + x + 1$

6.10 286.96°C

6.11 $f(x) = x^3 + x^2 - x + 2$

6.12 $y = -\frac{1}{30}(39x^2 + 123x - 252)$

6.14 $y = x^3 - 1$

6.16 $u(1.6, 0.33) = 1.8406$. Actual value from the function is 1.8350; Error = –0.006.

6.17 $y = x^3 - 4x^2 - 7x - 15$

6.18 Let the missing term be a, since we are assuming a polynomial of third degree, its third difference is constant and fourth difference is zero. From the forward difference table, we observe $\Delta^4 y = 6a - 2214 = 0$, which gives $a = 369$.

6.19 Form the forward difference table. Taking $x_0 = 3$, we find $y_0 = 0.205$, $\Delta y_0 = 0.035$, $\Delta^2 y_0 = -0.016$, $\Delta^3 y_0 = 0.0$. Newton's forward difference formula gives

$$y = y_0 + p\Delta y_0 + \frac{p(p-1)}{2}\Delta^2 y_0 + \cdots = 0.205 + 0.035p - 0.016\,\frac{(p^2 - p)}{2} + \cdots$$

Therefore, for minimum value of y,

$$\frac{dy}{dp} = 0.035 + \frac{2p - 1}{2}(-0.016) = 0$$

which gives $p = 2.6875 = (x - 3)/1$ and $x = 5.6875$. With this x, the minimum of y is 0.2628.

6.20 $\Delta^5 y_0 = \Delta^6 y_0 = 0$ gives $y_3 - y_2 = 56$ and $-4y_3 + 3y_2 = -236$, whose solution gives the missing values as $y_2 = 12$ and $y_3 = 68$.

6.21 For natural end boundary conditions, the equations are

$$4M_1 + M_2 = 6, M_1 + 4M_2 = -12$$

whose solution gives $M_1 = 2.4$, $M_2 = -3.6$. Natural end conditions imply $M_0 = M_3 = 0$.
The required cubic splines in each interval is given by

$S_0(x) = 0.4x^3 + 0.1x,$ for $0 \le x \le 1$

$S_1(x) = -(x-1)^3 + 1.2(x-1)^2 + 1.3(x-1) + 0.5,$ for $1 \le x \le 2$

$S_2(x) = 0.6(x-2)^3 - 1.8(x-2)^2 + 0.7(x-2) + 2.0,$ for $2 \le x \le 3$

6.22 $M_0 = -0.36$, $M_1 = 2.52$, $M_2 = -3.72$, $M_3 = 0.36$. The required piecewise cubic spline functions are

$S_0(x) = 0.48x^3 - 0.18x^2 + 0.2x,$ for $0 \le x \le 1$

$S_1(x) = -1.04(x-1)^3 + 1.26(x-1)^2 + 1.28(x-1) + 0.5,$
for $1 \le x \le 2$

$S_2(x) = 0.68(x-2)^3 - 1.86(x-2)^2 + 0.68(x-2) + 2.0,$
for $2 \le x \le 3$

6.23 For natural cubic spline, we have the equations

$$3M_1 + 0.5M_2 = 12.7020$$
$$0.5M_1 + 2.5M_2 = 30.2754$$

which gives $M_1 = 2.2920$, $M_2 = 11.6518$. Natural end conditions imply $M_0 = M_3 = 0$. We also observe $h_0 = 1$, $h_1 = 0.5$, $h_2 = 0.75$. The required piecewise cubic splines are

$S_0(x) = 0.382x^3 + 2.0546x + 2.0,$ for $0 \le x \le 1$

$S_1(x) = 3.1199(x-1)^3 + 1.146(x-1)^2 + 3.2005(x-1) + 4.4366,$
for $1 \le x \le 1.5$

$S_2(x) = -2.5893(x-1.5)^3 + 5.8259(x-1.5)^2 + 6.6866(x-1.5)$
$+ 6.7134,$ for $1.5 \le x \le 2.25.$

6.24 The divided difference table is

x	$f(x)$	Ist. d.d.	2nd d.d.	3rd d.d.	4th d.d.
4	48				
		52			
5	100		15		
		97		1	
7	294		21		0
		202		1	
10	900		27		0
		310		1	
11	1210		33		
		409			
13	2028				

Newton's divided difference interpolation polynomial for the given data is

$$f(x) = 48 + (x - 4)(52) + (x - 4)(x - 5)(15) + (x - 4)(x - 5)(x - 7)(1)$$
$$= x^3 - x^2$$

Therefore,

$$f(2) = 4, \qquad f(15) = 315.$$

6.25 Since the abscissa is equally-spaced with h = 1, Eq. (6.83) gives

$$M_{i-1} + 4M_i + M_{i+1} = 6[y_{i-1} - 2y_i + y_{i+1}]/h^2 \tag{1}$$

For i = 1, 2 we get at once

$$M_0 + 4M_1 + M_2 = 36, \; M_1 + 4M_2 + M_3 = 72 \tag{2}$$

Note that $M_0 = M_3 = 0$, for a natural cubic spline. The solution of (2) is found to be $M_1 = 24/5$, $M_2 = 84/5$. Let the natural cubic spline be

$$S(x) = a_i(x - x_i)^3 + b_i(x - x_i)^2 + c_i(x - x_i) + d_i$$

where, a_i, b_i, c_i and d_i are given in Eqs. (6.77) to (6.80). Since $x = 1/2$ is in the interval (0, 1), we set $i = 0$ and get $a_0 = 12/15$, $b_0 = 0$, $C_0 = 1/5$, $d_0 = y_0 = 1$. Hence $S(x) = 12(x - 0)^3/15 + (x - 0)/5 + 1$ which gives $S(1/2) = 6/5$. However, the actual value is $y(1/2) = 9/8$, which is obtained from the tabulated function $y = 1 + x^3$.

6.26 $y(1.05) = 0.8674$

6.27 $y(1.3) = 0.9634$

6.28 $y(0.6) = \ln(0.6) = -0.5108$

CHAPTER 7

7.3 $f'(0.4) = 1.4913$

7.4 $y'(1.10) = 0.4770, \quad y''(1.10) = -0.2$

7.5 Velocity $= \left(\dfrac{dx}{dt}\right)_{t=0.3} = 0.5416$ (by forward difference formula)

Acceleration $= \left(\dfrac{d^2x}{dt^2}\right)_{t=0.3} = -4.7$ cm/s^2 (by forward difference formula)

7.6 $y''(1.3) = -0.7050$

7.7 $y'(0.75) = -8 + 7.9998 - 1.7777 = -1.7779$ (quadratic fit)

$y'(0.75) = -8 + 7.9998 - 1.7777 - 0.3777 = -2.1556$ (cubic fit)

Actual value $= y'(0.75) = \left(-\dfrac{1}{x^2}\right)_{x=0.75} = -1.7778$

Therefore, quadratic fit gives the most accurate value.

7.8 Noting that $\cosh x = (e^x + e^{-x})/2$. Its table of values:

x	1.0	1.2	1.4	1.6	1.8
$\cosh x$	1.5431	1.8107	2.1509	2.5775	3.1075

using Simpson's 1/3 rule, the value of integral = 1.7670. Actual value = 1.76695 (accurate to four decimal places)

7.9 Values of the integral = 626

7.10 3.1416

7.11 $I = 0.6933, \dfrac{h^4}{120} \le E \le \dfrac{4h^4}{15}$

7.12 $I = 1.7183$

7.13 $y''(0.6) = 1.25766$

7.14 $\int_{1.0}^{1.8} y(x)\, dx = 1.7671$

7.15 $\int_1^2 \dfrac{dx}{x} = 0.693147$

7.16 $I = 0.5582$

7.17 $I = \int_1^2 \int_1^2 \dfrac{dx\, dy}{x^2 + y^2} = 0.2313$

7.18 Taking $h = k = 0.25$, we get $I = 2.6666$

7.19 $S = \int_0^{12} v\, dt = 552$ metres (using Simpson 1/3-rule). From Newton's formula,

$$v(a + ph) = v_0 + p\Delta v_0 + p\frac{p-1}{2}\Delta^2 v_0$$

$$\text{Acceleration} = v'(a + ph) = \frac{1}{h}[\Delta v_0 + \frac{2p-1}{2}\Delta^2 v_0]$$

Given $h = 2$, $a = 0$, $p = (2 - 0)/2 = 1$. From the forward difference table, we have $\Delta v_0 = 2$, $\Delta^2 v_0 = 8$ and all other differences are zero. Therefore, the required acceleration is

$$\frac{dv}{dt} = \frac{1}{2}\left(2 + \frac{8}{2}\right) = 3 \text{ m/s}^2$$

7.20 $y'(2.03) = -0.46708, y'' = -1.04167$.
Hint: Since $x = 2.03$ is not a tabulated value, use Newton's backward interpolation formulae (7.9b) and (7.9c) to get y' and y''.

7.21 Volume = 402.76 cubic cm

7.22 $I = -0.5876$

CHAPTER 8

8.2 $y(2.1) = 2.00238$

8.3 $y(1.3) = 0.4158$

8.4 $y = x + \dfrac{x^3}{6} + \dfrac{x^5}{120} + \cdots$

8.5

t	0.1	0.2
y	0.095	0.1810

8.6 $y(0.2) = 1.5067$

8.7 $y(0.1) = 0.525$ and $y(0.2) = 0.5536$

8.8 $y(0.1) = 1.0101345, y(0.2) = 1.0206776, y(0.3) = 1.0318418$ and $y(0.4) = 1.0438448$

8.9 $k_1 = 0.0, l_1 = -0.2, k_2 = -0.02, l_2 = -0.1998, k_3 = -0.01998, l_3\ -0.1958, k_4 = -0.0392, l_4 = -0.1906, y(0.2) = 0.9801, y'(0.2) = -0.1970$

8.10 $k_1 = 0.4, l_1 = 0.2, k_2 = 0.42, l_2 = 0.33, k_3 = 0.433, l_3 = 0.33612, k_4 = 0.4672, l_4 = 0.50414, y(0.2) = 1.4289, y'(0.2) = 2.3394$

8.11 $y(0.9) = 7.5772$

8.12 After first correction, the corrected value is $y(0.8) = 1.21808$

8.14 $y^{(p)}(4.4) = 1.01875, y^{(c)}(4.4) = 1.0187445.$

8.15 $y' = p,\ y(0) = 0$ and $p' = \dfrac{1}{EI}[M(x)\ (1 + p^2)^{3/2}],\ p(0) = 0$

8.16

t	0.0	0.1	0.2	0.3	0.4	0.5
y	0.0	0.0715	0.1983	0.2004	–0.0237	–0.3747

8.17

x	0.1	0.2	0.3	0.4	0.5
y	4.4	4.843	5.3393	5.9002	6.5383

8.18 $y(0.1) = 1.1448$

8.19 $y(0.2) = 1.0079$

8.20

x	0.5	1.0	1.5	2.0
y	1.3571	1.5837	1.7555	1.8956

8.21

x	0.1	0.2
y	1.0911	1.1678

8.22 $y^{(p)}(0.4) = 1.8344$, $y^{(c)}(0.4) = 1.8387$

8.23 $y(1.4) = 0.9493$

8.24

t	0.2	0.25	0.3	0.4	0.5
y	1.0203	1.0321	1.0468	1.0857	1.1395

8.25

t	0.02	0.04	0.06
x	6.2935	6.6156	6.9685
y	4.5393	5.1195	5.7440

CHAPTER 9

9.3

	x										
t	0.0	0.1	0.2	0.3	0.4	0.5	0.6	0.7	0.8	0.9	1.0
0.0	0.0	0.2	0.4	0.6	0.8	1.0	0.8	0.6	0.4	0.2	0.0
0.002	0.0	0.2	0.4	0.6	0.8	0.9200	0.8	0.6	0.4	0.2	0.0
0.004	0.0	0.2	0.4	0.6	0.7840	0.8720	0.7840	0.6	0.4	0.2	0.0
0.006	0.0	0.2	0.4	0.5968	0.7648	0.8368	0.7648	0.5968	0.4	0.2	0.0
0.008	0.0	0.2	0.3994	0.5910	0.7456	0.8080	0.7456	0.5910	0.3994	0.2	0.0
0.010	0.0	0.1999	0.3978	0.5836	0.7272	0.7830	0.7272	0.5836	0.3978	0.1999	0.0

9.4

	x								
t	0	1	2	3	4	5	6	7	8
0	0.0	3.5	6.0	7.5	8.0	7.5	6.0	3.5	0.0
1/8	0.0	3.0	5.5	7.0	7.5	7.0	5.5	3.0	0.0
2/8	0.0	2.75	5.0	6.5	7.0	6.5	5.0	2.75	0.0
3/8	0.0	2.5	4.625	6.0	6.5	6.0	4.625	2.5	0.0
4/8	0.0	2.3125	4.25	5.5625	6.0	5.5625	4.25	2.3125	0.0
5/8	0.0	2.125	3.9375	5.125	5.5625	5.125	3.9375	2.125	0.0

9.5

t	x										
	0.0	0.1	0.2	0.3	0.4	0.5	0.6	0.7	0.8	0.9	1.0
0.0	1.0	1.0	1.0	1.0	1.0	1.0	1.0	1.0	1.0	1.0	1.0
0.0025	0.8884	0.9773	1.0	1.0	1.0	1.0	1.0	1.0	1.0	0.9773	0.8884
0.005	0.8735	0.9608	0.9943	1.0	1.0	1.0	1.0	1.0	0.9943	0.9608	0.8735
0.0075	0.8612	0.9474	0.9874	0.9986	1.0	1.0	1.0	0.9986	0.9874	0.9474	0.8612
0.01	0.8508	0.9359	0.9802	0.9962	0.9997	1.0	0.9997	0.9962	0.9802	0.9359	0.8508
0.0125	0.8416	0.9257	0.9731	0.9931	0.9989	1.0	0.9989	0.9931	0.9731	0.9257	0.8416

9.6

t	x			
	0.0	1/3	2/3	1.0
0	1.0	0.8660	0.5	0.0
1/27	1.0	0.7887	0.4553	0.0
2/27	1.0	0.7553	0.4155	0.0

9.7 The solution to the transformed problem $u(x, t)$ is

t	x					
	0.0	0.2	0.4	0.6	0.8	1.0
0.0	1.0	1.0	1.0	1.0	1.0	1.0
0.0025	0.8246	0.9896	1.0	1.0	1.0	1.0
0.005	0.8166	0.9799	0.9993	1.0	1.0	1.0
0.0075	0.8091	0.9709	0.9981	0.9999	1.0	1.0

Using the transformation $T = \log u$, the solution $T(x, t)$ to the given problem is (of the original non-linear problem)

t	x					
	0.0	0.2	0.4	0.6	0.8	1.0
0.0	0.0	0.0	0.0	0.0	0.0	0.0
0.0025	–0.0837	–0.0454	0.0	0.0	0.0	0.0
0.005	–0.0879	–0.0882	–0.0030	0.0	0.0	0.0
0.0075	–0.0919	–0.0128	–0.0083	–0.0004	0.0	0.0

9.8 $T_0^0 = 0,\ T_1^0 = 1/\sqrt{2},\ T_2^0 = 1,\ T_3^0 = 1/\sqrt{2},\ T_4^0 = 0,\ r = 1/2$

$$\begin{bmatrix} 6 & -1 & 0 \\ -1 & 6 & -1 \\ 0 & -1 & 6 \end{bmatrix} \begin{pmatrix} T_1^1 \\ T_2^1 \\ T_3^1 \end{pmatrix} = \begin{bmatrix} 6 & 0 & 0 \\ -1 & \frac{35}{6} & 0 \\ 0 & -1 & \frac{204}{35} \end{bmatrix} \begin{bmatrix} 1 & -\frac{1}{6} & 0 \\ 0 & 1 & -\frac{6}{35} \\ 0 & 0 & 1 \end{bmatrix} \begin{pmatrix} T_1^1 \\ T_2^1 \\ T_3^1 \end{pmatrix} = \begin{pmatrix} \sqrt{2}+1 \\ \sqrt{2}+2 \\ \sqrt{2}+1 \end{pmatrix}$$

gives

$$(T_1^1 \; T_2^1 \; T_3^1) = (0.5265 \quad 0.7446 \quad 0.5265)$$

9.9 $r = 2$

$$\begin{bmatrix} 3 & -1 & 0 \\ -1 & 3 & -1 \\ 0 & -1 & 3 \end{bmatrix} \begin{pmatrix} T_1^1 \\ T_2^1 \\ T_3^1 \end{pmatrix} = \begin{bmatrix} 3 & 0 & 0 \\ -1 & \frac{8}{3} & 0 \\ 0 & -1 & \frac{21}{8} \end{bmatrix} \begin{bmatrix} 1 & -\frac{1}{3} & 0 \\ 0 & 1 & -\frac{3}{8} \\ 0 & 0 & 1 \end{bmatrix} \begin{pmatrix} T_1^1 \\ T_2^1 \\ T_3^1 \end{pmatrix} = \begin{pmatrix} 0 \\ 0 \\ \frac{1}{8} \end{pmatrix}$$

gives

$$(T_1^1 \; T_2^1 \; T_3^1) = (0.00595 \quad 0.01785 \quad 0.0476)$$

9.10 Durfort–Frankel finite difference scheme given by Eq. (9.26) is unconditionally stable for $r > 0$, that is, for all positive mesh ratios.

9.11 Crank–Nicolson scheme gives the following tridiagonal system.

$$\begin{bmatrix} 2 & -0.5 & 0 & 0 \\ -0.5 & 2 & -0.5 & 0 \\ 0 & -0.5 & 2 & -0.5 \\ 0 & 0 & -0.5 & 2 \end{bmatrix} \begin{pmatrix} T_1^1 \\ T_2^1 \\ T_3^1 \\ T_4^1 \end{pmatrix} = \begin{pmatrix} -0.12 \\ -0.2 \\ -0.2 \\ -0.12 \end{pmatrix}$$

Using LU decomposition, we get

$$\begin{bmatrix} 2 & 0 & 0 & 0 \\ -0.5 & 1.875 & 0 & 0 \\ 0 & -0.5 & 1.8667 & 0 \\ 0 & 0 & -0.5 & 1.8661 \end{bmatrix} \begin{bmatrix} 1 & -0.25 & 0 & 0 \\ 0 & 1 & -0.2667 & 0 \\ 0 & 0 & 1 & -0.2679 \\ 0 & 0 & 0 & 1 \end{bmatrix} \begin{pmatrix} T_1^1 \\ T_2^1 \\ T_3^1 \\ T_4^1 \end{pmatrix} = \begin{pmatrix} -0.12 \\ -0.2 \\ -0.2 \\ -0.12 \end{pmatrix}$$

Thus, heat loss is given by

t	x					
	0.0	0.2	0.4	0.6	0.8	1.0
0.0468	0.0	–0.1018	–0.1673	–0.1673	–0.1018	0.0
0.0936	0.0	–0.0860	–0.1402	–0.1402	0.0860	0.0

9.12 (i) Using explicit scheme, Eq. (9.22), and taking $\Delta x = 0.1$, the stability condition $\Delta t/\Delta x^2 < 1/2$ gives $\Delta t < 0.005$. Therefore, we take $\Delta t = 0.001$ and $r = 1/10$. The computed results are tabulated as follows:

t	x										
	0.0	0.1	0.2	0.3	0.4	0.5	0.6	0.7	0.8	0.9	1.0
0.0	0.0	0.2	0.4	0.6	0.8	1.0	0.8	0.6	0.4	0.2	0.0
0.001	0.0	0.2	0.4	0.6	0.8	0.96	0.8	0.6	0.4	0.2	0.0
0.002	0.0	0.2	0.4	0.6	0.796	0.928	0.796	0.6	0.4	0.2	0.0

(ii) The implicit scheme, Eq. (9.30), allows larger time step as it is stable for all mesh ratios r. Thus, choosing $\Delta x = 0.1$, $r = 1$ implying $\Delta t = 0.01$ and noting that the solution is symmetric about $x = 0.5$. It is enough to solve Eq. (9.30) for $i = 1, 2, \ldots, 5$ and $n = 1$, thus getting

$$\begin{bmatrix} 4 & -1 & & & \\ -1 & 4 & -1 & & \\ & -1 & 4 & -1 & \\ & & -1 & 4 & -1 \\ & & & -2 & 4 \end{bmatrix} \begin{pmatrix} T_1^1 \\ T_2^1 \\ T_3^1 \\ T_4^1 \\ T_5^1 \end{pmatrix} = \begin{pmatrix} 0.4 \\ 0.8 \\ 1.2 \\ 1.6 \\ 1.6 \end{pmatrix}$$

or

$$\begin{bmatrix} 4 & 0 & 0 & 0 & 0 \\ -1 & 3.75 & 0 & 0 & 0 \\ 0 & -1 & 3.7333 & 0 & 0 \\ 0 & 0 & -1 & 3.7321 & 0 \\ 0 & 0 & 0 & -2 & 3.4642 \end{bmatrix} \begin{bmatrix} 1 & -0.25 & 0 & 0 & 0 \\ 0 & 1 & -0.2667 & 0 & 0 \\ 0 & 0 & 1 & -0.2679 & 0 \\ 0 & 0 & 0 & 1 & -0.2679 \\ 0 & 0 & 0 & 0 & 1 \end{bmatrix} \begin{pmatrix} T_1^1 \\ T_2^1 \\ T_3^1 \\ T_4^1 \\ T_5^1 \end{pmatrix} = \begin{pmatrix} 0.4 \\ 0.8 \\ 1.2 \\ 1.6 \\ 1.6 \end{pmatrix}$$

whose solution at $T = 0.01$ is

$$\left(T_1' \; T_2' \; T_3' \; T_4' \; T_5'\right) = (0.1989 \;\; 0.3956 \;\; 0.5834 \;\; 0.7381 \;\; 0.7690)$$

CHAPTER 10

10.1 After discretization, Gauss–Siedel iterations are obtained from

$$u_{i,j}^{(r+1)} = \frac{1}{4}\left[u_{i-1,j}^{(r+1)} + u_{i+1,j}^{(r)} + u_{i,j-1}^{(r+1)} + u_{i,j+1}^{(r)} + 10\left(i^2 + j^2 + 10\right)\right]$$

The first few iterations are given below:

Iteration number, r	$u_{1,1}$	$u_{2,1}$	$u_{1,2}$	$u_{2,2}$
0	30.0	45.0	45.0	67.5
1	52.5	67.5	67.5	78.75
2	63.75	73.125	73.125	81.5625
3	66.5625	74.5313	74.5313	82.2657
4	67.2657	74.8829	74.8829	82.4414
5	67.4415	74.9707	74.9707	82.4854
6	67.4855	74.9928	74.9928	82.4964
7	67.4964	74.9982	74.9982	82.4991
8	67.4991	74.9996	74.9996	82.4998

10.2 After discretization, the Gauss–Siedel iteration for the given Poisson's equation can be written as

$$u_{i,j}^{(r+1)} = \frac{1}{4}\left[u_{i-1,j}^{(r+1)} + u_{i+1,j}^{(r)} + u_{i,j-1}^{(r+1)} + u_{i,j+1}^{(r)} + \frac{1}{8}\right]$$

The first few iterations are given in the following table:

Iteration number r	$u_{1,1}$	$u_{2,1}$	$u_{3,1}$	$u_{1,2}$	$u_{2,2}$	$u_{3,2}$	$u_{1,3}$	$u_{2,3}$	$u_{3,3}$
0	0.0313	0.0391	0.0410	0.0391	0.0508	0.0542	0.0410	0.0542	0.0584
1	0.0508	0.0669	0.0615	0.0669	0.0918	0.0842	0.0615	0.0842	0.0734
2	0.0647	0.0858	0.0736	0.0858	0.1163	0.0971	0.0738	0.0971	0.0798
3	0.0742	0.0973	0.0799	0.0973	0.1285	0.1033	0.0799	0.1033	0.0829
4	0.0799	0.1033	0.0829	0.1033	0.1346	0.1064	0.0829	0.1064	0.0844

10.3 The initial values at the internal nodes are obtained using diagonal and standard formulae (see the figure). Liebmann's iterations are obtained from

$$u_{i,j}^{(r+1)} = \frac{1}{4}\left[u_{i-1,j}^{(r+1)} + u_{i+1,j}^{(r)} + u_{i,j-1}^{(r+1)} + u_{i,j+1}^{(r)}\right]$$

and the computed solution at all the internal points of the region iteratively are shown in the figure.

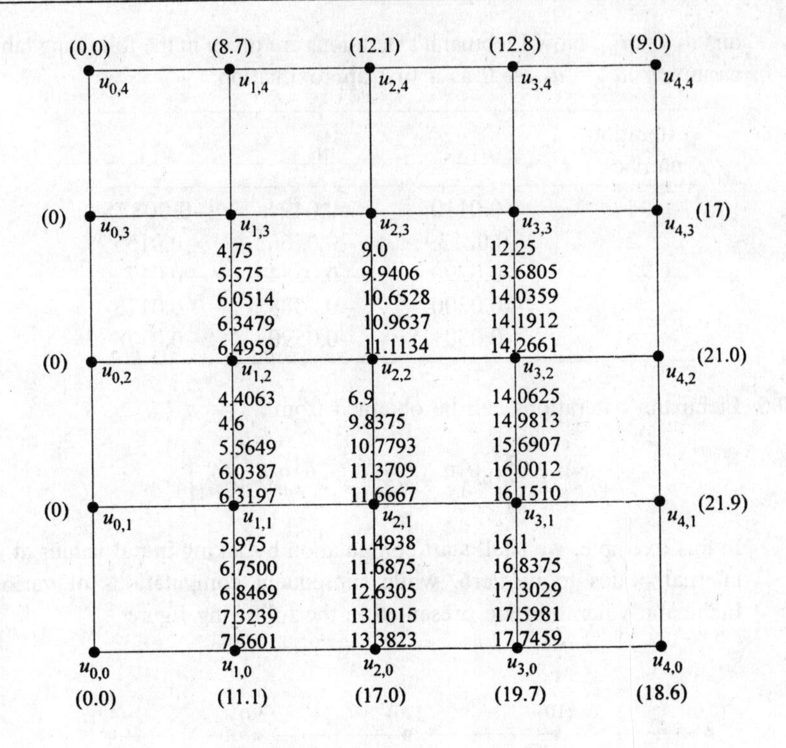

10.4 The standard five-point formula for the given Poisson's equation is

$$u_{i-1,j} + u_{i+1,j} + u_{i,j-1} + u_{i,j+1} - 4u_{i,j} = 8h^6 i^2 j^2 = \frac{8}{729} i^2 j^2$$

Thus,

$$u_{i,j} = \frac{1}{4}\left(u_{i-1,j} + u_{i+1,j} + u_{i,j-1} + u_{i,j+1} - \frac{8}{729} i^2 j^2 \right)$$

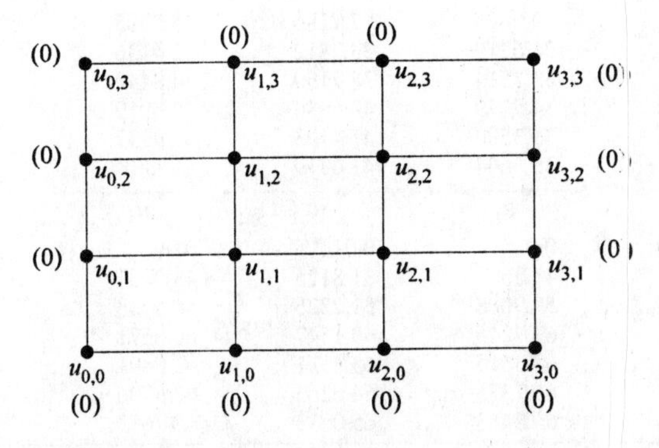

Now, from the figure, we observe that

$$u_{1,1} = \frac{1}{4}(u_{2,1} + u_{1,2} - 0.0110), \qquad u_{2,1} = \frac{1}{4}(u_{1,1} + u_{2,2} - 0.0439)$$

$$u_{1,2} = \frac{1}{4}(u_{1,1} + u_{2,2} - 0.0439), \qquad u_{2,2} = \frac{1}{4}(u_{1,2} + u_{2,1} - 0.1756)$$

and $u_{1,2} = u_{2,1}$. Now Liebmann's iterations are given in the following table, assuming $u_{1,1} = u_{2,2} = 0$ as a first approximation:

Iteration number, r	$u_{1,2}$	$u_{2,2}$	$u_{1,1}$
1	−0.0110	−0.0494	−0.00825
2	−0.02539	−0.0566	−0.0155
3	−0.0290	−0.0584	−0.0173
4	−0.0300	−0.0588	−0.0178
5	−0.0301	−0.0590	−0.0178

10.5 Liebmann's iterations can be obtained from

$$u_{i,j}^{(r+1)} = \frac{1}{4}\left[u_{i-1,j}^{(r+1)} + u_{i+1,j}^{(r)} + u_{i,j-1}^{(r+1)} + u_{i,j+1}^{(r)}\right]$$

In this example, we shall start computation by taking initial values at all internal nodes to be zero, while subsequent computations of various Liebmann's iteration are presented in the following figure:

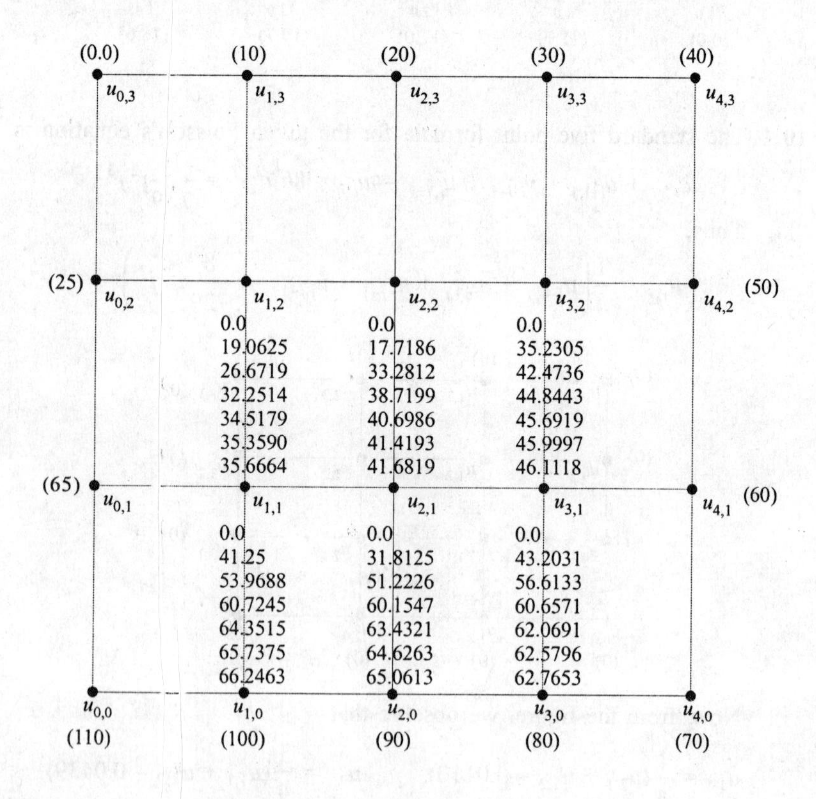

10.6 Applying Liebmann's over-relaxation method, the finite difference equivalent of the given Poisson's equation is

$$\phi_{i,j}^{(r+1)} = \frac{\omega}{4}\left[\phi_{i-1,j}^{(r+1)} + \phi_{i+1,j}^{(r)} + \phi_{i,j-1}^{(r+1)} + \phi_{i,j+1}^{(r)} + 8\right] + (1-\omega)\phi_{i,j}^{(r)}$$

Here

$$p = 3,\ q = 4,\ c = \cos\frac{\pi}{3} + \cos\frac{\pi}{4} = \frac{\sqrt{2}+1}{2\sqrt{2}}$$

Therefore, $c^2 = 0.7284$, $\omega_{\text{opt}} = 1.05$. Taking the initial values of ϕ at all internal nodes to be zero, the computed solution is shown in the following figure:

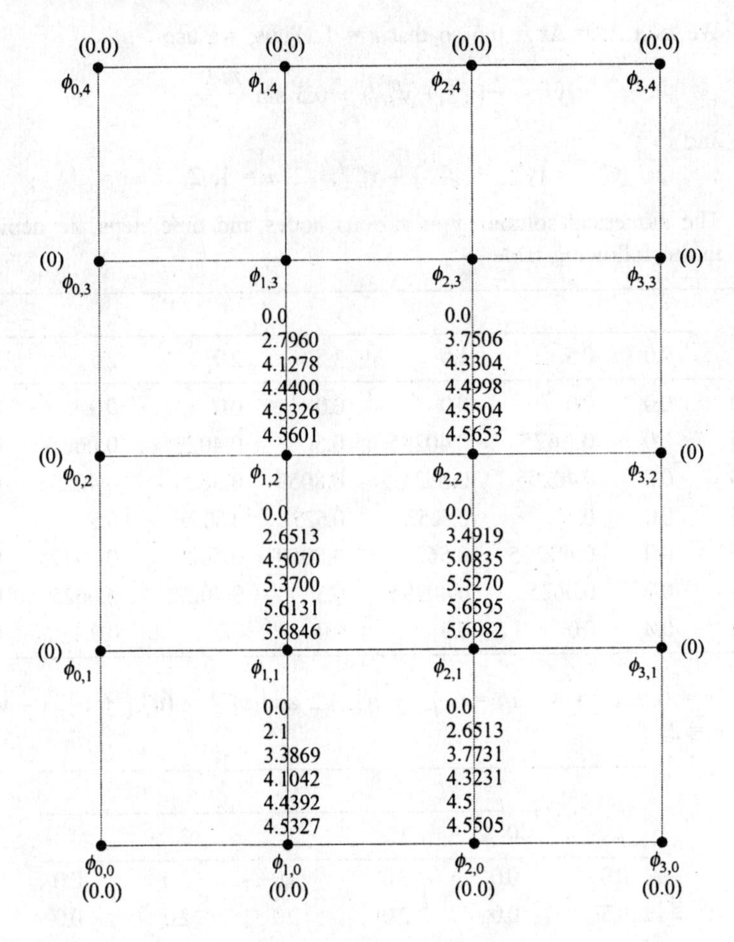

CHAPTER 11

11.1 $r = \Delta t/\Delta x = 0.1/0.1 = 1$. Therefore $u_i^1 = (u_{i-1}^0 + u_{i+1}^0)/2$ and $u_i^{n+1} = u_{i-1}^n + u_{i+1}^n - u_i^{n-1}$. Obviously, there is a symmetry about $x = 0.5$.

t	0.0	0.1	0.2	0.3	0.4	0.5	0.6
				x			
0.0	0.0000	0.0386	0.0735	0.1011	0.1189	0.1250	0.1189
0.1	0.0000	0.0368	0.0699	0.0962	0.1131	0.1189	0.1131
0.2	0.0000	0.0313	0.0595	0.0819	0.0962	0.1012	0.0962
0.3	0.0000	0.0227	0.0433	0.0595	0.0700	0.0735	0.0700
0.4	0.0000	0.012	0.0227	0.0314	0.0368	0.0388	0.0368
0.5	0.0000	0.0000	0.0001	0.0000	0.0002	0.0001	0.0002
	(0.0000)	(0.0000)	(0.0000)	(0.0000)	(0.0000)	(0.0000)	(0.0000)

11.2 Solution similar to that of Example 11.1

11.3 We take $\Delta t = \Delta x = 0.5$ so that $r = 1$. Thus, we use

$$y_i^1 = \frac{1}{2}(y_{i-1}^0 + y_{i+1}^0) + 0.5 \sin^3 \frac{\pi i \Delta x}{L}$$

and

$$y_i^{n+1} = (y_{i-1}^n + y_{i+1}^n) - y_i^{n-1}, \qquad n = 1, 2, \ldots$$

The numerical solution y at various nodes and time steps are depicted in the following table.

t	0.0	0.5	1.0	1.5	2.0	2.5	3.0
				x			
0.0	0.0	0.0	0.0	0.0	0.0	0.0	0.0
0.5	0.0	0.0625	0.40295	0.5	0.40295	0.0625	0.0
1.0	0.0	0.40295	0.5625	0.8059	0.5625	0.40295	0.0
1.5	0.0	0.5	0.8059	0.625	0.8059	0.5	0.0
2.0	0.0	0.40295	0.5625	0.8059	0.5625	0.40295	0.0
2.5	0.0	0.0625	0.40295	0.5	0.40295	0.0625	0.0
3.0	0.0	0.0	0.0	0.0	0.0	0.0	0.0

11.4 $r = 0.5$, $c^2 = 4$, $u_i^1 = (u_{i-1}^0 + u_{i+1}^0)/2$ and $u_i^{n+1} = (u_{i-1}^n + u_{i+1}^n) - u_i^{n-1}$, $n = 2, 3, \ldots$

t	0	1	2	3	4
			x		
0.0	0.0	3.0	4.0	3.0	0.0
0.5	0.0	2.0	3.0	2.0	0.0
1.0	0.0	0.0	0.0	0.0	0.0
1.5	0.0	−2.0	−3.0	−2.0	0.0
2.0	0.0	−3.0	−4.0	−3.0	0.0
2.5	0.0	−2.0	−3.0	−2.0	0.0

11.5 $r = \Delta t/\Delta x = 1.0$. Therefore, $u_i^1 = (u_{i-1}^0 + u_{i+1}^0)/2$ and $u_i^{n+1} = (u_{i-1}^n + u_{i+1}^n) - u_i^{n-1}$, $n = 1, 2, 3, 4$.

The values of u are tabulated as follows:

t	x										
	0.0	0.1	0.2	0.3	0.4	0.5	0.6	0.7	0.8	0.9	1.0
0	0.0	0.0	0.0	0.0	0.5	1.0	0.5	0.0	0.0	0.0	0.0
0.1	0.0	0.0	0.0	0.25	0.5	0.5	0.5	0.25	0.0	0.0	0.0
0.2	0.0	0.0	0.25	0.5	0.25	0.0	0.25	0.5	0.25	0.0	0.0
0.3	0.0	0.25	0.5	0.25	0.0	0.0	0.0	0.25	0.5	0.25	0.0
0.4	0.0	0.5	0.25	0.0	0.0	0.0	0.0	0.0	0.25	0.5	0.0
0.5	0.0	0.0	0.0	0.0	0.0	0.0	0.0	0.0	0.0	0.0	0.0

11.6

t	x										
	0.0	0.1	0.2	0.3	0.4	0.5	0.6	0.7	0.8	0.9	1.0
0.00	0.0000	0.8968	1.5388	1.7601	1.5388	1.0000	0.3633	−0.1420	−0.3633	−0.2788	0.0000
0.05	0.0000	0.7694	1.3284	1.5388	1.3800	0.9511	0.4290	0.0000	−0.2104	−0.1816	0.0000
0.10	0.0000	0.4316	0.7694	0.9484	0.9511	0.8090	0.5878	0.3606	0.1816	0.0684	0.0000
0.15	0.0000	0.0000	0.0516	0.1816	0.3774	0.5878	0.7407	0.7694	0.6394	0.3633	0.0000
0.20	0.0000	−0.3800	−0.5878	−0.5194	−0.1816	0.3090	0.7694	1.0194	0.9511	0.5710	0.0000
0.25	0.0000	−0.5878	−0.9511	−0.9511	−0.5878	0.0000	0.5878	0.9511	0.9511	0.5878	0.0000
0.30	0.0000	−0.5710	−0.9511	−1.0194	−0.7694	−0.3090	0.1817	0.5194	0.5878	0.3800	0.0000
0.35	0.0000	−0.3633	−0.6394	−0.7694	−0.7406	−0.5878	−0.3774	−0.1816	−0.0516	0.0000	0.0000
0.40	0.0000	−0.0684	−0.1816	−0.3606	−0.5878	−0.8090	−0.9511	−0.9484	−0.7694	−0.4316	0.0000
0.45	0.0000	0.1816	0.2104	0.0000	−0.4290	−0.9511	−1.3800	−1.5388	−1.3284	−0.7694	0.0000
0.50	0.0000	0.2788	0.3633	0.1420	−0.3633	−1.0000	−1.5388	−1.7601	−1.5388	−0.8968	0.0000

CHAPTER 12

12.1 $y(1/3) = 0.1222$, $y(2/3) = 0.1556$

12.2 $y(1.25) = 1.3513$, $y(1.5) = 1.6350$,

$y(1.75) = 1.8505$

12.3 $y(1/4) = -0.03488$, $y(1/2) = -0.05632$, $y(3/4) = -0.05003$

12.4 $y(0.2) = 1.3222$, $y(0.4) = 0.1968$

$y(0.6) = -0.9547$, $y(0.8) = -1.7812$

12.5 When $y'(1) = -3.5$, the solution is found to be

x	1.0	1.2	1.4	1.6	1.8	2.0	2.2	...
y	2.0	1.348	0.787	0.305	–0.104	–0.443	–0.712	...
y'	–3.5	–3.018	–2.599	–2.221	–1.867	–1.521	–1.167	...

...	2.4	2.6	2.8	3.0
...	–0.908	–1.026	–1.060	–1.0
...	–0.794	–0.391	0.054	0.547

12.6 $y(x) = 0.7829x^3 - 0.8070x^2 + 0.2325x$

12.8 *Hint:* Break the third order problem into a coupled set of first order equations, taking $f_1 = f''$,

$f_2 = f'$ and $f_3 = f$. That is solve

$$f_1' = -f_1 f_3$$

$$f_2' = f_1$$

$$f_3' = f_2$$

Take also, the initial guesses for $f_1''(0) = 1.0$ and $f_2''(0) = 0.5$ and use fourth order R–K method to march the solution from the wall to $\eta = 10$ in step of $\Delta\eta = 0.01$ etc... .

12.9 $y(1) = 0.3555$

CHAPTER 13

13.2 $-\dfrac{11}{32}T_0 + \dfrac{11}{8}T_1 - \dfrac{1}{8}T_2 + \dfrac{7}{48}T_3 - \dfrac{1}{32}T_4 + \dfrac{1}{80}T_5$

13.3 $\dfrac{81}{64}T_0 + \dfrac{9}{8}T_1 + \dfrac{13}{48}T_2 + \dfrac{1}{24}T_3 + \dfrac{1}{192}T_4$

13.6 $y = \dfrac{1}{2}(b-a)x + \dfrac{1}{2}(b+a)$

13.7 *Hint:* Start with $\tan^{-1}x = x - \dfrac{x^3}{3} + \dfrac{x^5}{5} - \dfrac{x^7}{7} + \dfrac{x^9}{9}$

$$\tan^{-1}x = \frac{\left(x + \frac{7}{9}x^3 + \frac{64}{945}x^5\right)}{\left(1 + \frac{10}{9}x^2 + \frac{5}{21}x^4\right)}$$

13.8 $e^x = \dfrac{24 + 18x + 6x^2 + x^3}{24 - 6x}$

13.9 $R_{22} = \dfrac{(6x + 3x^2)}{(6 + 6x + x^2)}$

13.10 $R_{54} = \dfrac{(945x - 105x^3 + x^5)}{(945 - 420x^2 + 15x^4)}$

Index